Craniofacial 3D Imaging

Onur Kadioglu · G. Fräns Currier
Editors

Craniofacial 3D Imaging

Current Concepts in Orthodontics and Oral and Maxillofacial Surgery

Springer

Editors
Onur Kadioglu
Division of Orthodontics, Department of Developmental Sciences
University of Oklahoma, College of Dentistry
Oklahoma City, OK
USA

G. Fräns Currier
Division of Orthodontics, Department of Developmental Sciences
University of Oklahoma, College of Dentistry
Oklahoma City, OK
USA

ISBN 978-3-030-00721-8 ISBN 978-3-030-00722-5 (eBook)
https://doi.org/10.1007/978-3-030-00722-5

Library of Congress Control Number: 2018965729

This Springer imprint is published by the registered company Springer Nature Switzerland AG
The registered company address is: Gewerbestrasse 11, 6330 Cham, Switzerland

We would like to dedicate this book to the many different individuals who have guided us along life's pathway, so we were able to share these developing concepts and changes with others.
There are our parents. They are Serap and Asif as well as Francis and George, who have unselfishly dedicated everything so we could be the best we can be. There are our brothers and sister who came along with us shouldering burdens and sharing joy. They are Aydin and Asli, Jeff and Barb, and Sue Ann and Dan.
There are loved ones. Sezin has enriched Onur's life beyond words with more purpose and happiness added by their sons, Arden and Aren.
We would also like to dedicate this book to our many previous faculty members and colleagues that have helped us at major times in our lives so we could arrive where we are today.
This book would not be possible without the invaluable contributions from our guest authors. We wholeheartedly thank each one along with Springer and their staff who trusted and supported us in this journey.
Last but not least, it has been our pleasure to help guide and direct the future members of our fine profession and specialty. Our current and past orthodontic graduate students/residents at the University of Oklahoma have given us great encouragement and a brightness that shines now and will continue in the years to come.
Cheers.

Onur Kadioglu and G. Fräns Currier

Preface

We have envisioned this textbook to be a current reference that highlights the use of 3D imaging through Cone Beam CT technology with emphasis on orthodontics and oral and maxillofacial surgery. We would like to recognize and present those areas that are impacted by 3D imaging by either changing the way one thinks about conventional orthodontic diagnosis and treatment planning or through the various types of tooth and surgical jaw movements. Our goal is to demonstrate planning and execution with emphasis on the limits of the alveolar bone, airway and temporomandibular joints. These areas of interest will be demonstrated not only by recalling the available science but also employing current translational research findings that our distinguished authors and editors have been executing in their respective disciplines, whether in departments, clinics and hospitals.

We have judiciously selected an outstanding group of authors whom we strongly believe will have a positive impact for any reader, directing their thought processes during diagnosing and treatment planning of their patients.

Oklahoma City, OK, USA

Onur Kadioglu
G. Fräns Currier

Contents

Part III Clinical Applications in Oral Surgery and Orthodontics

Part I

The Overview

1 History, Technique, and Safety

Farah Masood, Onur Kadioglu,
and G. Fräns Currier

Abstract

The scope of dental imaging has been greatly expanded with the invention of cone beam computed tomography (CBCT). The everyday functionality of dental practice, especially for dental specialties like orthodontics and oral surgery, has certainly changed with this radiographic technology for treatment planning and evaluation. The discovery of X-rays was made by Wilhelm Conrad Rontgen in 1895, who was a known physicist for his work. He won a Nobel Prize in 1901 for this discovery. Ever since this revolution, constant technologic advancements have been made in the field of dental radiology that had resulted in improving the diagnostic accuracy and reducing the radiation exposure in every day dental practice. In dentistry, conventional two-dimensional (2-D) radiographic imaging has been widely used. However, the conventional radiographic images have limitations like inherent magnification, distortion, superimposition of structures, and lack of depth for three-dimensional anatomical objects. Over the years, the technology has improved tremendously in terms of image quality and radiation dose. With CBCT, the visualization of structures is possible with much clarity and without superimposition. The technology has shown a profound impact on the dental practice with its widespread applications.

1.1 Introduction

Computed tomography (CT) imaging was originally called computed axial tomography (CAT) scan. Researchers started the development of medical CT scanners in the 1960s. Later around 1970–1972, Dr. Godfrey Hounsfield (electrical engineer at EMI Central Research Laboratories, England) and another physicist Allan Cormack of Tufts University (Boston, MA) introduced the CT imaging modality for clinical applications. The Nobel Prize was awarded to both of them in 1979 for the development of computerized tomography, as the technology had a profound impact in improving the diagnostic methodology. The technology was patented by Dr. Hounsfield in 1973 [1]. In the mid-1970s, a full-body scanner was developed by a dentist-physicist Dr. Robert Ledley of Georgetown University (Washington, DC).

F. Masood (✉)
Division of Oral Diagnosis and Radiology, Department of Diagnostic and Preventive Sciences, University of Oklahoma Health Sciences Center College of Dentistry, Oklahoma City, OK, USA
e-mail: Farah-Masood@ouhsc.edu

O. Kadioglu · G. F. Currier
Division of Orthodontics, Department of Developmental Sciences, University of Oklahoma Health Sciences Center College of Dentistry, Oklahoma City, OK, USA
e-mail: onur-kadioglu@uohsc.edu; frans-currier@ouhsc.edu

O. Kadioglu, G. F. Currier (eds.), *Craniofacial 3D Imaging*,
https://doi.org/10.1007/978-3-030-00722-5_1

To capture the data, the medical CT scanner uses an X-ray beam and image detectors, mounted and fixed on a rotating gantry, which rotates around the patient to capture the region of interest. During the rotational movement of the gantry, the X-ray beam passes through the patient, and the remnant X-ray photons remaining after the attenuation are captured by the image detectors. The "raw data" acquired by this process is reconstructed by a computer algorithm, and the end result is generation of cross-sectional images of the patient's tissues. The process uses a series of radiographic images to create sequential images, and virtual slices of body tissues are produced.

Initially, the first-generation CT scanners acquired the data in the axial plane by "slice-by-slice" scanning with a narrow fan-shaped X-ray beam and a single array of detectors. Eventually, the development of spiral CT (1989) and the multislice image detector systems (1988) leads to the acquisition of volumetric data [2]. Modern CT scanners are much faster as they use array of multiple detectors with the rotating fan-shaped X-ray beam and capture multiple slices of data simultaneously, in a short period of time. This has resulted in shorter scan times and lesser radiation dose to the patient as well [3].

With this technique, information about the internal structures is obtained by reformatting of the data and production of cross-sectional images, and the structures are visualized without superimposition. For image display, the components of the gray images are pixel and voxels. A voxel defines a point in three dimensions, whereas the pixel defines a point in two dimensions. Pixel or picture element represents the smallest single module of an image in a two-dimensional (2-D) framework. The attenuation of X-ray photons or signal by the patient's tissues determines the value and intensity of each individual pixel that is captured by the detector, and the information is displayed on the computer screen. Pixel size effects the image resolution. Voxel adds detail and third dimension (3-D) or depth to the image.

Medical multi-detector computed tomography (MDCT) units use Hounsfield units (HU) to display relative density values of various body structures according to a calibrated gray value scale.

For viewing, the reconstruction of data produces images in multiple imaging planes. During the CT scanning process, the data is captured in the axial or transverse plane. Axial plane is an imaginary plane that divides the structure or body into upper and lower portions. From this axial data set, the computer software programs can generate multiplanar reformatted images in axial, sagittal, and coronal planes by combining the information. The sagittal plane sections the structure or body into right and left, and the coronal plane divides the structure or body into anterior and posterior sections. Also, 3-D computer-generated models of the structures can be made. MDCT units have superior contrast resolution and can display soft tissues with more superior quality. Medical- or hospital-based CT units have large footprints and are supine gantry-style units with considerably high radiation exposure to the patient. Before the introduction of CBCT for dental needs, MDCT units were utilized for diagnosis and treatment planning for only limited cases in dentistry. Factors like lower radiation dose and ease of use in dental setting lead to the development of cone beam computed tomography.

1.2 What Is Cone Beam Computed Tomography (CBCT)?

CBCT was introduced in the early 2000s to the main market. As stated earlier, before CBCT was introduced to the main market, the conventional CT and MDCT scanners were used by the dental specialties to obtain cross-sectional views for pathology, maxillofacial trauma, and in limited number of dental implant cases. However, the utilization of MDCT was very limited due to higher radiation doses as compared to CBCT. Cost of the procedure was also very high. Many CBCT systems are available (Fig. 1.1).

Cone beam computed tomography (CT) has the potential to reduce the size and cost of CT scanners. Because this emerging technology

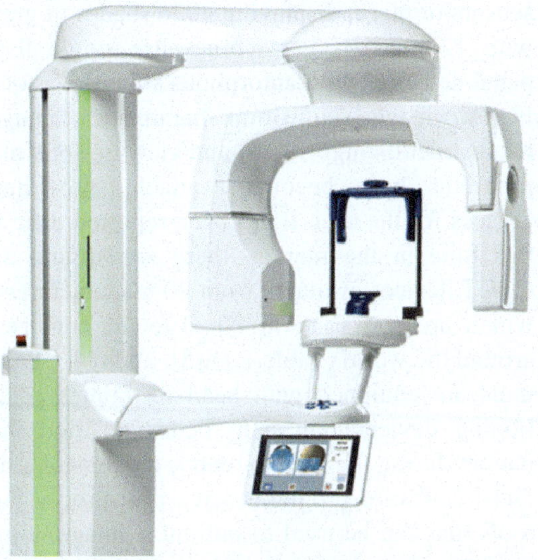

Fig. 1.1 Picture of currently available ProMax 3D CBCT scanner (Courtesy of Planmeca Oy, Helsinki, Finland)

produces images with isotropic submillimeter spatial resolution, it is ideally suited for dedicated dentomaxillofacial CT scanning. When combined with application-specific software tools, cone beam computed tomography can provide dentomaxillofacial practitioners with a complete solution for performing specific diagnostic and surgical tasks, such as dental implant planning.

The other terms used to describe this technology include cone beam volumetric imaging (CBVI) and cone beam volumetric tomography (CBVT).

The introduction of low-dose CBCT scanning systems has changed this approach for everyday dental practice. It is specifically designed to produce three-dimensional images of the maxillofacial region. The computer software programs are designed for dental needs.

CBCT scanners are connected to a computer, and the data or the region of interest is acquired with a single full 360° or partial rotation of the cone-shaped X-ray beam and reciprocal rotating single image detector around the patient's head. The scan times are usually less than 15–20 s. The system uses back-projection reconstruction tomographic technique. MDCT acquires image data using multiple rows of detectors, where multiple slices must be stacked to obtain a complete image [3].

The CBCT technology exposes the whole region of interest or the head of the patient with one flat-panel detector. This baseline data are then used to generate individual image slices in different planes. In CBCT image acquisition, there is no additional mechanism needed to move the patient during the scanning, and also the use of cone-shaped beam in CBCT increases the utilization of the X-ray energy by lowering the X-ray tube heat capacity required for volumetric scanning as compared to a fan-shaped beam in MDCT [4]. CBCT units have isotropic (equal in all three dimensions) voxel resolution in which images with isotropic submillimeter spatial resolution are produced [3]. Image detectors with smaller pixels tend to capture fewer X-ray photon per voxel and thus result in more noise. Higher radiation doses are required for a reasonable signal-to-noise ratio, which improves the image quality. The following factors also effect the spatial resolution and image quality: focal spot of the X-ray generator, patient-to-detector distance, X-ray source-to-patient distance, and patient movement. Smaller focal size, reduced patient-to-detector distance, and increased X-ray source-to-patient distance minimize the geometric unsharpness of the images. In practice, movement of the patient's head is a big factor that will deteriorate the image quality [4]. Many CBCT machines have artifact reduction tools that allow minimization of the noise due to metal streaking after acquiring the images. One company offering a tool which helps reduce movement related artifact after acquisition of the data. The rotation of the cone-shaped X-ray beam and the detector around the patient's head generates large amount of data that is rapidly transferred from the rotating scanning system to the external computers for further processing for visualization in axial, sagittal, and coronal planes, and 3-D reconstruction is done. Images are produced with isotropic submillimeter spatial resolution, and the application-specific software tools are available for use by the dentists.

With advancements, the modern CBCT systems have integrated very well in the dental practice. Smaller footprint of the machine, simplicity of operator training procedures, ease of use, short exposure time, easy patient positioning for scanning, integration into the workflow of the practice, accuracy of information, and availability of relatively simple viewing software have led to the popularity of CBCT systems. However, the CBCT cost and radiation dose are considered to be higher as compared to the conventional 2-D dental imaging procedures. As with any other radiographic technique, the aim is to achieve optimal image quality with the lowest possible radiation dose, which could be challenging. CBCT units are smaller in size and can fit in a dental office with some modification. In majority of the CBCT machines, the patient sits in the chair for the short exposure time.

CBCT digital imaging produces 3-D data of the area of interest with diagnostically acceptable spatial resolution, much lower radiation dose, and cost as compared to the MDCT. For dental practice the first CBCT machine, NewTom 9000 (Quantitative Radiology, Verona, Italy), was developed and introduced in the European market in the late 1990s.

This technology was brought to the market in the United States in 2001. For scanning in the NewTom 9000, the patient had to be in supine position and the X-ray tube and the detector rotated 360° around the patient's head to obtain a relatively larger field of view (FOV) 15 cm × 15 cm volume. The system utilized an image intensifier and a charge-coupled device. The sensor was 8-bit displaying 256 shades of gray. Later developments were made to fabricate CBCT machines with smaller, adjustable FOVs. Later Ortho-CT based on Scanora stand (Soredex Corporation, Helsinki, Finland) made it possible where patient would sit in a chair during the scan. In 2002, 3D Accuitomo unit (J. Morita Corporation, Japan) became available in the European market. In this scanner the patient sat in a chair for exposure, and the FOV size was reduced to 3 cm × 4 cm cylinder [5].

Since then, tremendous improvements have been made in image quality. Currently the CBCT systems offer different sizes of FOVs and 12-bit sensors or more, displaying 4096 shades of gray with 12-bit. CBCT machines use single flat-panel detectors with amorphous silicon. Various FOVs, image acquisition parameters, image reconstruction algorithms, and viewing software programs have become available, providing choices for the user. It has been reported that, at this time in the market, there are around 50 CBCT devices available from 20 manufacturers, which are operating in 20 different countries around the world (Table 1.1) [5]. Today's CBCT units are equipped with head restraining/positioning devices that help better position the patient during the scan to reduce movement artifacts. Software programs have post-processing tools that can be used to minimize image noise artifacts after data acquisition. It is important that the whole dental team be knowledgeable about the availability of these tools in the software. This will potentially reduce the number of re-exposures.

1.3 Acquisition of CBCT Volume

CBCT devices consist of X-ray source (cone-shaped divergent beam) and a 2-D image detector. The X-ray source and the image detector are connected by an arm that rotates around the patient's head during the scan. Rotation varies from 180° to 360°. Typically, the volumetric data is captured with single rotation around the patient's head as the transmitted beam of radiation is aimed at the image detector. Beam collimators either match the size of the detector and the beam size or can be used to further reduce or collimate the field of view.

With most CBCT units, a series of 2-D raw base images or projections are captured. The number of raw images varies from 180 to 600 or up to 1000 in some machines. Exposure times vary from 6 s to 40 s. Many machines have pulsating radiation, which helps reduce the patient dose. The ranges for tube current (mA) and peak voltage (kVp) are 1–15 mA and 85–120 mA, respectively. After processing axial, coronal, and sagittal planes, images appear on the computer monitor as an Explore screen (Fig. 1.2).

Table 1.1 Selected CBCT systems available with larger fields of view (FOVs)

Model	Manufacturer	Voxel mm^3	Detector size/field of view cm
3-D Accuitomo 170	J. Morita	0.125–0.2	4 × 4–17 × 12
Galileos Comfort Plus	Sirona Dental systems	0.25/0.125	15 × 15
I-CAT FLX	Imaging sciences	0.125–0.4	8 × 8–17 × 23
CS 9300	Carestream	90 to 500 μm	5 × 5–17 × 13.5
NewTom 3D	Quantitative radiology	0.08	6 × 6–10 × 10
i3D-Premium	Vatech	0.2, 0.3, 0.4	8 × 8 × 21 × 19
PaX-i3D Pano + CBCT + Ceph	Vatech	0.12–0.3	8 × 8–12 × 9
Picasso Pro	Vatech	0.2–0.3	5 × 5–12 × 9
ProMax3D Max	Planmeca	0.1, 0.2, 0.4	5.5 × 5–17 × 22
KaVo OP 3D Vision	KaVo Dental	0.125–0.4	5 × 8–17 × 23
SCANORA 3D	Soredex	0.13–0.35	5 × 5–24 × 16.5

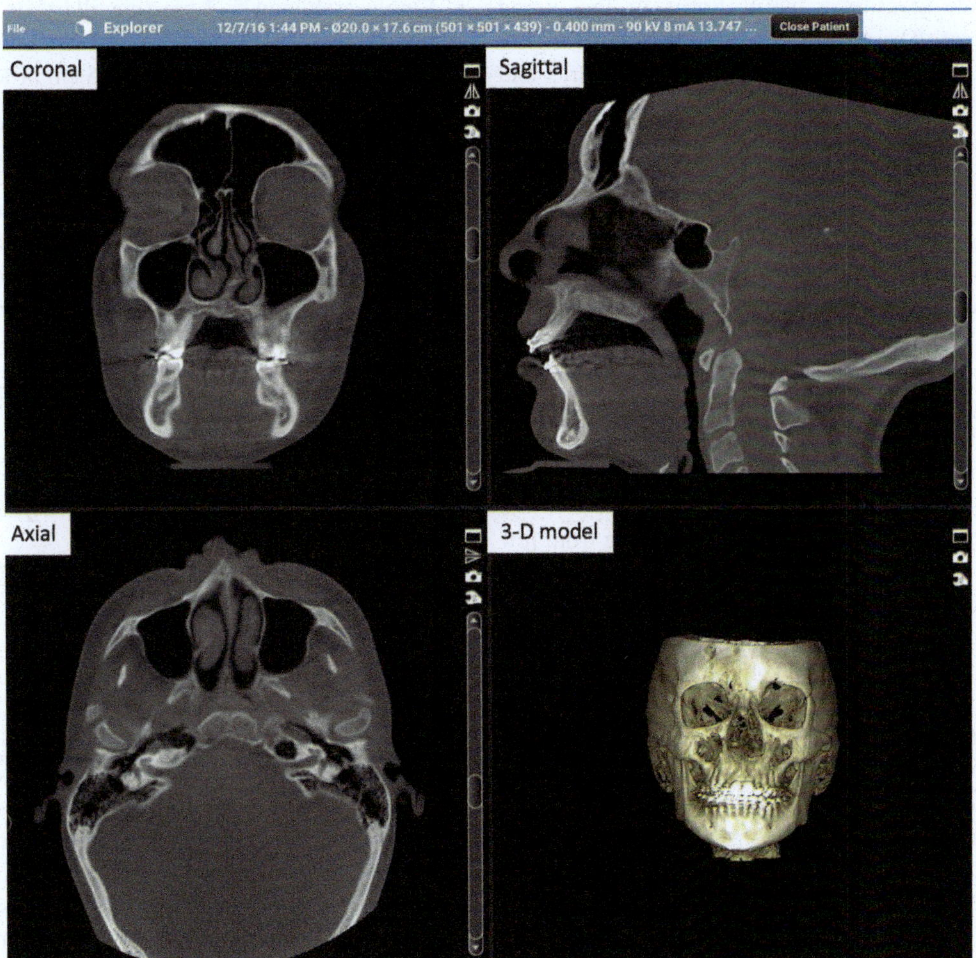

Fig. 1.2 Explore screen from a CBCT machine with large field of view is shown. This is a typical image display in coronal, sagittal, and axial planes

1.4 Image Detectors Used in CBCT Units

The image detector or receptor converts the incoming remnant X-ray photons from the patient into electrical signals. Later computer processing converts these signals into visible images. CBCT machines are equipped with either image intensifier tubes/charge-coupled device (II-CCD) or a flat-panel detector (FPD). II-CCD units are usually bulkier as compared to the FPD. FPD is made up of scintillation crystal screen on a matrix of photodiodes embedded in a solid-state amorphous silicon layer with thin-film transistors. The signal intensity is proportional to the stored charges. The advantages of FPD include higher radiosensitivity, lesser radiation exposure, and better image quality.

1.5 Field of View (FOV)

FOV is the anatomical volume that can be captured by the detector. FOV varies in size. The machines come with various detector sizes. Machines with larger detectors offer larger FOV, with ability to collimate the X-ray beam to a small area or FOV. With collimation of the X-ray beam, the FOV can be reduced to suit the needs, and this reduces the amount of exposure to the patient. Multiple FOV options, ranging from few centimeters to full head size, are available for various clinical scenarios.

Larger detectors tend to be more expensive. Due to the cost factor, some CBCT systems offer smaller detectors with limited FOV. When there is need to acquire the larger FOV, two or more adjacent scans can be made, and the volumes can be stitched together by the computer software to produce a larger FOV.

Decreasing the size of FOV or beam collimation improves image quality by decreasing scatter artifacts in the image. The extent of anatomic coverage should be based on clinical evaluation by the treating clinician. Over collimation or too narrow collimation to achieve smaller FOV may result in excluding essential anatomic structures needed for evaluation, and thus a "not needed" retake of CBCT may be needed. Scarfe and Farman [6] published a FOV categorization of the different CBCT systems according to the CBCT volume height and provided examples of coverage as follows:

- Craniofacial region: Height > 15 cm (extending from the head vertex to the inferior mandibular border).
- Maxillofacial region: 10–15 cm in height (nasion to inferior mandibular border).
- Interarch region: 7–10 cm (extending from the inferior nasal concha to the mandible).
- Single arch/jaw: 5–7 cm (maxillary or mandibular arch only).
- Localized to region of interest: 5 cm or less in height (1–2 teeth and surrounding bone, temporomandibular joints).

FOV can also be classified large for craniofacial coverage (>10 cm in height) and small to medium for dentoalveolar coverage (variable depending on the region of interest <10 cm in height). Smaller volume or FOV should be considered if it addresses the diagnostic needs [7].

1.6 Reconstruction Process and Display of CBCT Images

With a single rotational movement of the CBCT machine used for exposure of 20 s or less, approximately 100 to more than 600 individual frames may be captured by the acquisition computer. A volumetric data set is created with the individual basic frames by a series of algorithms or reconstruction process at the processing computer or the workstation. Both computers are connected via an Ethernet connection for transfer of the acquired individual frames from acquisition computer to the workstation for processing.

For image display, the CBCT units also use HU units. However, in CBCT, the measured density numbers correspond to the grayscale values and do not directly represent HU units. Smaller field of view scans have more discrepancies related to the density values as the contributing structure may be located outside the area of interest. After exposure, the raw data set is reconstructed with the computer software to produce cross-sectional images in axial, coronal, and sagittal planes (orthogonal planes). Sectional or reconstructed panoramic image can also be obtained with multiplanar reconstruction from the same data set. Furthermore, cross-sectional images of a region of interest can also be obtained perpendicular to the curve of the dental arch, which are widely used for implant treatment planning.

The operator can change the cross section or slice thickness. With this data, lateral cephalometric images can also be generated. Other advantages of CBCT images include no image magnification or distortion, and the software designed for dentistry is equipped with tools such as ruler for accurate 1:1 linear and angular measurements.

The CBCT data can be displayed in various formats such as volume rendering and maximum intensity projection (MIP). Spatial relationship between structures can be visualized by volume rendering, which gives a three-dimensional impression of the volume with different colors and transparency levels, based on attenuation or gray values. In a MIP, only the highest voxel value is displayed within the selected thickness in area of interest.

The contrast and brightness can be easily adjusted (changing the window width and window level) to improve the display. Bit depth of the system determines the number of shades of gray available to display the attenuation. Display of the images depends on the ability of the system to display the variations in attenuation and the capability of the image detector to show the subtle contrast variances.

Most newer machines offer 14-bit or 16-bit image detectors, which translates to 2^{14} (16,384) and 2^{16} (65,536) shades of gray or contrast display, respectively. Higher bit-depth systems require more processing time and considerably larger data set files, which may require greater storage capability.

Segmentation of an area of interest is a very useful tool when separation of certain structures is desired from the volume for in-depth analysis.

CBCT machines from the most major manufacturers have the capability to export data in standard DICOM (digital imaging and communications in medicine) format. With DICOM data, the user can use a third-party software to view and analyze the data set.

Some CBCT machines offer "volume stitching," where a larger field of view can be obtained by stitching the data sets obtained by a smaller detector.

1.7 Factors Influencing Image Quality in CBCT

In order to understand the image quality, the noise, scatter, spatial, and contrast resolution of an image must be discussed. Image noise and scatter are factors that are often encountered. Image noise occurs due to inconsistent distribution of signal and inconsistent attenuation or gray values and is visualized as "grainy appearance" of an image. Image noise can potentially degrade the image display and obscure the structure of interest. In order to decrease noise, the exposure time has to be increased.

Scatter in the images is produced by diffraction of X-ray photons from the original pathways, and upon interaction with the image detector, these photons end up producing nonuniformed increased intensities of structures. This interaction results in inferior contrast resolution on the resultant image. Changes of scatter are more with larger image detector or field of view. As CBCT machines use a single 2-D image detector with a cone-shaped beam, scatter is more seen with CBCT as compared to medical CT scanners [5]. Scatter is also produced by the patient from anatomical structures or existing restorations. Current CBCT systems have software tools to minimize the scatter after running the artifact reduction algorithm.

The spatial resolution is measured in line pairs per millimeter (lp/mm). It is the ability to distinguish fine detail or structures that are located very close together. Higher spatial resolution means clear or sharp distinction between the shades of gray or structures on the image. Higher spatial resolution can be attained with smaller voxel size. However, this requires higher exposure time. Scans with smaller field of view also have superior spatial resolution.

Ability of an image to display subtle differences between tissues of different radiodensities is the contrast resolution. In other words, clear distinction in various shades of gray is seen on the image. Following factors tend to decrease the contrast resolution: image noise, scatter, larger fields of view, reduced milliamperes, and kilovoltage settings of the X-ray generator.

1.8 Imaging Protocols for CBCT and Indications

Utilization of CBCT imaging has certainly increased over the last decade, and many research studies have been published that document that its use enhances the diagnosis in a significant number of clinical cases and thus has shown to improve treatment outcomes [7]. The clinician may choose to this technology if it is believed that there will be benefit in the patient care. However, the factors like higher radiation exposure, greater cost, and cost of interpretation must be considered.

Only after detailed clinical examination, the need for any radiographic imaging must be determined. Following main parameters influence the imaging protocol and image quality: exposure settings, voxel size, scan time, and field of view. The operator must understand that change in the exposure setting will effect the image quality and radiation dose to the patient. Spatial resolution is the ability of an image to display detail. Different voxel sizes are offered with CBCT machines. Voxel size should be specified for acquisition or reconstruction stage.

Longer scan times acquired more basic frames. The advantage is fewer artifacts and better image quality. One must remember that using the longer scan time protocol will result in longer reconstruction times and more radiation dose to the patient [8].

Some CBCT machines are equipped with low-dose or ultralow-dose exposure modes.

CBCT may be advocated for patients in situations where plain conventional radiographic images are not adequate for addressing the diagnostic issues [9].

General common reasons for obtaining CBCT scan in orthodontics would include localization of unerupted teeth, resorption of root, bone grafting, and assessment of cleft palate [10–12].

Depending on the indication or the region of interest, the CBCT field of view (small, medium, or large) may be selected. The fields of view and possible indications are listed in Table 1.3. Once the data set is acquired, the clinician should screen through the full volume systematically in all three dimensions. The region of interest should be evaluated in axial, sagittal, and coronal planes. It must be kept in mind that the CBCT data must be evaluated in entirety or the complete volume by the clinician prescribing the scan. The scan can be read by an oral radiologist to get an additional report on the region of interest and incidental findings of significance and to rule out pathologic conditions.

After gathering this information, the practitioner can refine or change the initial diagnosis and the proposed treatment plan, as needed.

The CBCT volumetric data is usually backed up in the proprietary format. The data export is usually done as DICOM v3 (digital imaging and communications in medicine standard version v3) format, so that the data can be imported and viewed in third-party software applications as needed.

The following guidelines for the use of CBCT imaging can be recommended for the orthodontic clinical practice:

- Anomalies of the teeth, especially marked oligodontia and supernumerary teeth, impactions

(esp. permanent incisors and canines), transpositions.
- Anomalies of the craniofacial complex, especially craniofacial syndromes, including cleft lip and palate. Class II division 2 malocclusions.
- Orthodontic treatment involving functional orthopedics (fixed or removable functions), maxillary orthopedics, self-ligating non-extraction protocols, extractions of permanent teeth, multiple TADs (temporary anchorage devices) and orthognathic surgery.
- Airway obstruction, including constant mouth breathing, snoring, obstructive sleep apnea.
- Class II division 1 malocclusions with mandibular retrognathia Class III malocclusions with mandibular prognathia and/or maxillary hypoplasia.
- Early treatment cases involving facebow headgear, facemask, lip bumpers.
- Cases prone to root resorption, especially those involving trauma, previous history of root resorption, marked intrusion/extrusion cases.
- Advanced periodontitis.

A task force was formed by the American Association of Oral and Maxillofacial Surgeons to study the possible indications for CBCT, clinical use, radiation exposure, safety, and legal issues in oral and maxillofacial surgery practices. The published applications and usage of CBCT at the clinical practitioner level were reviewed. They identified the current position of academic leaders in the field. A nationwide survey was done to determine how CBCT was being used and adopted by institutions and private practices. According to this published paper, the best practices supported evaluation of the entire CBCT volume with a written report with the findings, patient exposure, and FOV. The use of ALARA ("as low as reasonably achievable") principle was emphasized. They reported that the third-party patterns for reimbursements varied widely and seem to lack consistency [13].

As there has been a marked increase in the use of CBCT imaging in dentistry and especially in specialties like orthodontics, the American Academy of Oral and Maxillofacial Radiology published a paper [14] supporting the safe use of CBCT in practice. The paper summarized the potential benefits and risks of this technology in orthodontic diagnosis, treatment planning, and outcomes to aid the clinicians. The recommendation was to use the principle of justification based on clinical exam for each individual patient. The benefits must overweigh the potential risks associated with radiation exposure. The position paper provides the following guidelines for using CBCT in orthodontics: (a) image selection criteria and recommendation should be used, (b) radiation risks and dose to the patient must be taken into account, and (c) the clinician must maintain professional competency in acquisition and interpretation.

1.9 Benefits and Risks of CBCT

CBCT imaging offers many obvious advantages over the conventional 2-D extraoral imaging modalities like lateral cephalometric imaging, which is widely used. It must be emphasized that CBCT utilizes ionizing radiation and thus involves known potential risks that are associated with the use of radiation. The use of radiographic imaging should not be considered routine for practice. Like with any other radiographic imaging modality, the clinician must weigh in potential benefits that will be gained by the CBCT exposure versus risks to the patient.

According to the known principle of ALARA (as low as reasonable achievable), the radiation exposure should be kept to the minimum to protect the patient from any unnecessary radiation exposure. Based on guidelines for CBCT in Orthodontics[14], it has been recommended that following viewpoints must be taken into consideration in determining the need and potential benefits of a radiographic modality such as CBCT for an orthodontic patient:

1. Is the operator obtaining the images trained adequately and is using the best exposure parameters under the circumstances?

2. Will this procedure provide additional useful information that will further aid in clinical diagnosis and treatment planning?
3. Is the practitioner competent in the interpretation of CBCT images?

Before prescribing any radiographic images, it should be kept in mind that ionizing radiation can potentially lead to genetic mutations and carcinogenesis. The risk of potential side effect is more with greater exposure. Unnecessary radiographic imaging such as CBCT may contribute to an increased risk in the patients. The three guiding principles of justification, optimization, and dose limitation must be considered when ordering radiographic images [8].

Clinician's professional judgment is essential to justify every radiographic exposure. The decision to expose a patient must be made after a detailed clinical examination and only if the clinician thinks the findings from the acquired images will provide additional information that will benefit the patient. It is important for the clinician to understand the functioning of the chosen imaging modality including the advantages and the limitation. This will result in gaining the adequate and best quality images with minimum exposure to the patient, using the optimal exposure parameters. After the images are attained, the suitable formatting should be done.

The interpretation should not be limited to the region of interest only. The entire CBCT volume must be evaluated.

CBCT imaging is an excellent tool for the orthodontists and oral surgeons and helps to improve diagnosis, treatment planning, and outcome assessment in appropriate cases. Literature shows that CBCT is a powerful imaging modality and provides orthodontists with 3-D images of the craniofacial osseous structures, dentition, and soft tissue (Figs. 1.3 and 1.4). A virtual cephalometric image can also be made (Fig. 1.5). This information is important for diagnosing malocclusion. "While orthodontists await the American Association of Orthodontists' position paper on identifying appropriate cases for CBCT imaging, case selection using current evidence-based criteria suggest that complex craniofacial and surgical cases and cases of missing or impacted teeth may be the most suitable candidates for CBCT imaging (Fig. 1.6) although the absolute need for CBCT imaging must be determined on a case-by-case basis." [15]

Other common uses for CBCT evaluation of anatomical structures such as maxillary sinuses (Fig. 1.7), inferior alveolar canal and foramen (Fig. 1.8), and evaluation of periapical pathology (Fig. 1.9). Advantages of CBCT technology include lower cost, faster acquisition, and lower radiation dose to the patient as compared to MDCT. A personal computer is utilized for data reconstruction and viewing with an interactive software designed for dental applications. As compared to MDCT, the limiting factors for CBCT technology may include greater image noise and noticeably poor soft tissue contrast.

1.10 CBCT Radiation Doses

A variety of CBCT machines are available in the market (Table 1.1). The radiation doses with each machine are going to be different due to variability in device type, imaging protocols including the field of view, and exposure parameters (mA, Kv). Due to this, the effective dose, which is used to estimate the risk in humans, from CBCT may be similar or greater as compared to the conventional intraoral full-mouth radiographic survey or panoramic image. The effective dose from medical CT is more substantial to the patient as compared to CBCT. Variability in doses has been found in the previous printed literature. Therefore, effective doses for the various modalities are listed in (Tables 1.2 and 1.3) from two different sources.

Ludlow et al. [16] published with meta-analysis of the published data analyzing the effective doses of nine dental CBCT units. Dentists prefer high-resolution images and high signal-to-noise ratios. This technique is associated with higher radiation dose. In many clinical situations, such absolute high-quality image may

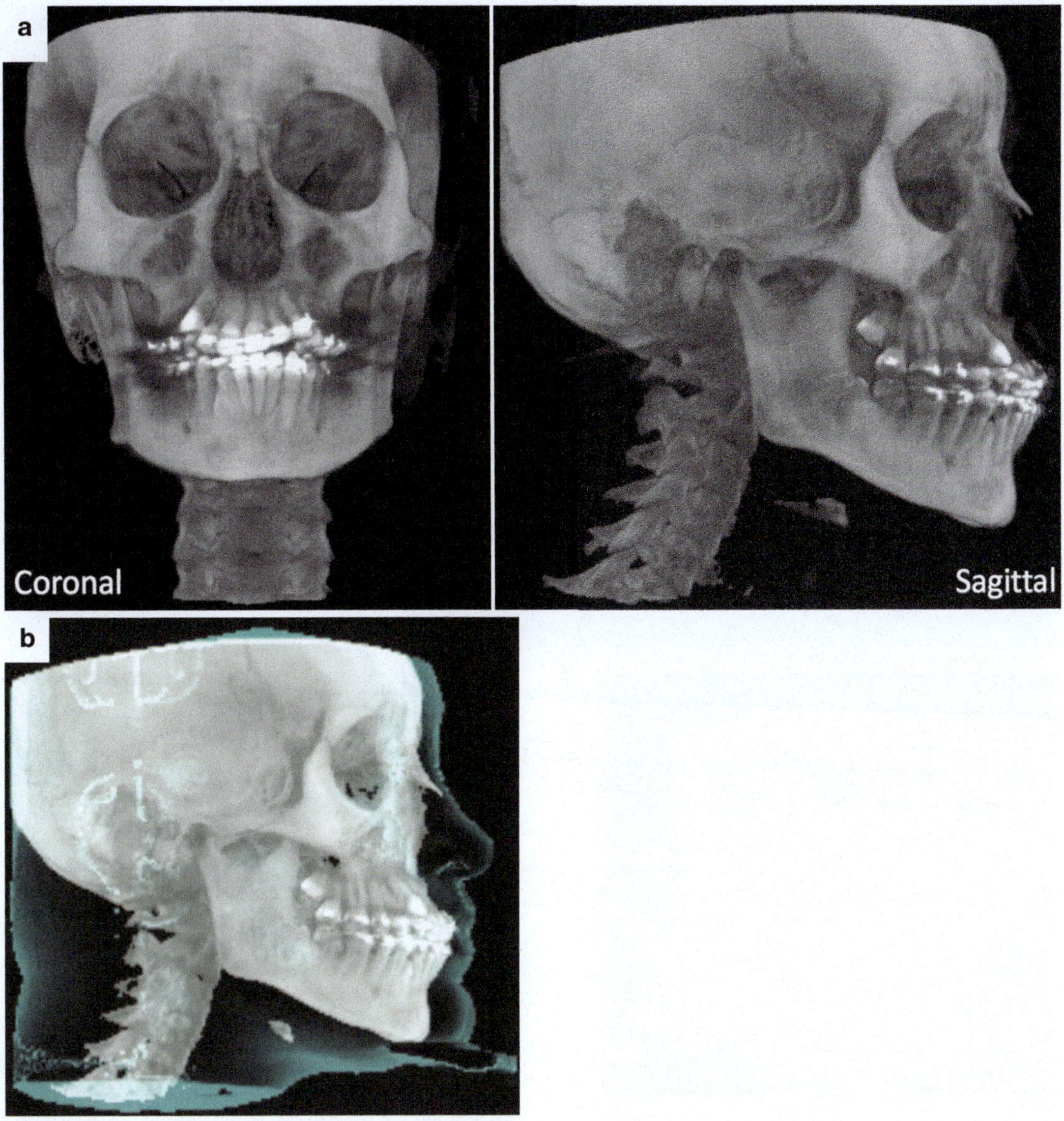

Fig. 1.3 (**a**) Hard tissue renderings from CBCT data are shown. (**b**) Example of another type of rendering with soft tissue is shown

not be needed as lower dose data may provide sufficient information for diagnostic tasks (Fig. 1.10).

1.11 Artifacts on CBCT Images

An artifact can be described as any distortion in the image which is unrelated to the patient. These are volumetric data set errors and do not represent the corresponding region of the patient's tissues. It has been reported that CBCT images inherently have more artifacts due to the use of lower energy spectrum and beam geometry as compared to the medical CT units [8]. Many artifacts result from discrepancies between the physical imaging process and mathematical assumptions in the data reconstruction algorithms. One has to be careful during the interpretation process. The noise level is found to

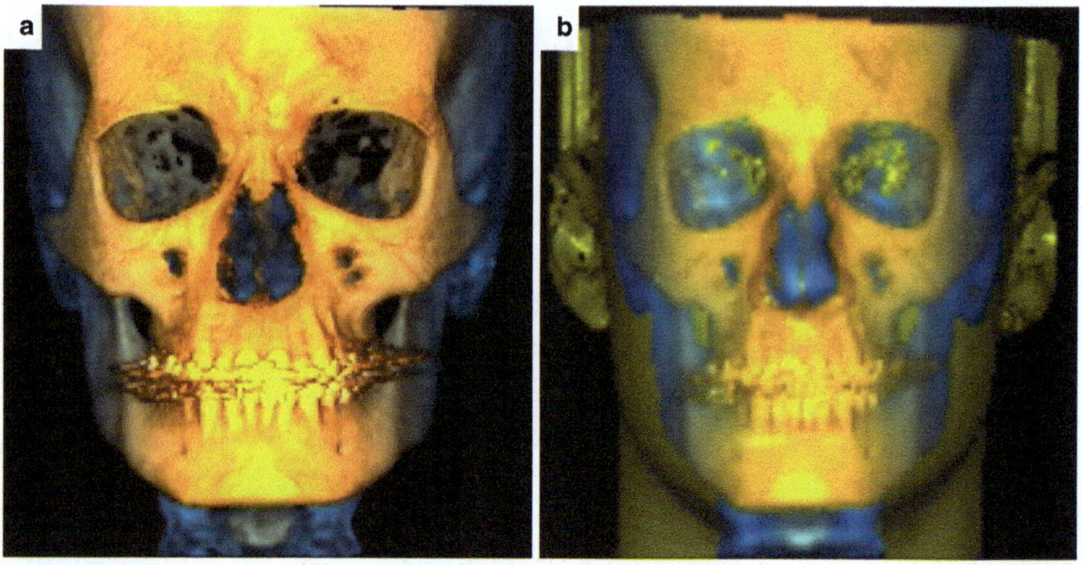

Fig. 1.4 (**a**) Enhanced depth 3-D model. (**b**) Enhanced depth with soft tissue

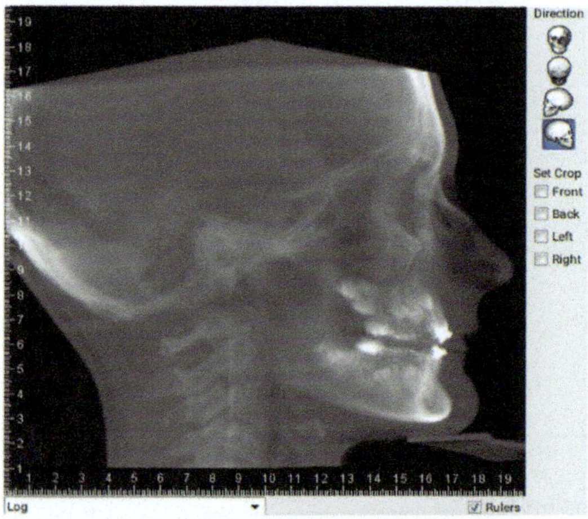

Fig. 1.5 Virtual cephalometric image from a CBCT unit

be more in CBCT. Artifacts or errors on CBCT images are often seen as dark or white areas or streaks. The following artifacts may be encountered in the CBCT images:

Beam hardening artifact occurs when there is more attenuation in the center of the structure as compared to the edges. Structures located adjacent to high-density metallic objects like amalgam restorations, dental implants, and orthodontic brackets may appear missing or burnt-out. Streak artifacts are commonly seen around the amalgam restorations and cast metal restorations, as these metals have a higher atomic number. Radiographically, they appear as dark or white streaks (Fig. 1.11).

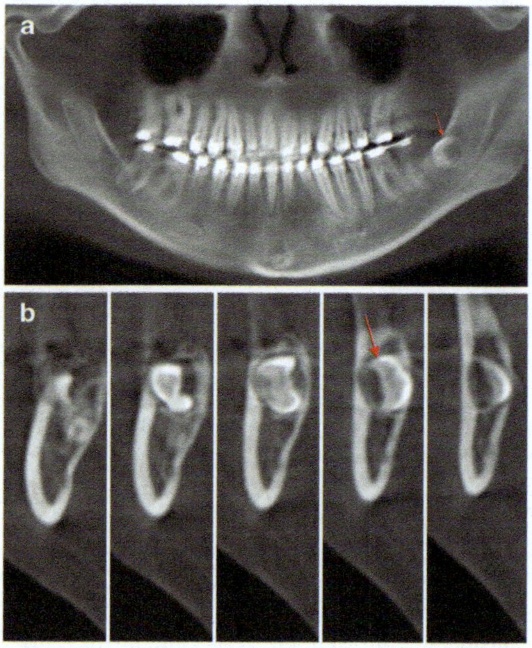

Fig. 1.6 Reconstructed panoramic image (**a**) and cross-sectional images (**b**) from CBCT data showing the relationships of the forming tooth with the surrounding structures

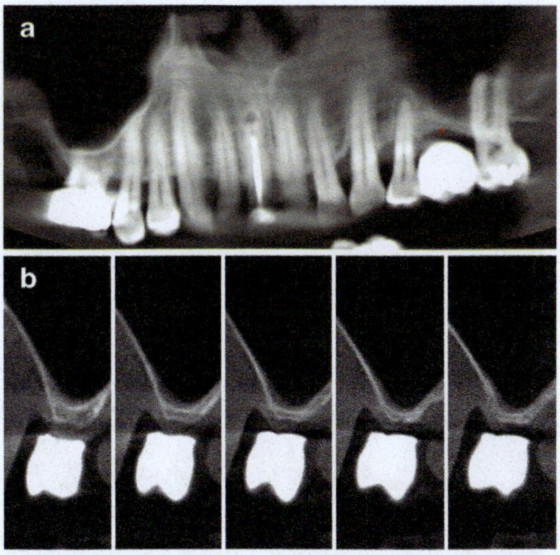

Fig. 1.7 CBCT maxillary images produced for dental implant treatment planning with an opaque radiographic marker in the left posterior maxilla. (**a**) Reconstructed panoramic image. (**b**) Cross-sectional images. Pneumatization of the maxillary sinus is also visible in the region of interest (red arrow)

Scatter may also cause streak artifacts. Scatter occurs when X-ray photons are diffracted from the original path after interaction with the matter or patient's tissues (Table 1.4).

Partial volume averaging artifacts are seen when an image voxel contains more than one type of tissue. Thus, after attenuation, the gray value is not representative of any specific tissue. It appears as hazy or blurred tissue outlines. This artifact is more associated with larger-sized voxels, where the object being imaged is smaller and the voxel is larger in size. If both hard and soft tissues are included in one voxel, the end result will be an average of brightness values of different tissues. The region may have a "step appearance," and the displayed pixel may both be representative of either tissue.

Ring artifacts are seen on the displayed images due to a faculty pixel in the detector.

Motion artifact is usually caused by the movement of the patient during the scanning process. Movement of the patient can be easily recognized by double margins or blurry cortical outlines (Fig. 1.12). In case of severe movement, the resultant scan is of no diagnostic value and must be acquired again. To reduce the chance of patient movement during the exposure, the operator should educate the patient about the procedure, use head-restraining devices available with the CBCT unit, and use shorter exposure time.

Scanner-related artifacts have been described as round or ring artifacts, which result from problems associated with improper scanner calibration. Example of another type of machine calibration-related artifact is shown as black line in the middle of the scan (Fig. 1.13). After discussion with the manufacturer, it was concluded that either the detector or beam was off-centered. Realignment and calibration was needed as the corrective action. Some of the machines with smaller detector operate by acquiring two scans to cover the full skull, and thus the data sets have to be overlapped correctly and "stitched" by the computer program. Imperfections in overlap before stitching can also appear as stitching artifacts that show a step formation (Fig. 1.14).

1.12 Training for Using CBCT

The practicing dentist should be aware of the state laws and requirements before purchasing the unit. It is best to check with the governing agency directly to rule out any restrictions that may be in place for purchasing the CBCT unit. In the United States, certain states do require a "certificate of need." Legal ramifications should also be understood. Radiographic images may be acquired in a dental office or independent dental radiology imaging centers. In the United States, any licensed dentist may own and operate a CBCT unit. Non-dentists, who own an imaging center, can also own and operate these units. In both cases, adequate training is required, which is typically provided by the manufacturer [17].

Before operating the unit for the patient care, the training is critical for the whole dental office. After the installation of the unit, the manufac-

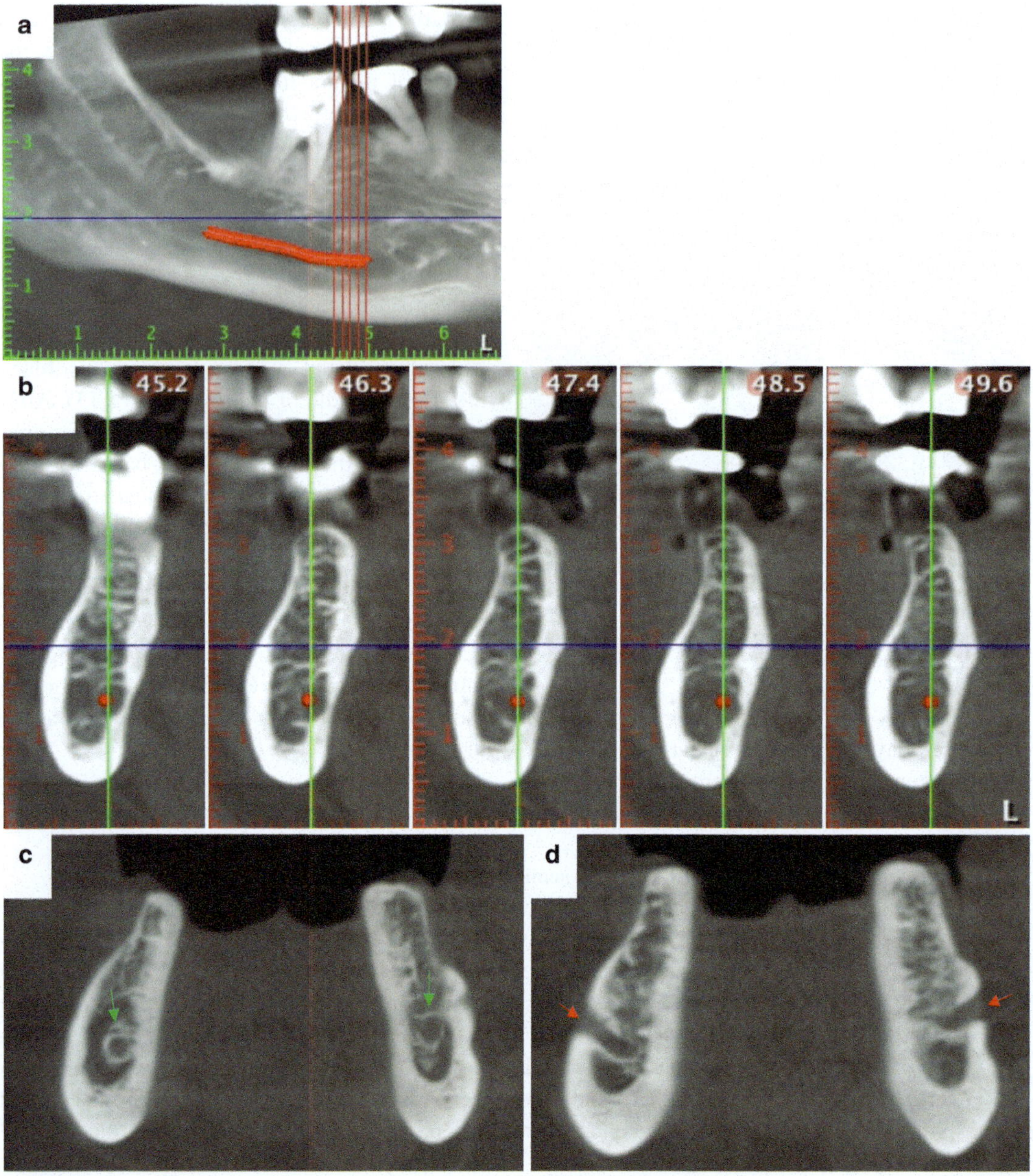

Fig. 1.8 CBCT images produced for dental implant treatment planning. (**a**) reconstructed panoramic and (**b**) cross-sectional views. Inferior alveolar nerve marking is also shown by red line. (**c** and **d**) Coronal CBCT views show corticated inferior alveolar canal (green arrows) and the mental foramen (red arrows)

turer is responsible for hands-on training on how to use the machines, proper patient positioning, and the various modes of image capture. This will enable the operator and the referring dentist to acquire the best quality images to fulfill the diagnostic aim and also will keep the radiation dose as low as possible for the patient.

Referring Dentist: The referring dentist has to clearly indicate why the CBCT images are needed and justify the use. Other conventional

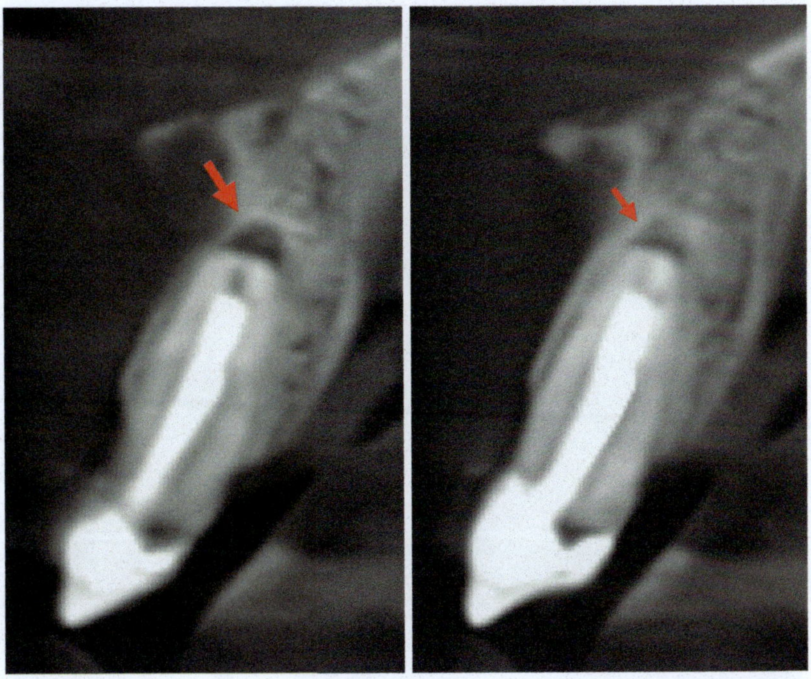

Fig. 1.9 Cross-sectional views show root resorption with periapical low-density lesion with corticated borders (red arrow)

Table 1.2 Effective doses from CBCT systems [5]

Field of view (FOV)	Effective dose range
Small and medium FOV Volumes <10 cm	11–674 μSv
Large FOV Volumes >10 cm	30–1073 μSv

imaging alternates may be considered before prescribing CBCT. It is important to select the appropriate cases and understand the associated risks with the use of radiation. The region of interest or field of view must be identified before exposure. The dentist must be trained in the proper use of the viewing software program for manipulating the CBCT data. Different software tools are provided to help the clinician in the process of visualization of the region of interest in different enhancement modes.

He or she should have sufficient knowledge of anatomy, variation of anatomy, incidental findings, and other pathologic conditions, in order to correctly interpret the CBCT data. The referring dentist is responsible for evaluation of complete CBCT data set to rule out abnormalities. Finding of the significance should be reported. Many practitioners choose to send CBCT for interpretation by an oral and maxillofacial radiologist.

Table 1.3 Effective doses from different imaging modalities [4]

Modality	Effective dose range (μSv)	Equivalent background exposure (days)
Multislice CT (conventional head protocol)	860–1500	101–177
CBCT		
Large FOV Medium FOV Smaller FOV	68–1073 45–860 19–652	8–126 5–101 2–77
Panoramic image	9–24	1–3
FMX Round collimation:		
FMX: CCD sensor (estimated) FMX: PSP plates/F-speed film	85 171	10 20
Full-mouth intraoral survey FMX Rectangular collimation:		
FMX: CCD sensor (estimated) FMX: PSP plates or F-speed film	17 35	2 4
Cephalometric	2–6	0.3–0.7

FMX full-mouth intraoral survey, *CCD* charge-coupled device, *PSP* photostimulable phosphor plate, *FOV* field of view

Source of information [4]

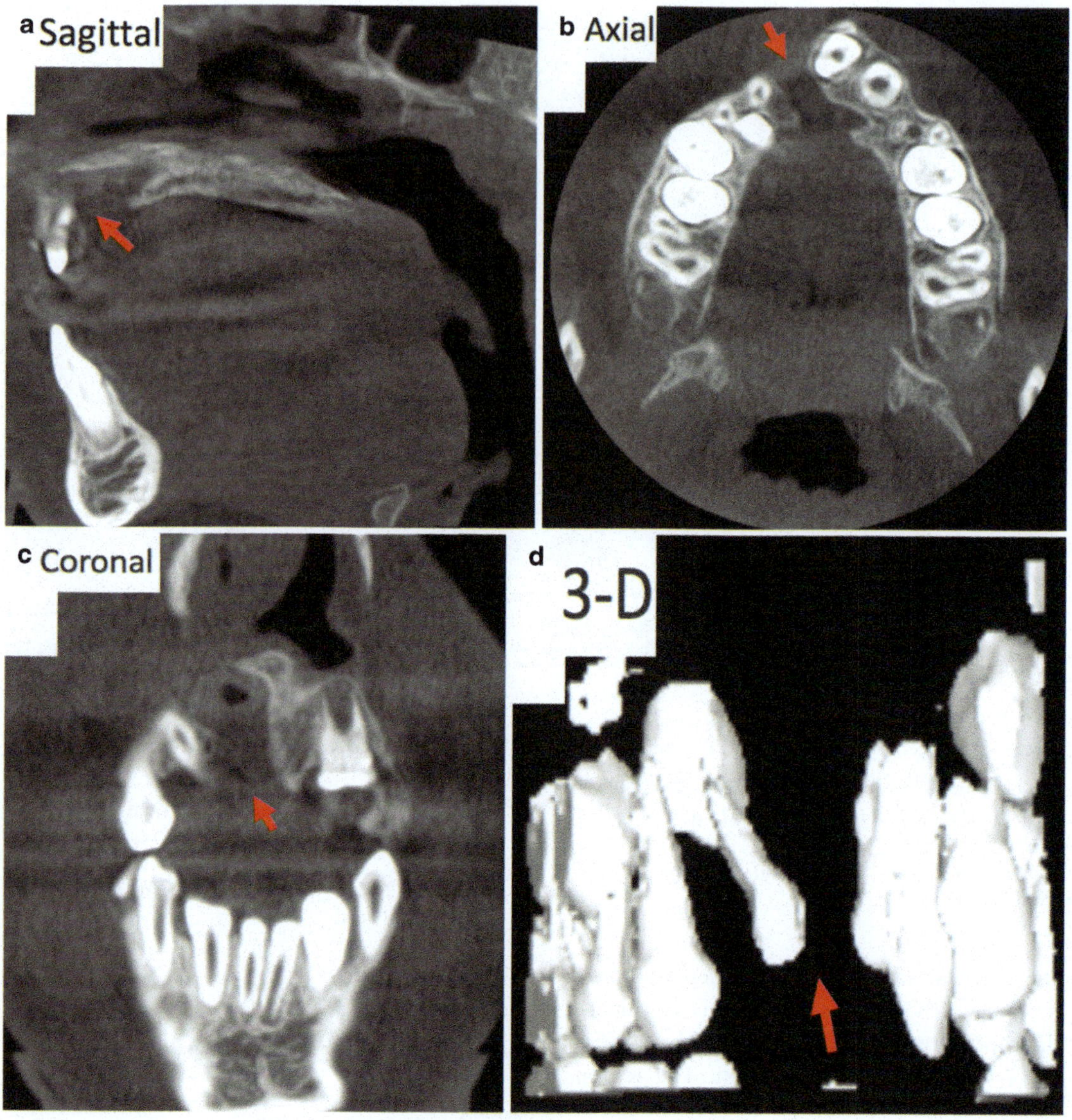

Fig. 1.10 Ultralow-dose CBCT scan: Images appear slightly grainy. CBCT images also show cleft palate (red arrows)

This is similar to other procedures, where a procedure may be referred to a specialist if the general dentist or the specialist does not feel competent to perform the task or to seek a second opinion.

Machine Operator: It is the responsibility of the referring dentist to convey the region of interest and the diagnostic aim to the person exposing the patient or the operator of the CBCT unit. A written prescription from the referring dentist is advised for clear directions to the operator and documentation. This will result in dose optimization and will reduce or eliminate the need for retaking the images.

Quality assurance is very important. If any errors are noted on the images, corrective actions must be taken. It is recommended that the machines should be calibrated annually by the factor certified maintenance staff to maintain the proper functioning of the machine and the image

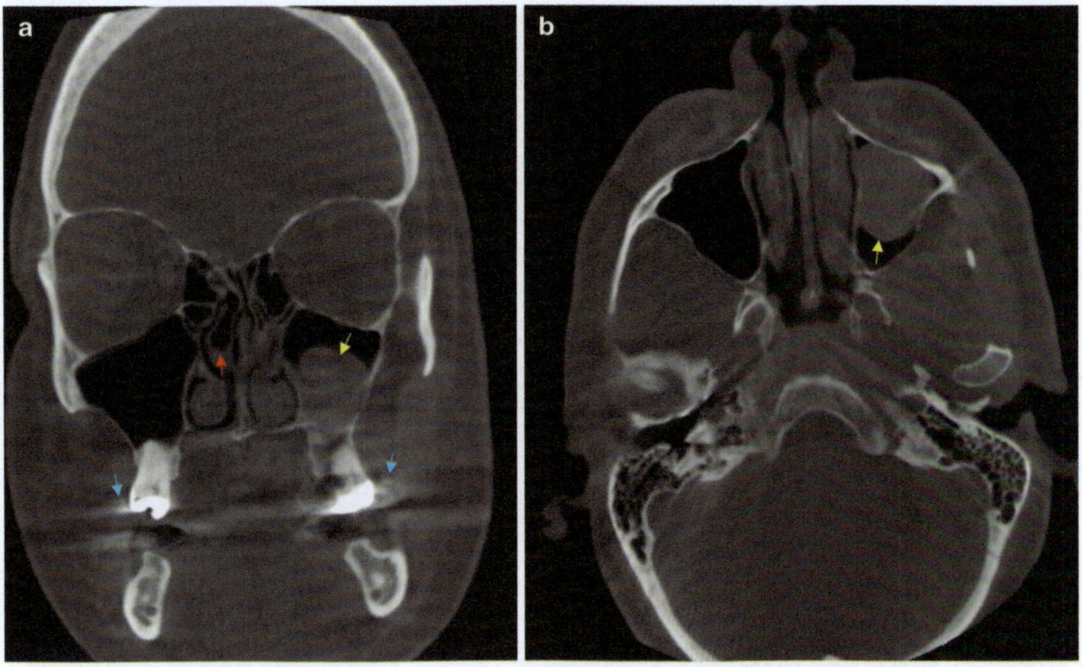

Fig. 1.11 Scatter artifacts (blue arrows) due to metal are seen on large field of view CBCT image. Other findings include unilateral concha bullosa (red arrow) and mucus retention pseudocyst (yellow arrow) on the floor of the left maxillary sinus

Table 1.4 Indication for different CBCT fields of view (FOVs)

Small FOV	1–2 impacted teeth. Localized area for bone quality and quantity assessment. Morphology of tooth crown and root, localized supernumerary teeth. Single quadrant
Medium FOV	Partial or complete maxillary or mandibular arches. Limited TMJ region. Partial view of the maxillary sinuses
Large FOV	Whole head for both skeletal and dental structures. Orthognathic surgery. Craniofacial anomalies. Facial deformities

quality. Therefore, training should be done for the whole dental team. Dawood et al. [1] stated that the comprehensive training should include training from the manufacturer by a training specialist on how to operate the particular CBCT machine, an update on radiation risks and imaging pitfalls, and the selection criteria. Training should also include the use of the viewing software tools, to help in the interpretation process of cross-sectional and three-dimensional CBCT images.

1.13 Summary

CBCT technology has come a long way. Technically, the CBCT technology has evolved momentously over the past decade and thus is considered indispensable for many diagnostic scenarios. The image quality has improved tremendously; the cost of the CBCT machines and the scanning time has decreased. Various companies have developed task-specific software tools which are aimed for orthodontists and oral surgeons. As the technology is being integrated in the dental practice more and more, it is important that the principles of patient selection for imaging and radiation protection guidelines must be followed.

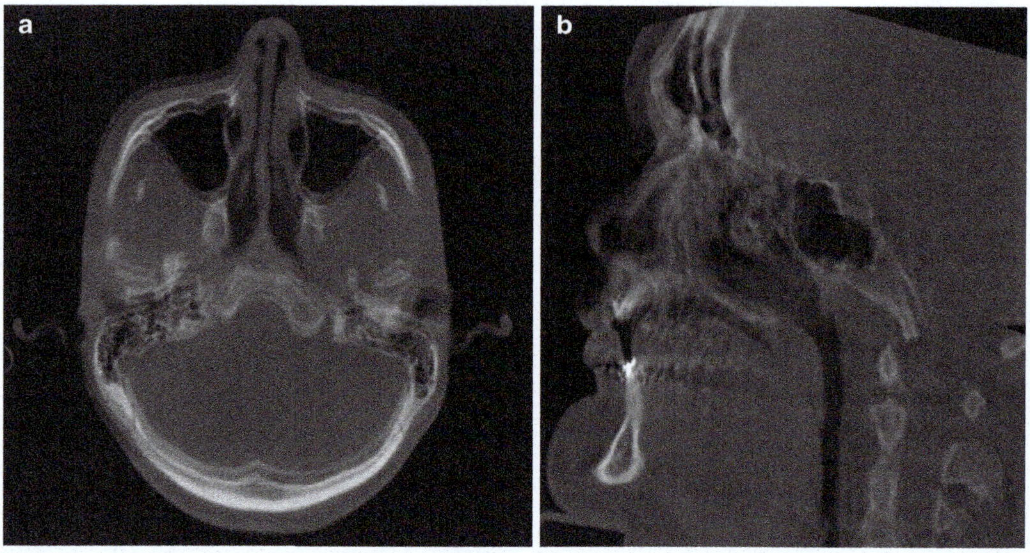

Fig. 1.12 Bony outlines appear double or blurry due to movement artifacts

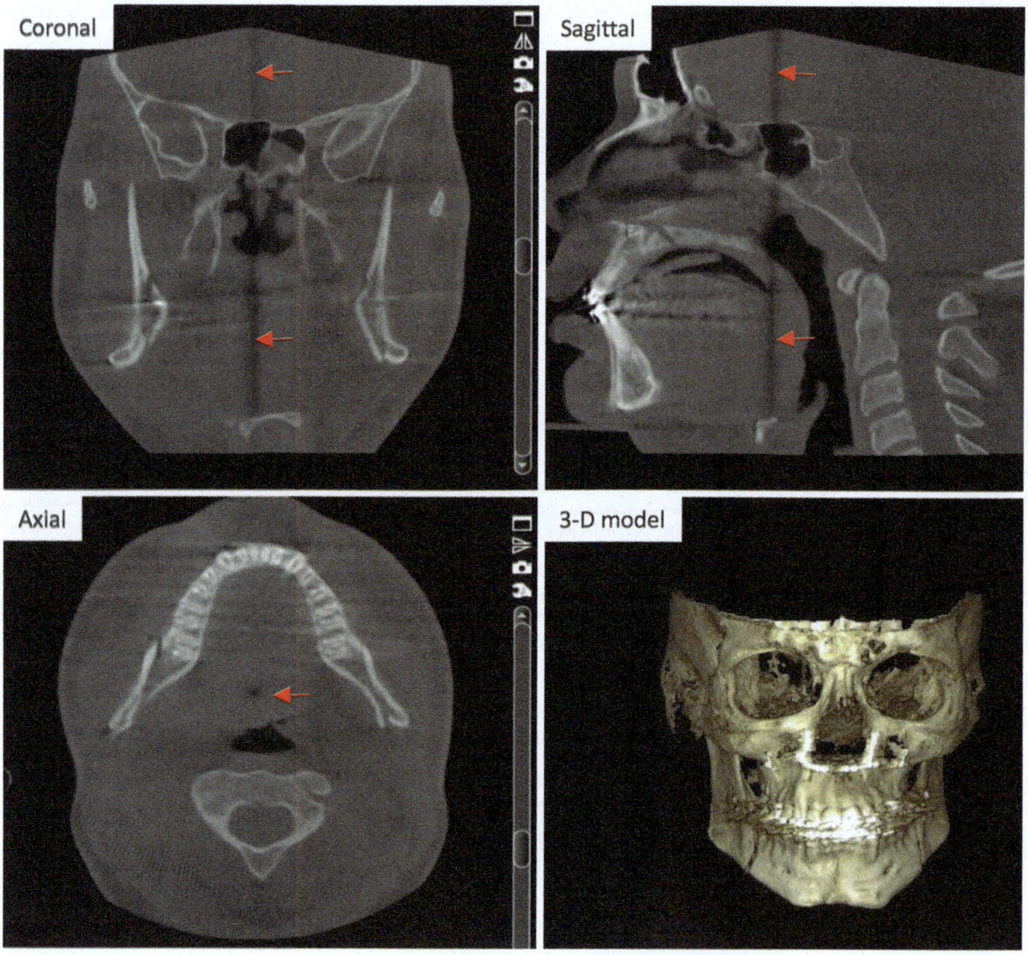

Fig. 1.13 Images show a solid black line in the midline (red arrows). It was reported as calibration error of the machine

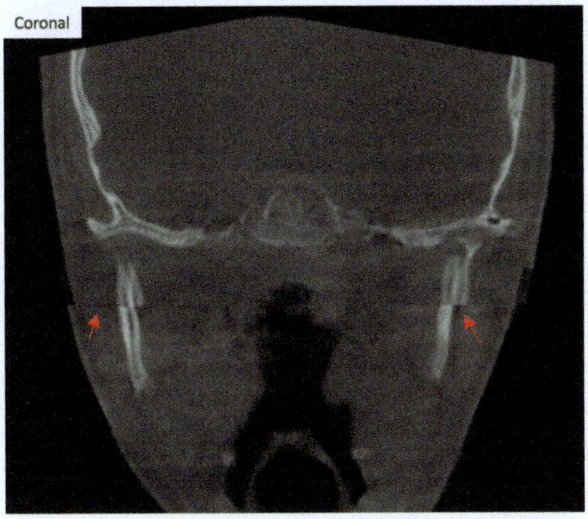

Fig. 1.14 Coronal image shows irregularity of the bony margins at the mid-ramus level. This error was attributed to problems in alignment of the two volumes before stitching

References

1. Dawood A, Patel S, Brown J. Cone beam CT in dental practice. Br Dent J. 2009;207:23–8. https://doi.org/10.1038/sj.bdj.2009.560.
2. Zoller JE, Neugebauer J. Cone-beam volumetric imaging in dental, oral and maxillofacial medicine: fundamentals, diagnostics and treatment planning. Batavia , IL: Quintessence Publishing Co. Ltd; 2008.
3. Sukovic P. Cone beam computed tomography in craniofacial imaging. Orthod Craniofac Res. 2003;6:31–6. https://doi.org/10.1034/j.1600-0544.2003.259.x.
4. Jaffray DA, Siewerdsen JH. Cone-beam computed tomography with a flat-panel imager: initial performance characterization. Med Phys. 2000;27(6):1311–23.
5. Tamimi DF, Koenig A-E, Rathi S, Bajunaid S, Angle C. Specialty imaging TM: dental implants. Salt Lake City, UT: Amirsys Publishing; 2014.
6. Scarfe WC, Farman AG. What is cone-beam CT and how does it work? Dent Clin N Am. 2008;52:707–30.
7. SEDENTEXCT Guideline Development Panel. Radiation protection: cone beam CT for dental and maxillofacial radiology provisional guidelines 2009. 1. 2009. http://www.sedentexct.eu. Accessed 10 Jun 2013.
8. White SC, Pharoah MJ. Oral radiology: principles and interpretation. 7th ed; 2014.
9. SEDENTEX CT. Radiation protection: cone beam CT for dental and maxillofacial radiology. Evidence based guidelines (v2·0 final). 2011. http://www.sedentexct.eu/files/guidelines_final.pdf
10. Kau CH, Pan P, Gallerano RL, English JD. A novel 3D classification system for canine impactions-the KPG index. Int J Med Robot. 2009;5:291–6.
11. Liu D, Zhang W, Zhang Z, Wu Y, Ma X. Localization of impacted maxillary canines and observation of adjacent incisor resorption with cone-beam computed tomography. Oral Surg Oral Med Oral Pathol Oral Radiol Endod. 2008;105:91–8.
12. Oberoi S, Chigurupati R, Gill P, Hoffman WY, Vargervik K. Volumetric assessment of secondary alveolar bone grafting using cone beam computed tomography. Cleft Palate Craniofac J. 2009;46:503–11.
13. JB C, JD S, RS C, Mercer JE. Applications of cone-beam computed tomography in oral and maxillofacial surgery: an overview of published indications and clinical usage in United States Academic Centers and Oral and Maxillofacial Surgery practices. J Oral Maxillofac Surg. 2016;74(4):668–79. https://doi.org/10.1016/j.joms.2015.10.018.
14. American Academy of Oral and Maxillofacial Radiology. Clinical recommendations regarding use of cone beam computed tomography in orthodontics. Position statement by the American Academy of Oral and Maxillofacial Radiology. Am Acad Oral Maxillofac Radiol. 2013;116(2):661.
15. Nervina J. Cone beam computed tomography use in orthodontics. Aust Dent J. 2012;57:95–102. https://doi.org/10.1111/j.1834-7819.2011.01662.x.
16. Ludlow JB, Timothy R, Walker C, Hunter R, Benavides E, Samuelson DB, et al. Effective dose of dental CBCT—a meta analysis of published data and additional data for nine CBCT units. Dentomaxillofac Radiol. 2015;44:20140197. https://doi.org/10.1259/dmfr.20140197.
17. Friedland B. Conebeam computed tomography: legal considerations. Alpha Omegan. 2010;103(2):57–61.

Current Applications

2

Farah Masood, Onur Kadioglu,
and G. Fräns Currier

Abstract

With the introduction of cone beam computed tomography (CBCT), the practice of dentistry has taken a new approach. Before the emergence of this technology, most of the dental professionals depended on the conventional two-dimensional (2-D) radiographic imaging for treatment planning and evaluation. Previously, the multi-detector computed tomography or medical CT scanners were utilized for assessment of pathology and trauma cases in dentistry. CBCT technology has found its way into the dental offices and offers many advantages and specific clinical applications for both specialist and general dentists. CBCT image quality is superior as compared to 2-D as structures can be viewed without superimposition and distortion, in three dimensions.

F. Masood (✉)
Division of Oral Diagnosis and Radiology,
Department of Diagnostic and Preventive Sciences,
University of Oklahoma Health Sciences Center
College of Dentistry, Oklahoma City, OK, USA
e-mail: Farah-Masood@ouhsc.edu

O. Kadioglu · G. F. Currier
Division of Orthodontics, Department of
Developmental Sciences, University of Oklahoma
Health Sciences Center College of Dentistry,
Oklahoma City, OK, USA
e-mail: onur-kadioglu@uohsc.edu;
frans-currier@ouhsc.edu

2.1 Utilization of Cone Beam Computed Tomography in Orthodontics and Oral Surgery

Since the discovery of X-rays, without a doubt, the conventional two-dimensional (2-D) radiographic imaging has remained an integral part of the diagnostic process in dentistry and also in its specialties like orthodontics, oral surgery, periodontics, and implantology. For the practice, depending on the need and the treatment stage of the patient, the appropriate radiographic imaging modality should be selected and used, if there is adequate reason to believe that this exposure will effectively aid the clinician in the initial diagnosis, better treatment planning, on-going evaluation, and also with the posttreatment assessment of the cases. Imaging serves as an important adjunctive tool and provides baseline information about the patient. Pretreatment understanding of the relationships of underlying osseous structures, soft tissues, and dentoalveolar components is essential in order to form a treatment plan of the various craniofacial abnormalities, malocclusion, and other dental anomalies. During the treatment phase, the follow-up imaging allows to evaluate the effectiveness of the treatment administered. After the completion of the treatment, with appropriate radiographic imaging, the clinician is able to assess the outcome.

O. Kadioglu, G. F. Currier (eds.), *Craniofacial 3D Imaging*,
https://doi.org/10.1007/978-3-030-00722-5_2

With the introduction of cone beam computed tomography (CBCT), the clinicians were faced with new challenges in terms of usage, effectiveness, benefits, and financial issues. It was especially difficult as initially no evidence-based systemic guidelines or position papers were available.

The American Academy of Oral and Maxillofacial Radiology (AAOMR) [1] published a position paper in 2013. Both board-certified orthodontists and oral and maxillofacial radiologists contributed in development of this paper to establish orthodontic-specific clinical guidelines for practice. According to this published position paper, the utilization of CBCT in different phases of orthodontic treatment should be justified on an individual basis and should be based on clinical signs and presentation of the patient. The panel established that "there was no clear indication to support the routine use of ionizing radiation in standard orthodontic diagnosis and treatment planning, including the use of CBCT." This position paper by AAOMR supported the position of the American Dental Association Council of Scientific Affairs [2] in the selection of CBCT imaging, which suggested that imaging should be based on clinical examination and must be decided on the individual patient needs.

Hodges et al. [3] evaluated the impact of CBCT on the orthodontic diagnosis and treatment planning. They reported that changes in the diagnosis and treatment plan varied widely with patient characteristics. The results supported obtaining a CBCT scan before orthodontic diagnosis and treatment planning when a patient had an unerupted tooth with delayed eruption or a questionable location, suspected severe root resorption, or a severe skeletal discrepancy. They also concluded that "CBCT scans should be ordered only when there was clear, specific, individual clinical justification." No advantage was found in terms of changes in treatment plan for patients when the reason for obtaining a CBCT scan was to assess the temporomandibular joint abnormalities or airway analysis. However, the participating orthodontists in the study who used the CBCT imaging frequently in practice were more confident in the diagnostic process and in forming a treatment plan after viewing the CBCT scans during the study [3].

This new 3-D technology made it possible, in a dental office setting, to have superior quality structural images in three planes (axial, sagittal, and coronal) without superimposition and with a radiation dose much less then medical CT units at a lesser expenditure.

According to some practicing orthodontists, most of the orthodontic practices are no longer using full-mouth intraoral radiographic surveys. Even the conventional extraoral posterior-anterior cephalometric views are not made as CBCT provides all the needed information required for outcome assessments for orthodontic and oral surgery procedures. Thus, it has been suggested that conventional 2-D images may not be needed, if CBCT imaging is available and the radiation dose of CBCT is similar to conventional imaging, as CBCT provides more in-depth information.

For the needs of the oral surgery and the orthodontic procedures, typically CBCT machines with a larger sensor or image detector are used to capture the craniofacial region. A smaller sensor or a more collimated smaller region of interest is sometimes utilized for localized problems such as impacted teeth. Most common uses of CBCT would include diagnosis and treatment planning, skeletal evaluation, tooth localization for impacted teeth, assessment of root shape and condition in suspected external apical root resorption, evaluation of alveolar bone thicknesses, treatment planning for alveolar bone grafting in cleft lip and palate, pre-orthognathic surgery, and evaluation of airway patency and size.

2.2 Tooth Impactions

Most common impacted teeth are third molars and permanent maxillary canines. CBCT imaging is often done to localize the posi-

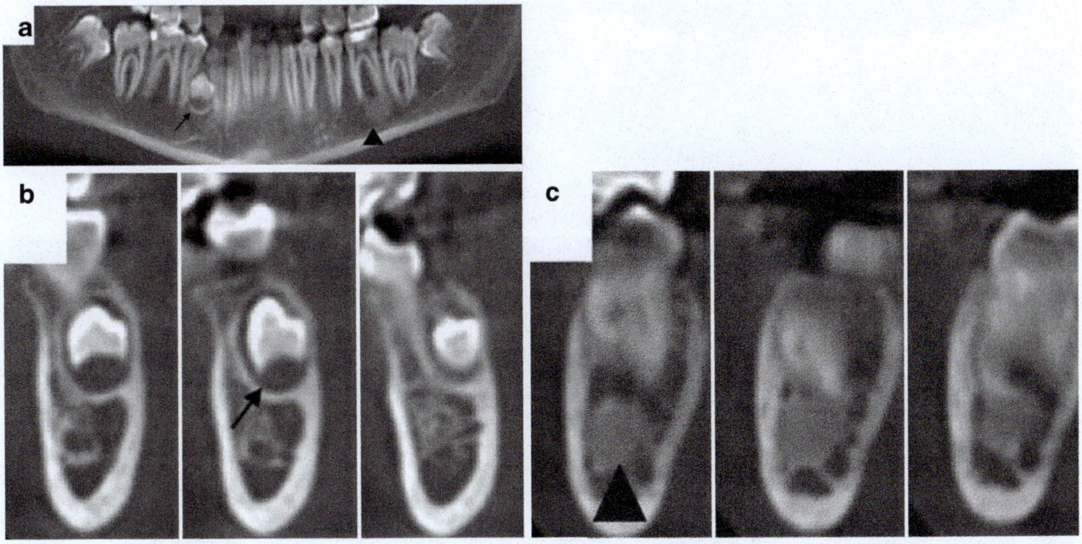

Fig. 2.1 CBCT-reconstructed panoramic image (**a**) and cross-sectional views (**b** and **c**). Impacted tooth is noted on the right mandibular premolar region (black arrow). Idiopathic osteosclerosis also noted periapical to tooth # 19 (arrowhead)

tion, angulation, and effect of the impacted teeth on the surrounding structures, as the technology has been shown to improve diagnosis and contribute in treatment modifications in such cases in a significant number of subjects [4, 5].

CBCT is considered very helpful in planning surgical access and assessing the direction of extrusion of the impacted canines in the oral cavity and provides a 3-D insight for proximity of these impacted canines to adjacent teeth and structures, extent of resorption of adjacent teeth, size of the follicular space, and the presence of pathology [6, 7].

Visualization of 3-D root structure of a tooth with CBCT is substantially superior as compared to the conventional 2-D radiographic imaging (Fig. 2.1). It has been suggested that the small field of view may be used for CBCT imaging of impacted maxillary canines if the canine inclination in the arch on a conventional 2-D panoramic radiograph exceeds 30° relative to a perpendicular midline and also in cases where adjacent root resorption and/or dilaceration of the root is in question [8].

2.3 Osseous or Bony Evaluation

Condition of the buccal and lingual alveolar bone and thickness is determined by the dentoalveolar anatomy prior to start of the treatment and by the bone's morphology and adaptability during tooth movement during the treatment and its morphology following the final positioning of teeth after completion of the process. Kapila et al. [9] described alveolar boundary conditions in orthodontics, which included the depth, height, and morphology of alveolar bone relative to tooth root dimensions, angulation, and spatial position. They stated that for orthodontic tooth movements, alveolar boundary conditions can be considered dynamic and determined by the patient's pretreatment bone condition and gingival biotype as well as bone physiology (see Chap. 10 and 11).

Alveolar bone is not static in shape as remodeling of the alveolar bone occurs, without which the orthodontic tooth movement would not be possible. However, use of excessive orthodontic forces to a tooth can affect alveolar boundary conditions unfavorably and

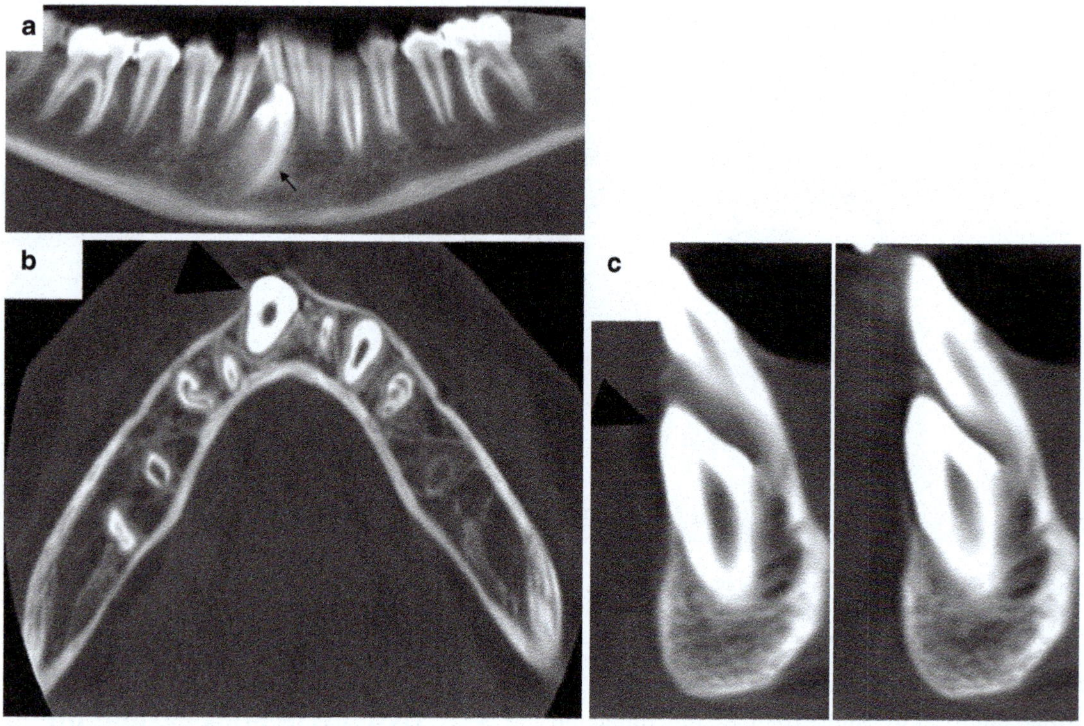

Fig. 2.2 CBCT-reconstructed panoramic image (**a**) and axial (**b**) and cross-sectional view (**c**) images show buccally located impacted mandibular right canine (black arrow). Thinning of the buccal cortex is evident (black arrowhead)

may result in dehiscences and fenestrations. The cross-sectional views from the CBCT are very useful in verifying the thickness of the buccal and lingual cortex (Fig. 2.2) not visualized on conventional 2-D radiographic images, both before and after the treatment.

CBCT can also be a useful tool for evaluation of the bone quantity, quality, the underlying trabecular bone pattern, and thus stability of the bone [10]. Temporary Anchorage Devices (TADs) have been used in orthodontic procedures to provide a stable anchor for the application of orthodontic forces. TADs can be placed nearly anywhere in the oral cavity, but it is important that there is no impingement on the complex surrounding anatomical structures, such as roots or vessels and nerves. CBCT may be used to determine the optimal site and treatment plan for the placement of TADs, as the proximity and relationship to the surrounding structures such as roots, nasal fossa, maxillary sinuses, and vasculature can be visualized beforehand to avoid complications.

2.4 Orthognathic Surgery

CBCT 3-D volumetric reconstructions provide detailed information for treatment planning of orthognathic surgery (Fig. 2.3). Volumetric analysis can help predict the procedure. CBCT data can be used to create stereolithic models of the area of interest as well (Fig. 2.4). One cannot emphasize enough the usefulness of this 3-D technology in orthognathic surgery to visualize the relationship between hard and soft tissues [11]. To a large extent, CBCT has replaced lateral cephalometric imaging for diagnosing skeletal and dental deformities like hemifacial macrosomia and Treacher Collins syndrome.

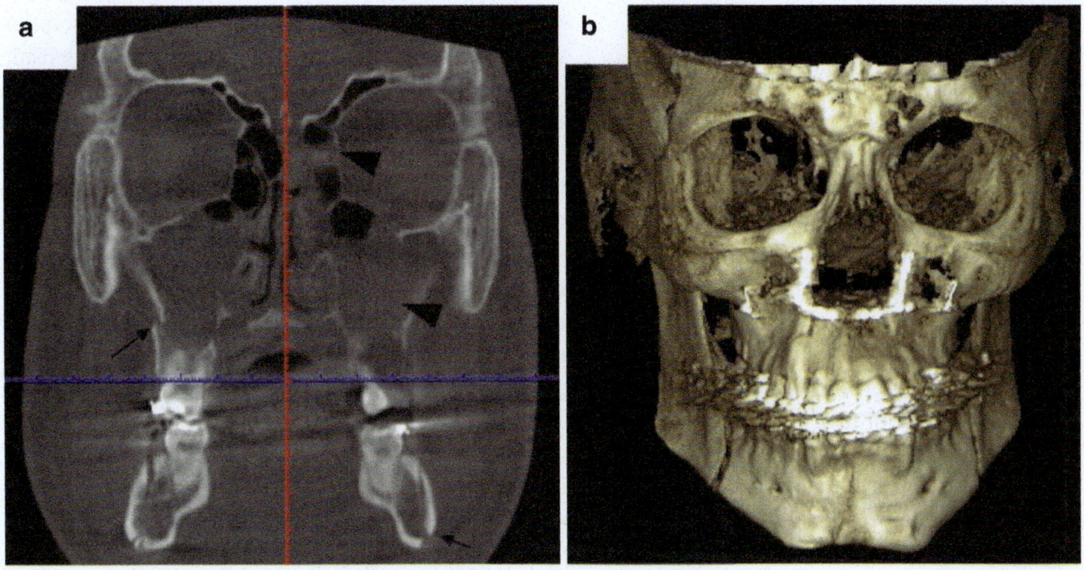

Fig. 2.3 (**a**) Large field-of-view CBCT coronal image (**a**) and a 3-D view (**b**). Changes after the orthognathic surgery are seen. Images show disruption of the bone due to surgery (black arrow) and opacification of the maxillary and ethmoid sinuses (black arrowhead). Surgical pins are also noted in the maxilla in the 3-D view

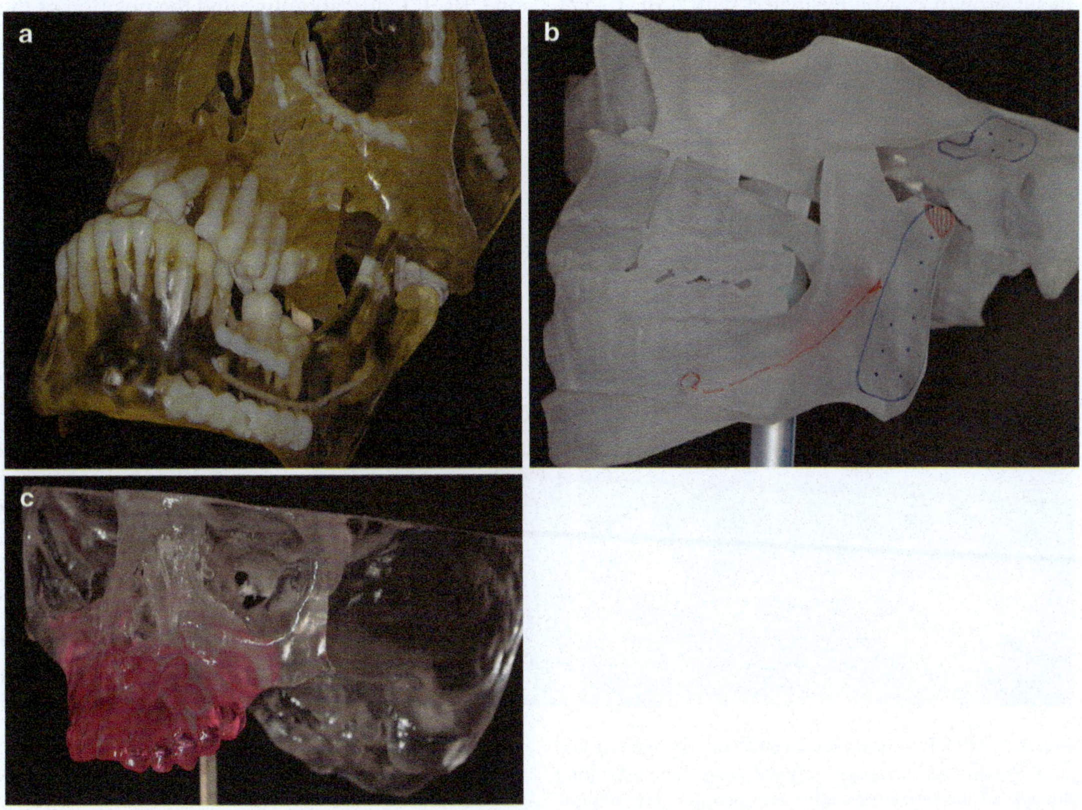

Fig. 2.4 Three different stereolithic models (**a–c**) are shown, which were fabricated to simulate the surgical treatment planning. Courtesy Dr. Kevin Smith and Dr. Steven Sullivan

2.4.1 Cleft Lip and Palate

This anomaly is commonly encountered and adversely effects the involved human beings. CBCT provides unique useful information for patients with cleft lip and palate. It is very useful in pretreatment and posttreatment planning phases, providing information about the cleft defect site, eruption status and position of the canines in the involved sites, and pre- and post-graft bone width and height. Timing of alveolar cleft repair is often determined based on conventional panoramic and occlusal imaging. In such cases, CBCT allows better evaluation of dental age, arch segment positioning, and cleft size compared with traditional radiography. Volumetric analysis with CBCT provides better prediction in terms of the cleft defect morphology (Fig. 2.5) as well as the volume of graft material needed for repair. After the surgery, the stability of the arch after grafting, the quality of the bone graft over time, and the effect on overall facial growth can be evaluated with CBCT [12]. Other uses include evaluation of impacted teeth for potential complications such as root resorption of adjacent roots. With CBCT the relationship between the impacted and supernumerary teeth and the surrounding structures such as the walls of the maxillary sinuses, cortical borders of the inferior alveolar canals, and mandibular cortices can be studied before the actual procedures to avoid potential postsurgical complications. Surgical prediction and treatment planning have become easier. However, it is important to understand the data manipulation, software tools along with normal anatomy, and anatomical variations for maximum treatment planning and surgical accuracy (see Chap. 13).

2.4.2 Temporomandibular Joints (TMJ)

If included in the field of view, TMJ region can be visualized in detail, without superimposition on CBCT. Cortical outline and the position of the condyles, glenoid fossae, articular eminence, and

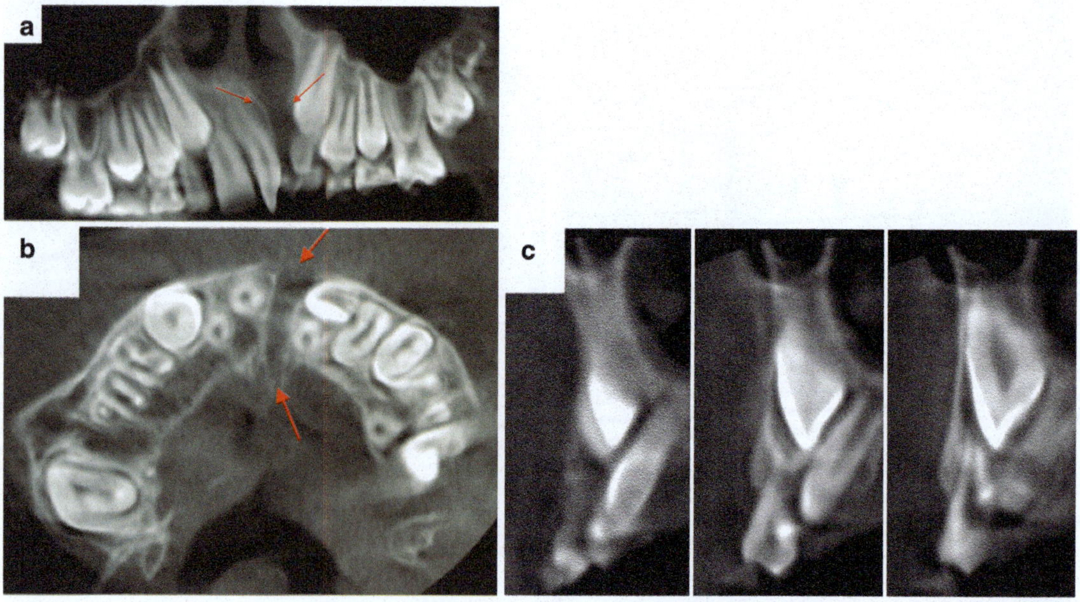

Fig. 2.5 CBCT-reconstructed panoramic image (**a**) and axial (**b**) and sagittal cross-sectional views (**c**) made along the axis of the impacted canine are shown. Cleft palate is visible on the left anterior maxilla on reconstructed panoramic image and axial image (red arrows). Buccal-lingual position of the impacted canine within the arch is visible on the images

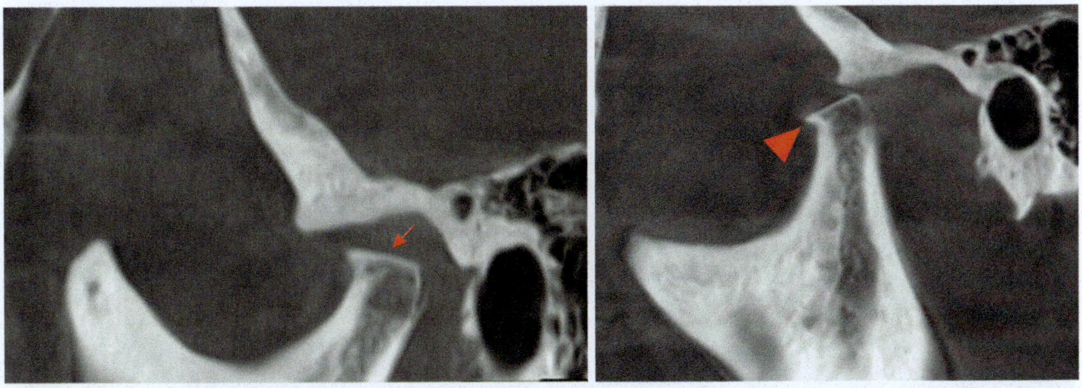

Fig. 2.6 CBCT sagittal cross sections show degenerative joint disease in a 52-year-old female. Severe flattening of the condylar head (red arrow) and possibly a small osteophyte formation are observed on the anterior aspect (red arrowhead)

joint spaces can be evaluated. Radiographic progressive changes include condylar flattening (Fig. 2.6), irregular and/or thickened cortical outlines, osteosclerosis, cortical erosions, osteophyte formation, subchondral cysts, and narrowing of the joint space [13]. Referring the patients to the appropriate specialists prior to commencing orthodontic treatment is recommended [7].

It has been emphasized that although CBCT provides diagnostic information about the TMJ disorders, it does not reveal if the disease process is active or not. Kapila et al. [7] stated that "CBCT images allow the concurrent visualization of the TMJs and assessment of the maxillo-mandibular-spatial relationships and occlusion and provide the opportunity to visualize and quantify the local and regional effects associated with the TMJ abnormalities" [7].

2.4.3 Airway Analysis

Factors like mandibular growth, function of the soft tissues and the jaw musculature, dentoalveolar development, and airway morphology affect development of vertical malocclusions. It has been reported that in children with mouth breathing issues, vertical malocclusions may develop with a constricted pharyngeal airway considered a potential contributing factor [14]. Although constricted airways, especially in children with enlarged adenoids and tonsils, are often diagnosed clinically with conventional 2-D lateral cephalometric images [15], the volume or cross-sectional area without superimposition may be a better measure of airway narrowing, which requires CBCT, rather than conventional 2-D images [16].

Earlier it was suggested that a constricted pharyngeal airway may contribute to mouth breathing and to the development of a steep mandibular plane angle with anterior open bite tendency, [14] but later studies have generated conflicting results with one study showing no relationship between facial pattern and airway volume, while the other study demonstrated the existence of such a relationship [16, 17]. Kapila et al. [18] stated that the discrepancies in the findings of the two studies highlight the need to use a standardized protocol for measuring airway volumes. An example of visualization of narrow airway is shown from a CBCT scan in Fig. 2.7.

2.5 Incidental Findings

In addition to the diagnostic information from the region of interest, CBCT scans can present with a variety of incidental findings. A thorough

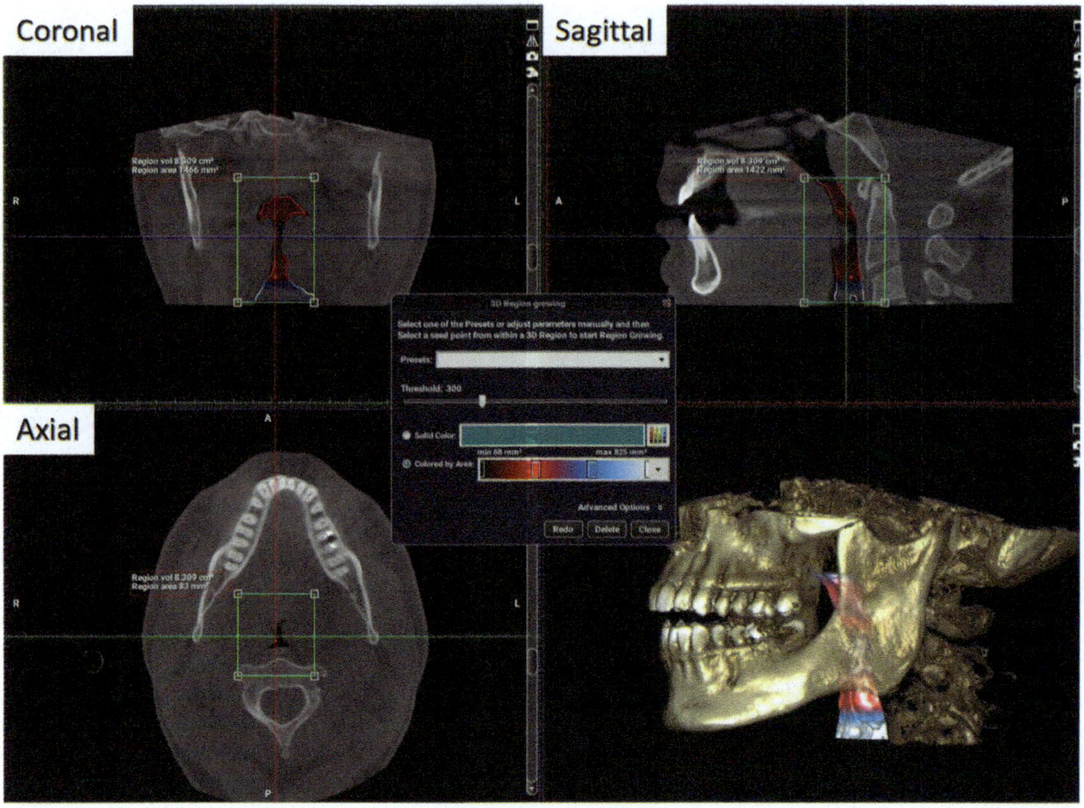

Fig. 2.7 Airway analysis shows narrow airway (region outlined in red on coronal and sagittal images). Axial image also shows the narrow airway. Outline of the airway is also observed in the 3-D model

knowledge of anatomical structures and their variations is of utmost importance. It is the responsibility of the clinician to evaluate and interpret the complete CBCT data set to rule out any abnormalities and potentially pathology. Clinician must also recognize incidental findings encountered in these images. Incidental findings are abnormal findings, unrelated to the problem in question, encountered in images unintentionally as the image was not made for that purpose. Findings should be reported and discussed with the patient. Appropriate actions or recommendations should be made as needed.

The frequency of incidental findings on CBCT images has been reported in several research papers with a high range between 25% and 54% by Cha et al. [19] They evaluated the location, nature, and occurrence of incidental findings in maxillofacial structures on 500 CBCT scans done for various diagnostic reasons. They also assessed association between these findings and symptoms in orthodontic patients. They reported the overall rate of incidental findings as 24.6%, and the highest was in the airway area (18.2%), followed by TMJ findings (3.4%), endodontic-related findings (1.8%), and others (1.2%). Specifically in the orthodontics, the airway-related incidental findings were 21.4%, TMJ findings 5.6%, and endodontic lesions 2.3%. However, only 22% of the airway findings, such as mucosal thickness, polyps, and retention cysts, were correlated with clinical signs and symptoms. It was recommended that for clinical diagnosis, the CBCT data should be interpreted with a full history of clinical signs and symptoms and with detailed communications with specialists to comprehensively evaluate possible underlying diseases.

Another study [20] reported the incidental findings in CBCT scans done for orthodontics. They reported at least one such finding in 66% of the patients; most common were retained primary root tips, followed by periapical disease. According to the results of this study, the overall orthodontic treatment was not altered. However, a high proportion of these cases required further follow-up or intervention (72.5%). Orthodontic treatment was altered in two cases. The first case involved root resorption of a premolar due to an ectopic maxillary permanent canine, which changed the proposed extraction plan. Dilaceration of the poorly positioned central incisor was also detected. In the other case, resorption and pulpal involvement were observed that changed the prognosis of the tooth and thus the extraction pattern.

Avserver et al. [21] evaluated 691 CBCT scans for incidental findings outside the primary region of interest. They reported 1109 incidental findings in the paranasal sinuses in 79.3% of the scans. The majority of the findings were in the maxillary sinus (mucosal thickening, polypoid mucosal thickening, air-fluid level, partial to complete opacification, hypoplasia mucus retention pseudocyst, aplasia, and tooth in the sinus), followed by the nasal cavity (deviated nasal septum, concha bullosa, and onodi cells). Most of the incidental findings required no treatment, but the authors recommended that the clinicians should be aware of the incidental findings and possible anatomic variations. Corrective action should be taken if needed to avoid future complications.

Edwards et al. [22] evaluated the rater agreement between the orthodontic clinicians in the assessments of reported incidental findings with regard to both the need for additional follow-up and the impact on future orthodontic treatment in large-field maxillofacial CBCT scans. Raters demonstrated higher levels of agreement for dentoalveolar findings as compared with all other extragnathic regions when assessing clinical significance of the findings. Fair to excellent rater agreements were discovered for the need for further follow-up and their potential impact on future orthodontic treatment.

Allareddy et al. [23] assessed the number of incidental findings on CBCT scans inside and outside the primary region of interest. The review of 1000 scans showed that 943 (94.3%) scans had findings within and outside the primary regions of interest. They reported 77 different conditions that were observed in these scans, both in the primary region of interest and outside the area. Larger study samples of this paper have provided a better clarification of the importance of analyzing the CBCT data completely to rule out any significant disease.

Edwards et al. [24] reported a higher frequency of incidental findings in large field of view maxillofacial CBCT scans of an orthodontic sample. The majority of the finding may be outside the regions of interest of many dental clinicians. Specifically, incidental findings in the airway and paranasal air sinuses were the most frequent. Other findings were found in the dentoalveolar region and the surrounding hard and soft tissues. This study underscores the importance for comprehensive review of the entire CBCT volume and the requisite to properly document all findings, regardless of the region of interest. The authors emphasized the importance of comprehensive review of the entire CBCT volume and documentation of the findings, regardless of the area of interest.

In orthodontics and oral surgery practices, often a larger field of view is used, and thus the probability of the incidental findings is somewhat more. Price et al. [25] also evaluated the type and prevalence of incidental findings from CBCT of the maxillofacial region. For reporting, the findings were divided into the following groups: (1) needed intervention/referral, (2) monitoring only, and (3) no further evaluation. Assessment of 300 CBCT revealed findings that were categorized into airway, soft tissue calcification, bone, temporomandibular joint (TMJ), endodontic, dental developmental, and pathological findings. A total of 272 scans revealed 881 incidental findings, and the most prevalent were airway findings (35%) followed by soft tissue calcification (20%), bone related (17.5%), TMJ (15.4%), endodontic (11.3%), dental developmental (0.7%), and pathological findings (0.1%). Intervention/referral was needed for 16.1% cases, 15.6% required monitoring. and the remaining (68.3%) required neither. This study also underscored the need to thoroughly examine all CBCT volume for significant findings within and beyond the area of interest.

Mutalik and Tadinada [26] reported a high prevalence (58%) of pineal gland calcifications in patients who were referred for CBCT for implant therapy. The pineal gland is located between the two cerebral hemispheres and produces a hormone called melatonin that affects sleep patterns. With age, the pineal gland calcifications increase. However, calcifications in the pineal gland have been reported in younger population as well. Most studies have considered these calcifications as physiologic, but a thorough medical history and clinical exam are recommended to rule out neurodegenerative disorders.

Incidental findings are listed according to the region where they are more commonly detected (Table 2.1). Paranasal sinuses and nasal fossae: Very common incidental findings in the paranasal sinuses are the mucosal thickening (Fig. 2.8) and mucus retention pseudocyst (Fig. 2.9). These

Table 2.1 Region of occurrence and common incidental findings

Maxillary and mandibular arches	Impacted and supernumerary teeth, periapical inflammatory lesions, incisive canal cysts, idiopathic osteosclerosis, Stafne bone defect, periapical cemental dysplasia. Retained primary roots, crown and root anomalies
Paranasal sinuses	Mucosal thickening, mucus retention pseudocysts, polyps, antroliths, dystrophic calcifications, osteomas, hypoplasia of the sinuses, irregularity of cortical outline
Nasal fossa	Anatomical variations, concha bullosa, septum deviation, inflammatory changes
Airway	Narrow airway, enlarged adenoids
TMJ	Abnormal shape, flattening, sclerosis, erosion, subcondylar pseudocysts, osteophytes, and bifid condyle
Intracranial calcifications	Intracranial pineal gland calcifications
Soft tissue calcifications	Sialoliths, tonsilloliths, calcified carotid artery atheroma, ossification of stylohyoid ligament
Cervical spine	Developmental variations and anomalies such as clefts in the arch of the atlas and degenerative changes
Ear	Debris or wax in the external auditory canals, opacification of the middle ear, cholesteatoma/keratosis, and dehiscence of the jugular bulb, soft tissue lesions
Skull	Opacification of the mastoid air cells, large jugular foramen, possible jugular diverticulum, and underdeveloped mastoid air cells

Source: Drage et al. [20]

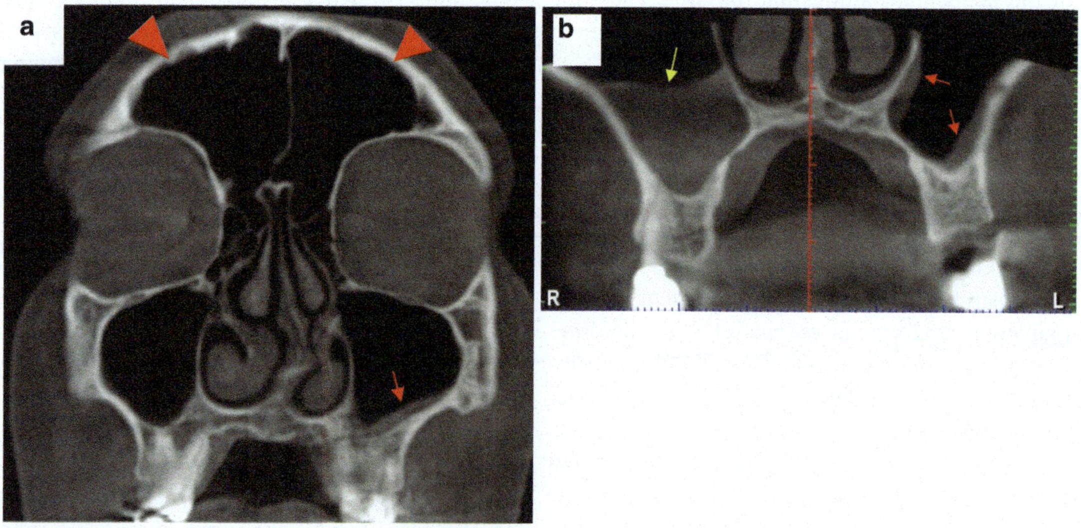

Fig. 2.8 Coronal CBCT views. (**a**) shows mild unilateral mucosal thickening on the left floor of the left maxillary sinus (red arrow) and large frontal sinuses (red arrowheads). Some deviation of the nasal septum is also noted. (**b**) shows opacification/air-fluid level in the right maxillary sinus (yellow arrow)

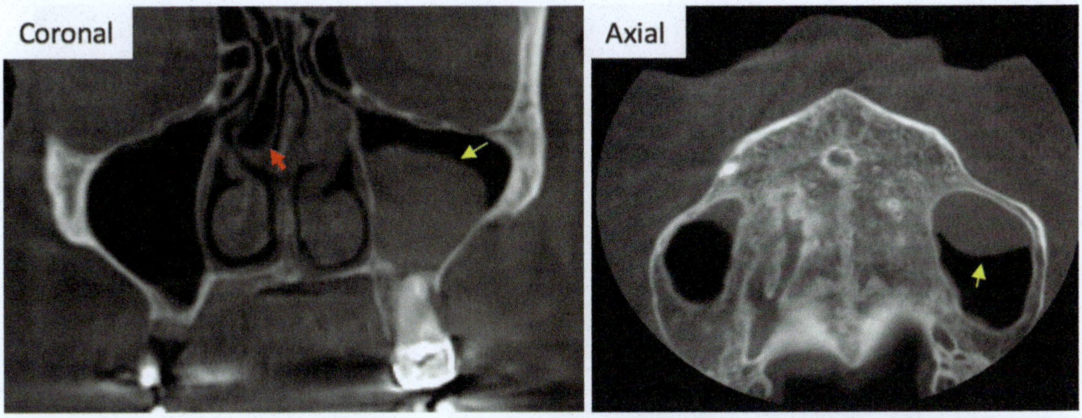

Fig. 2.9 CBCT images show a relatively large homogeneous rounded opacity, possibly a mucus retention pseudocyst, in the left maxillary sinus (yellow arrows). Concha bullosa (pneumatization of the middle concha) is noted on the right side (red arrow)

changes can occur due to chronic inflammation. These findings may be suggestive of chronic sinusitis. When only sinuses are involved, the term sinusitis may be used. The term rhinosinusitis is used when the changes also extend to the nasal cavity. Inflammation can be viral, bacterial, or fungal. Clinical symptoms associated with chronic sinusitis are nasal congestion, discharge, and pain and discomfort. The diagnosis of chronic sinusitis is based on endoscopy or if the radiographic findings have been present for longer than 12 weeks.

Mucosal thickening can lead to obstruction of the passages between the paranasal sinuses and nasal cavity, and this causes a blockade. If other findings such as moderate to severe opacification within the paranasal sinuses and air-fluid levels are noted, acute sinusitis may be suspected. Mucosal thickening can be noted in any of the sinuses. Attention should be paid to the frontal sinus due to its proximity to the brain. While interpreting the images, one should look for any signs of bone changes such as sclerosis or erosion. Rosenfeld et al. [27] have recommended that acute bacterial rhinosinusitis must be distinguished from acute rhinosinusitis caused by viral upper respiratory infections and noninfectious conditions. Clinician should confirm a clinical diagnosis of acute bacterial rhinosinusitis with objective documentation of sinonasal inflammation, which may be accomplished using anterior rhinoscopy, nasal endoscopy, or computed tomography. Although rare, complications may arise like osteomyelitis, orbital and periorbital cellulitis, and intracranial abscesses. Other abnormal radiographic findings associated with the sinus disease include air-fluid level and nonhomogeneous opacification (Fig. 2.10). Less common incidental finding associated with the maxillary sinus is hypoplasia of the maxillary sinus (Fig. 2.11).

Other incidental findings include concha bullosa (Fig. 2.9), asymmetry of the nasal structures (Fig. 2.12), opacity in the ethmoid sinus such as osteoma and mucosal thickening (Fig. 2.13), and mucosal thickening in the sphenoid sinus (Fig. 2.14). While using larger fields of view in CBCT images, the cervical spine is often captured. Degenerative changes in the cervical spine can be noted as osteosclerosis, pseudocysts, and flattening and ligament calcifications (Fig. 2.15).

Maxillary and mandibular arches can also have pathologic conditions not related to the primary region of interest. A case of incisive canal cyst vs. large incisive or nasopalatine foramen is shown in Fig. 2.16. The presence of incisive canal cyst is presumed if the width of the foramen is greater than 1 cm or enlargement is noted on

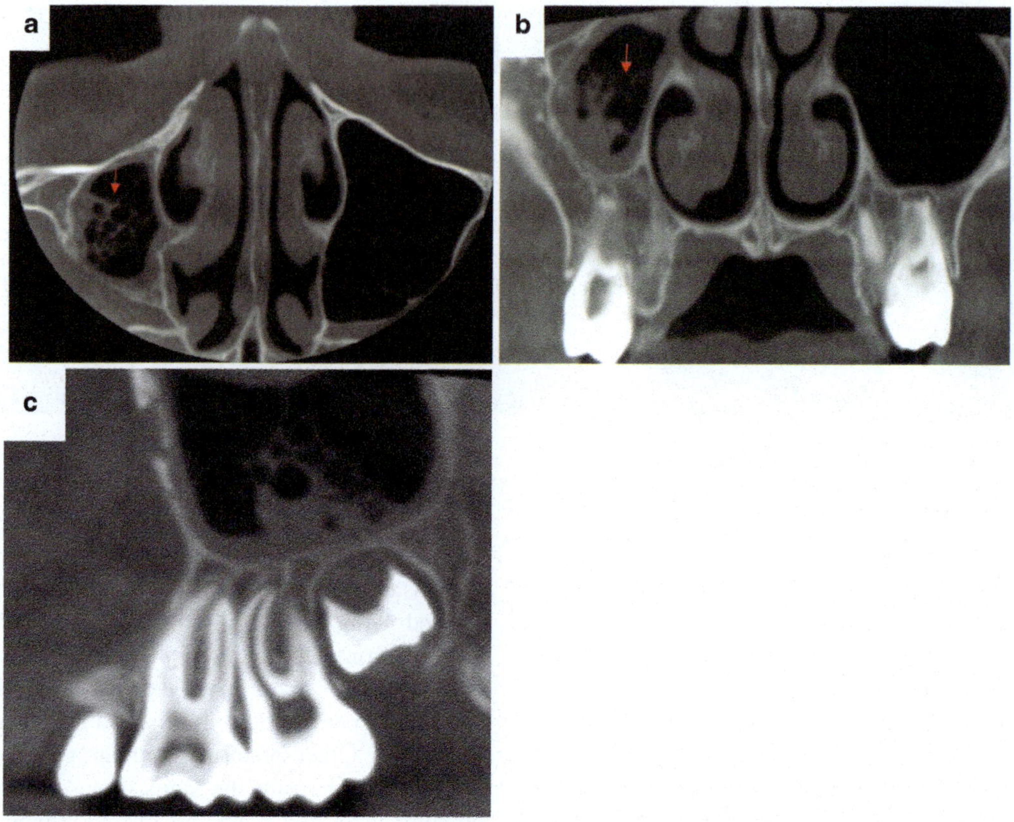

Fig. 2.10 CBCT axial (**a**), coronal (**b**), and sagittal (**c**) images show inflammatory changes in the right maxillary sinus (red arrow). Note the bubbly-appearing (dark) fluid in the right maxillary sinus. No inflammatory changes are visible in the left maxillary sinus. Cortical outline of the maxillary sinuses appear to be within normal limits. Normal nasal septum is also noted

successive radiographic images. Oral-antral communication or fistulae can also be found (Fig. 2.17). Incidental calcifications in the maxillary sinus have also been reported (Fig. 2.18). Intrasinus calcifications can be idiopathic in nature or due to chronic inflammatory or fungal diseases. Calcifications may appear as dense and well-defined masses, with irregular, nodular or linear shapes. Differential diagnosis may include dystrophic calcifications, anthrolith, osteoma, polyp or foreign material. Stafne's bone defect or lingual salivary gland depression may also be visualized. This is extraosseous and is located often below the inferior alveolar canal and anterior to the angle of the mandible (Fig. 2.19).

Airway: Narrowing and asymmetry of the airway or the pharyngeal space can be noted on the CBCT images. Hypertrophy of the adenoids can lead to narrowing (Fig. 2.20). Causes may include sleep apnea, asymmetry of structures, and tumors. One must keep in mind, like other scenarios, imaging findings should be correlated with the clinical evaluation for a more definitive diagnosis.

Carotid artery calcifications: Plaque formation can occur within the artery due to disease. These calcifications with the vessel lumen can diminish in the size, causing reduction of the blood flow. Loose plaque deposits can cause conditions such as pulmonary embolism. The end result can be life-threatening and debilitating conditions, such as myocardial infarction or

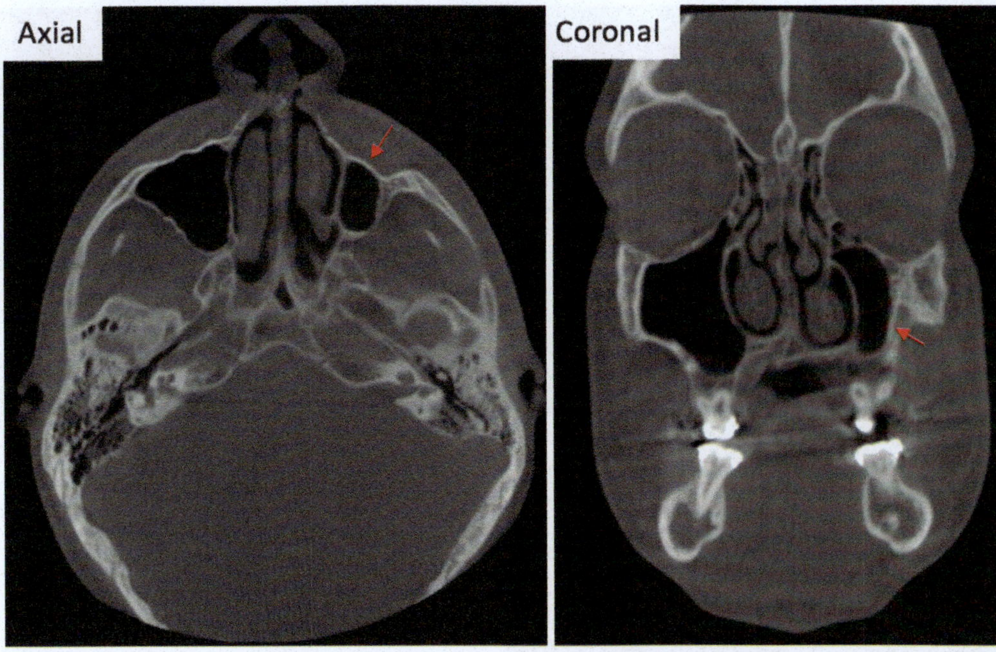

Fig. 2.11 A comparison of the right and left maxillary sinuses shows unilateral hypoplasia of left maxillary sinus (red arrows)

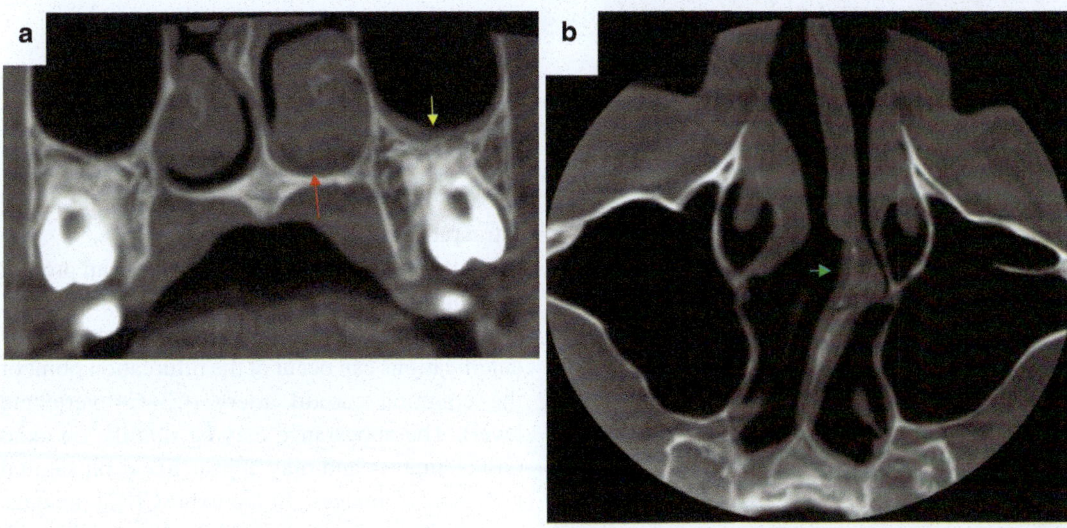

Fig. 2.12 Coronal view (**a**) shows mild mucosal thickening on the floor of the left maxillary sinus (yellow arrow), hyperplastic left inferior nasal turbinate (red arrow). Left inferior nasal meatus appears minimized. Axial (**b**) view shows deviation of the nasal septum to the left side (green arrow)

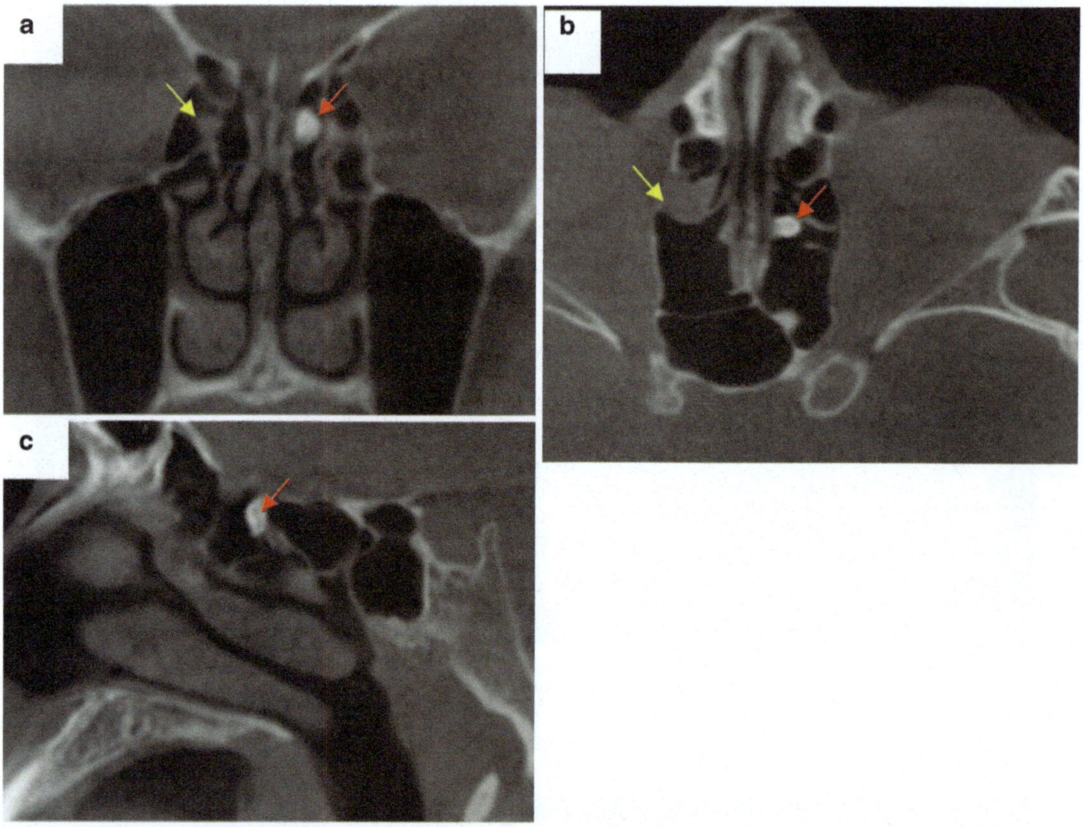

Fig. 2.13 CBCT coronal (**a**), axial (**b**), and sagittal (**c**) images show a well-defined small opacity, possibly an osteoma (red arrows) and mucosal thickening (yellow arrows) within the ethmoid sinuses

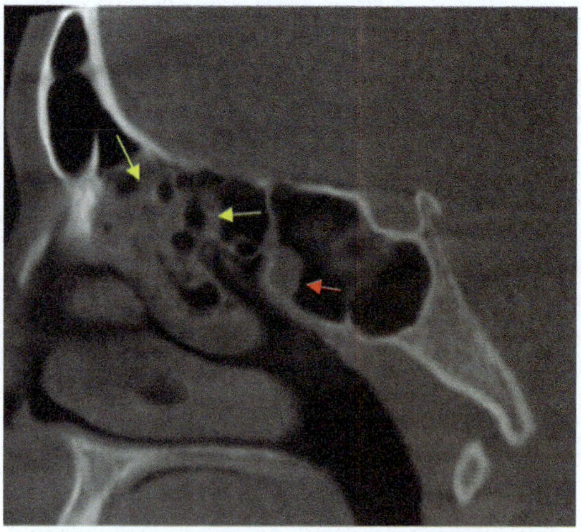

Fig. 2.14 Sagittal view shows mucosal thickening in the ethmoid (yellow arrows) and sphenoid sinuses (red arrow)

stroke. This can occur extracranially and intracranially.

Calcifications of the carotid artery can present as single or multiple, high-density structures, with generally defined outline. Extracranially, the calcifications can occur at the bifurcation point of the common carotid artery (C3–C4 vertebrae level). The appearance may be ringlike on axial CBCT images and may appear linear on sagittal and coronal images. In the axial CBCT images, the calcifications are located medial and anterior to the sternocleidomastoid muscle. Within the cranium, these calcifications may be located on either sides of the sella turcica or the sphenoid sinus area. Other structures that may be confused with calcified carotid atheromas may include

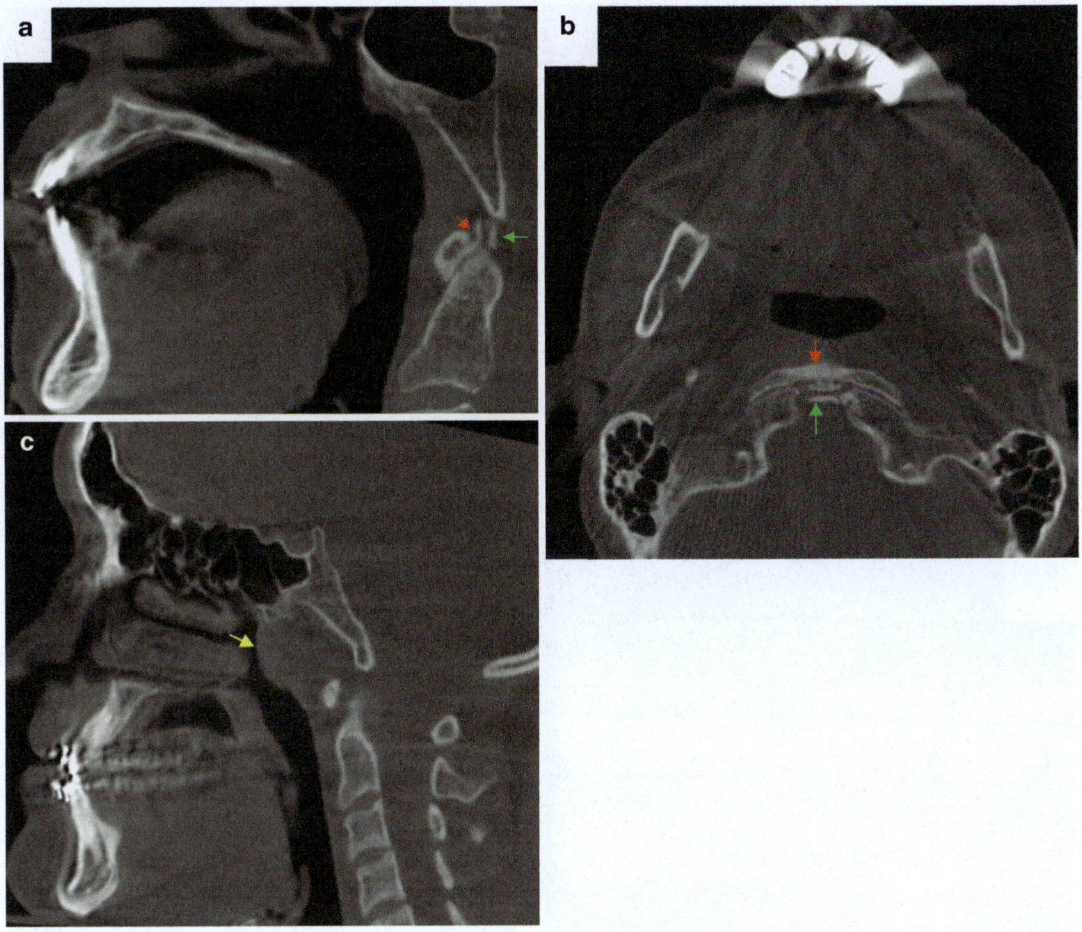

Fig. 2.15 CBCT views show normal unobstructed airway on sagittal view (**a**) and axial view (**b**). Enlarged adenoids causing narrowing of the airway are shown by the yellow arrow on sagittal view (**c**). Cortical changes (red arrow) are also noted in the cervical vertebrae (C1 and C2) with possibly a calcification of a ligament (green arrow)

calcified triticeous cartilage, superior cornu of calcified thyroid cartilage, and greater cornua of the hyoid bone due to the location of these structures. Further medical evaluation is recommended as this may be an indicator of arterial stenosis and stroke (Fig. 2.21).

Tonsilloliths: Dystrophic calcifications, possibly due to previous inflammation and infection, are present in the crevices of the palatine and pharyngeal tonsils and are often seen radiographically. On CBCT images, these calcifications may appear as single or multiple small high-density somewhat rounded structures (Fig. 2.22).

Sialoliths: Calcification or mineralization can occur in the salivary glands. On CBCT images, single or multiple, unilateral or bilateral high-density calcifications can be noted within the salivary glands (Fig. 2.23). Further evaluation is recommended.

Pineal gland calcifications: The pineal gland is located in the center intracranially between the two hemispheres of the brain. It is also known as pineal body or pineal organ. This small gland produces melatonin hormone that regulates sleep patterns and body metabolism. In large field of view CBCT scans, calcifications may be noted in the pineal gland region. These calcifications may present as

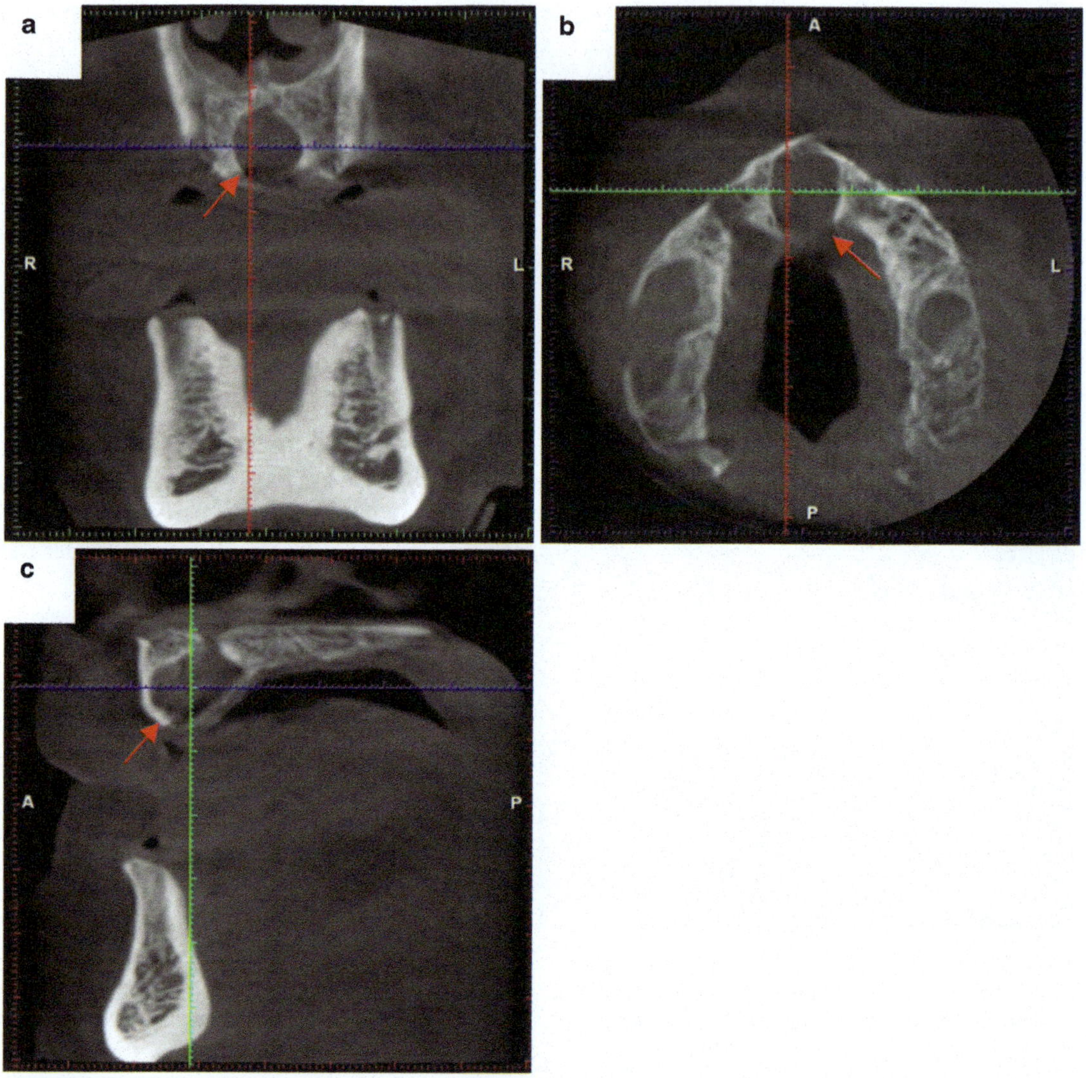

Fig. 2.16 Coronal (**a**), axial (**b**), and sagittal (**c**) images show midline incisive canal cyst vs. enlarged incisive foramen in a 62-year-old male. CBCT scan was taken for implant treatment planning. Diameter of the well-defined corticated entity was more than 5 mm. Lingual cortex was partially missing (red arrow). Recommendation was a clinical examination to determine the need for biopsy and to rule out pathology

single opacification or a group of small higher density rounded to irregular structures (Fig. 2.24). The size is variable. However, if the calcifications appear larger than 1 cm in size, further evaluation for pathology is recommended. Presence of sleep disorders has been also linked to presence of calcifications, especially in very young children.

Extradermal and intradermal opacifications and calcifications: Facial jewelry, soft tissue esthetic implants, cosmetic surgery, and foreign bodies may be seen on CBCT images. Obtaining a clinical history would certainly guide the clinician in better radiographic interpretation. Calcifications within the skin may be seen on

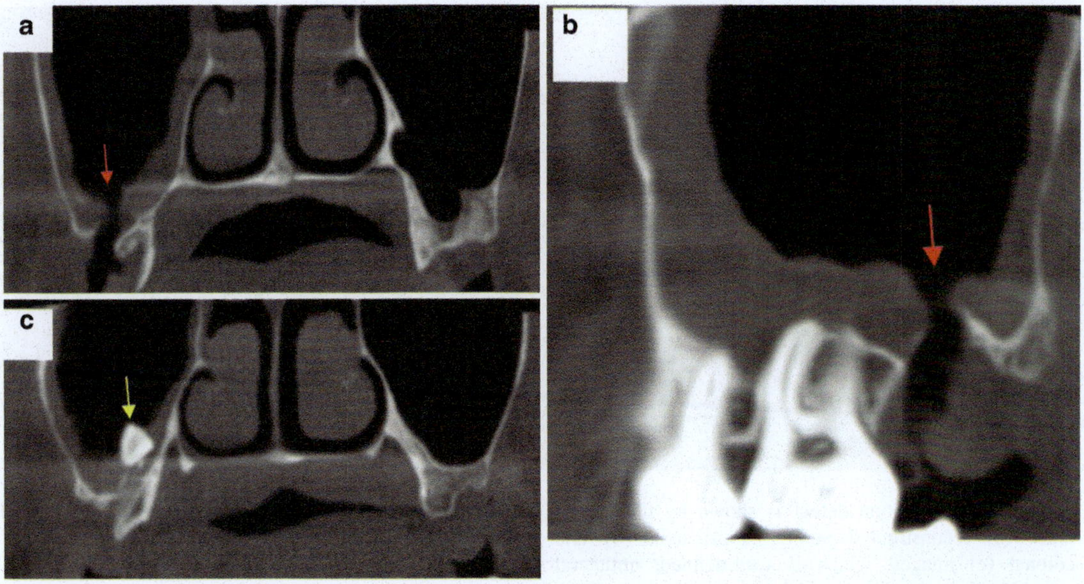

Fig. 2.17 Coronal (**a**) and sagittal (**b**) images show an oroantral communication (red arrows) with mucosal thickening on the floor of the right maxillary sinus. Coronal image (**c**) also shows displaced root into the maxillary sinus

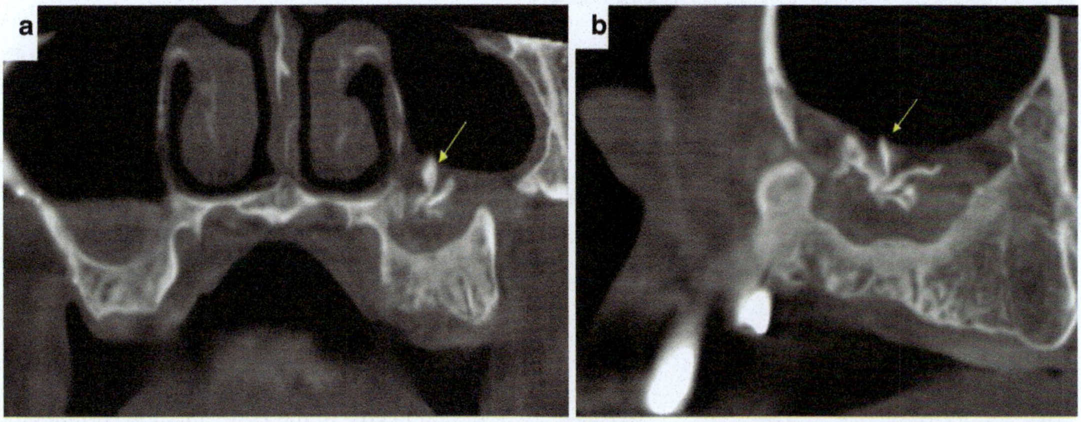

Fig. 2.18 Linear calcifications / foreign material is shown by yellow arrows in the maxillary sinus on coronal (**a**) and sagittal (**b**) CBCT images. Irregularly shaped high-density structure appears to be embedded in the thickened mucosal covering on the floor of the left maxillary sinus. Discontinuity of the bone is also noted in the left lateral aspect of the maxilla maybe due to previous surgical procedure

CBCT images due to various reasons such as idiopathic or dystrophic conditions, trauma, previous surgical procedure, systemic diseases, or metastatic condition. Elevation of serum calcium or phosphate levels should be considered. On CBCT images, high-density single or multiple calcifications of various shapes may be observed. Examples of smaller dispersed extradermal calcification are seen in Fig. 2.25.

Intradermal shunts and catheters: For management of various systemic diseases, shunt systems and catheters are used. Shunts provide alternative pathways through which cerebral-spinal fluids bypass obstructions and may run from the sub-

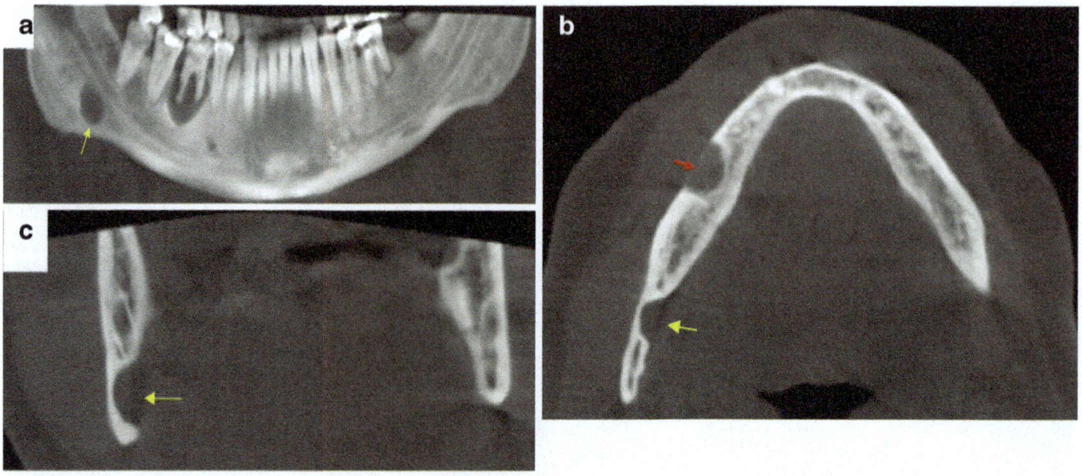

Fig. 2.19 Stafne bone defect is shown by the yellow arrows on reconstructed panoramic (**a**), axial (**b**), and coronal (**c**) images, which is well-defined, unilateral rounded area extending from the medial surface of the mandible or on the lingual aspect of the mandible. Red arrow on image **b** shows loss of mandibular buccal cortex due to an osseous inflammatory lesion

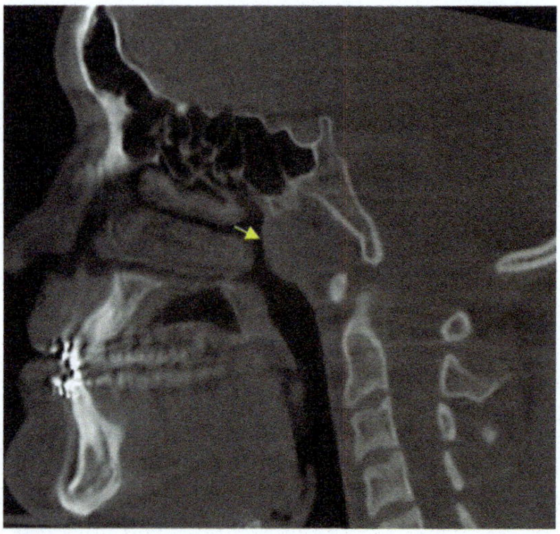

Fig. 2.20 Sagittal CBCT view shows the narrowing of airway due to enlarged adenoids (yellow arrow)

arachnoid spaces or the ventricles within the brain. Shunts and catheters divert CSF to another body region where it will be absorbed to restore the physiological balance between CSF production, flow, and absorption when one or more of these functions have been impaired. These are used to relieve the pressure on brain due to fluid accumulation. These tubes appear hyperdense on CBCT images (Fig. 2.26).

Soft tissue calcification of external auditory canal (EAC): The EAC is an important part of the temporal bone and is involved in conduction of the sound waves. It is approximately a 1-in.-long, slightly (S-shaped) curved dermal-lined passageway from the outside of the head or auricle toward the tympanic membrane or the eardrum, which separates it from the middle ear; the outer one-third is cartilaginous and inner two-third osseous. Soft tissue abnormalities or growths that may be incidentally seen on CBCT images, taken for dental needs, in the EAC, include cerumen or earwax, atresia (narrowing), posttraumatic or infection-caused keloid, external otitis (infection), hemangioma, lymphangioma, papilloma, keratosis obturans, acquired cholesteatoma, adenoma, fibroma, mixed tumor, and carcinomas. The most common lesion is congenital atresia. Wax accumulation is considered a physiological process unless clinical symptoms are reported. Cholesteatomas are not common in EAC but arise as a result of ingrowth of the stratified squamous epithelium of the EAC into the middle ear. Cholesteatoma can involve the tympanic mem-

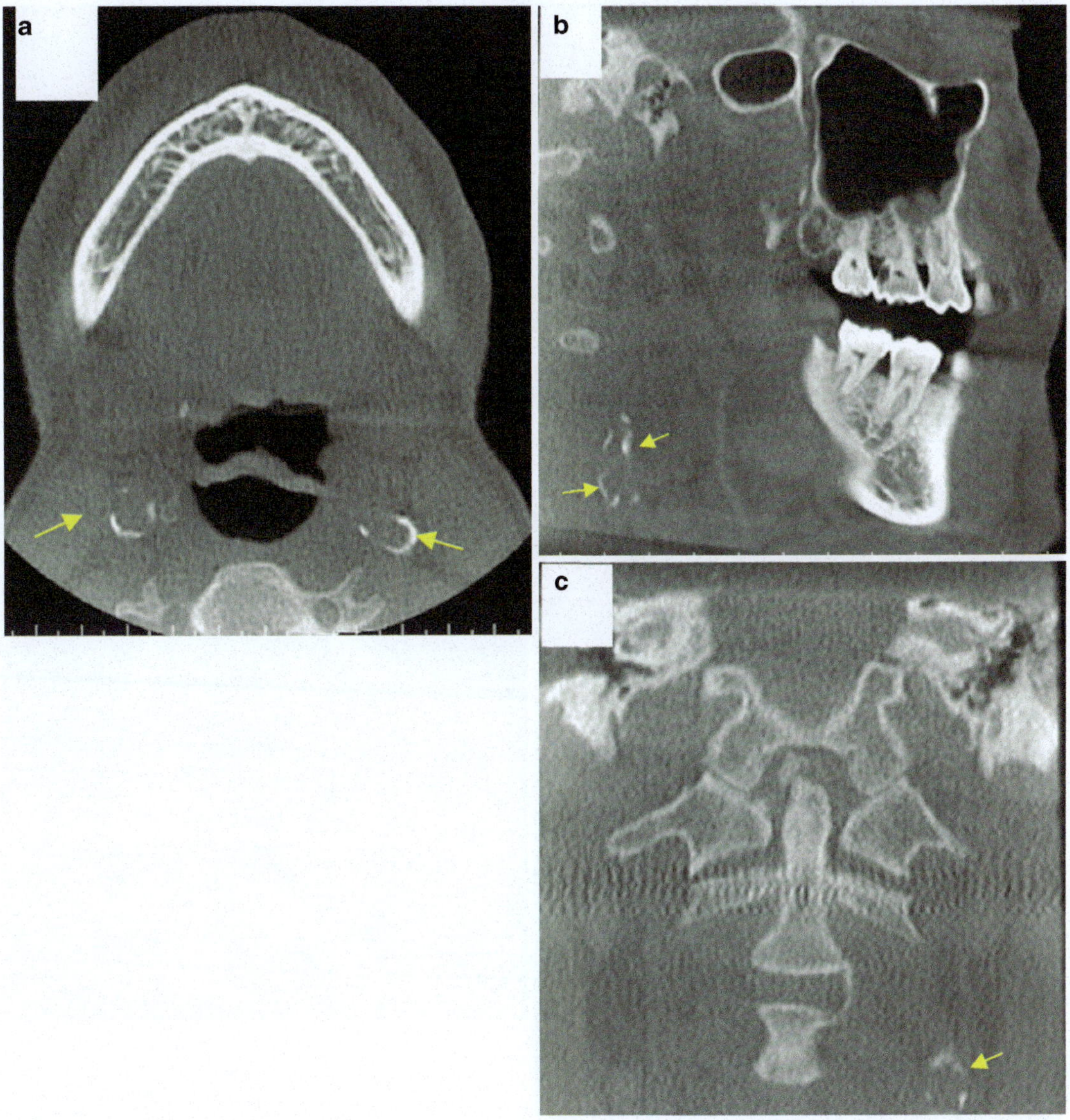

Fig. 2.21 Carotid artery calcifications are noted bilaterally (yellow arrows) as circular high densities on the axial image (**a**). Sagittal view (**b**) shows somewhat linear appearance of the calcifications. Coronal image (**c**) also shows irregular calcification at the level of C3–C4 (yellow arrow) on the left side of the partially visible cervical spine. Courtesy: Dr. Hui Liang

brane, the middle ear, and mastoid process. On the CBCT, the soft tissue lesion within EAC will show as hypodense asymmetric growth of variable size. Differential diagnosis must be made to avoid complications, and thus a consultation with otolaryngologist is recommended (Fig. 2.27). As emphasized earlier, a thorough medical history will aid in radiographic interpretation. If unsure, communication with the medical team is essential.

Elongated styloid process: The styloid process projects down and forward from the inferior aspect of the temporal bone, below the ear. Elongated styloid process is a common

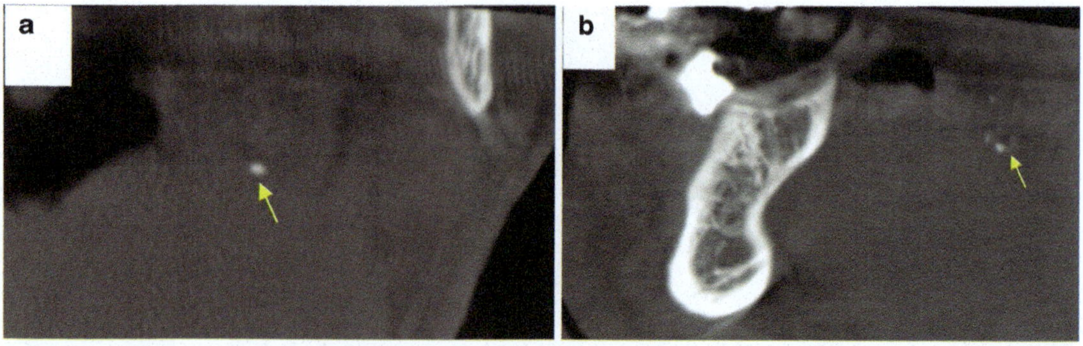

Fig. 2.22 CBCT images show smaller multiple tonsilloliths (yellow arrows)

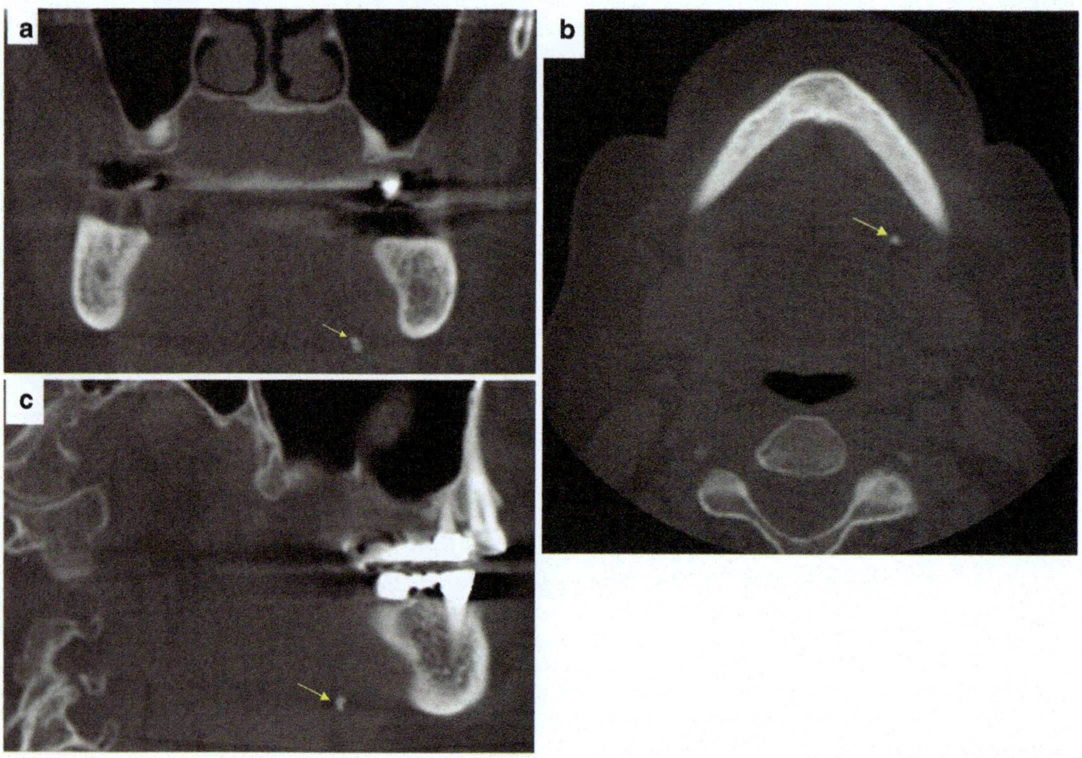

Fig. 2.23 CBCT coronal (**a**), axial (**b**), and sagittal (**c**) images show sialoliths (yellow arrows). Courtesy: Dr. Hui Liang

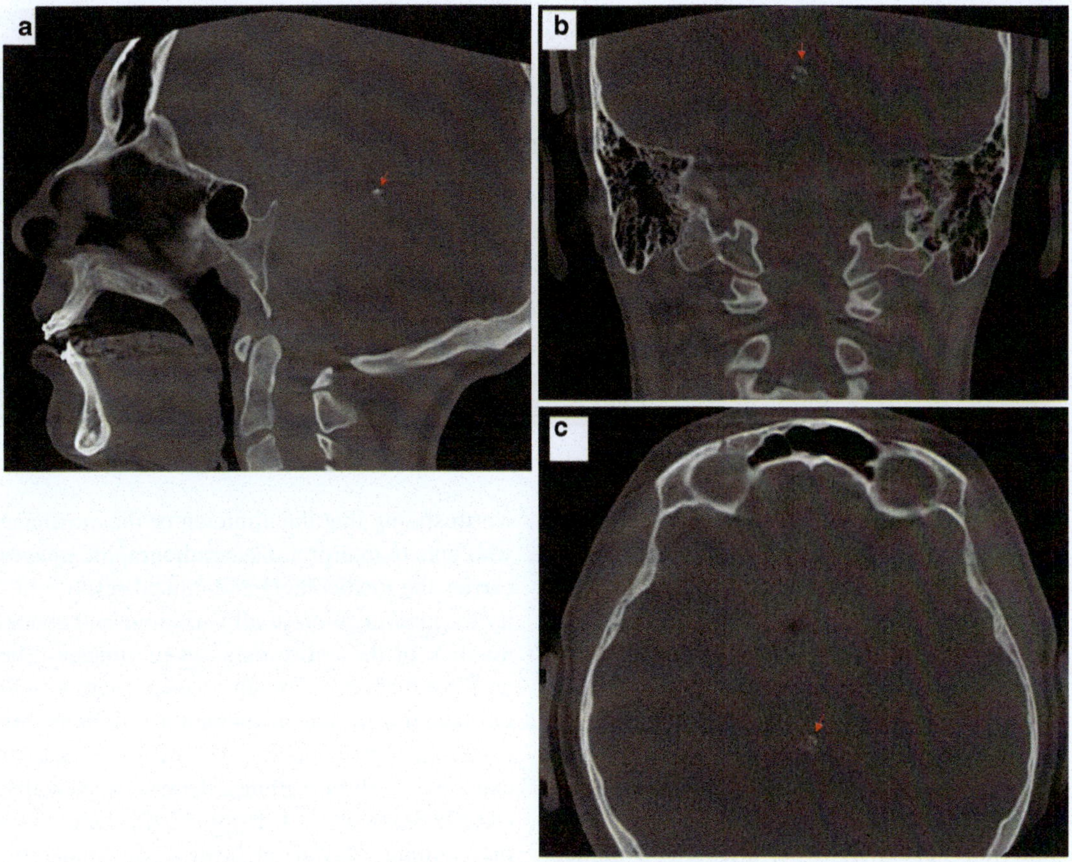

Fig. 2.24 Pineal gland calcifications are noted on sagittal (**a**), coronal (**b**), and axial (**c**) views by the red arrows. Narrow airway is also noted on the sagittal view (**a**)

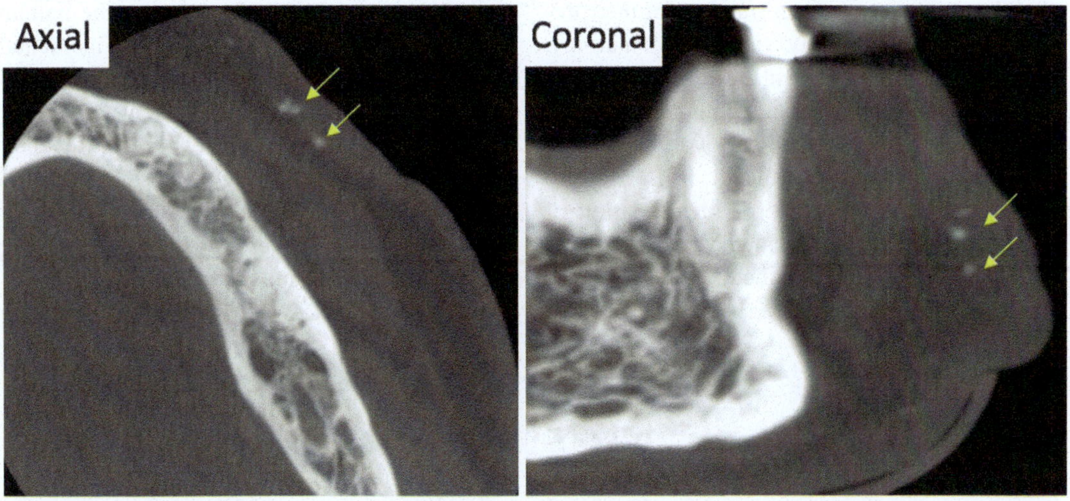

Fig. 2.25 Extradermal calcifications are shown on the CBCT images (yellow arrows). Axial, coronal and sagittal images show multiple smaller rounded calcifications on the buccal aspect of the mandible within the soft tissue

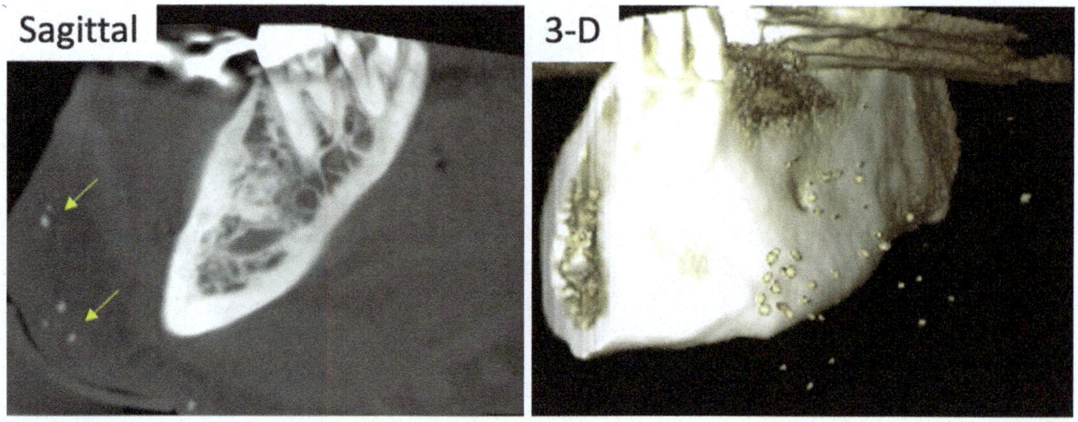

Fig. 2.25 (continued)

radiographic finding. The normal length of styloid process ranges from 20 to 30 mm. Elongation may be unilateral or bilateral. Example of an elongated styloid process is shown in Fig. 2.28. Foreign material: Nowadays, several injectable midface augumentation materials are available in the market. Facial filler may be visualized on radiographic images incidentally and should not be mistaken as a disease process. These materials appear as hyper-attentuated numerous rounded foci or as linear opaque structures dispersed within subcutaneous facial tissues (Fig. 2.29).

It is essential for the clinician, obtaining a CBCT scan, to have proper training for interpretation of normal anatomy, variation of anatomy, and other abnormalities in these images. Most incidental findings are encountered in larger field of view scans. Entire data volume should be comprehensively evaluated. It is vital to understand the importance of identification of the incidental findings, frequency of occurrence, and the medicolegal implications. It is also worth stating that the clinician or diagnostician who opts to interpret those radiographic images carries the medicolegal and ethical responsibility for identification of all variations and abnormalities in the entire data set of images. The need for further follow-up or assessment should be recognized when incidental findings are encountered that appear to be outside the area of expertise of the practicing dentist or specialist (Kapila SD book [28], Turpin [29]). Some cases may require referral for further clinical assessment and follow-up imaging to confirm the diagnosis or to rule out pathology. Although the type and frequency of the follow-up imaging vary, many clinicians advise a 6- to 12-month period. It is not justified to expose the patient for the purpose of identifying incidental findings. It should be noted that while dental clinicians are not expected to treat conditions outside of their professional expertise, it is their responsibility to identify abnormalities and deviations in the complete CBCT data set. If there are concerns, then the patient should be referred to the relevant specialist [30].

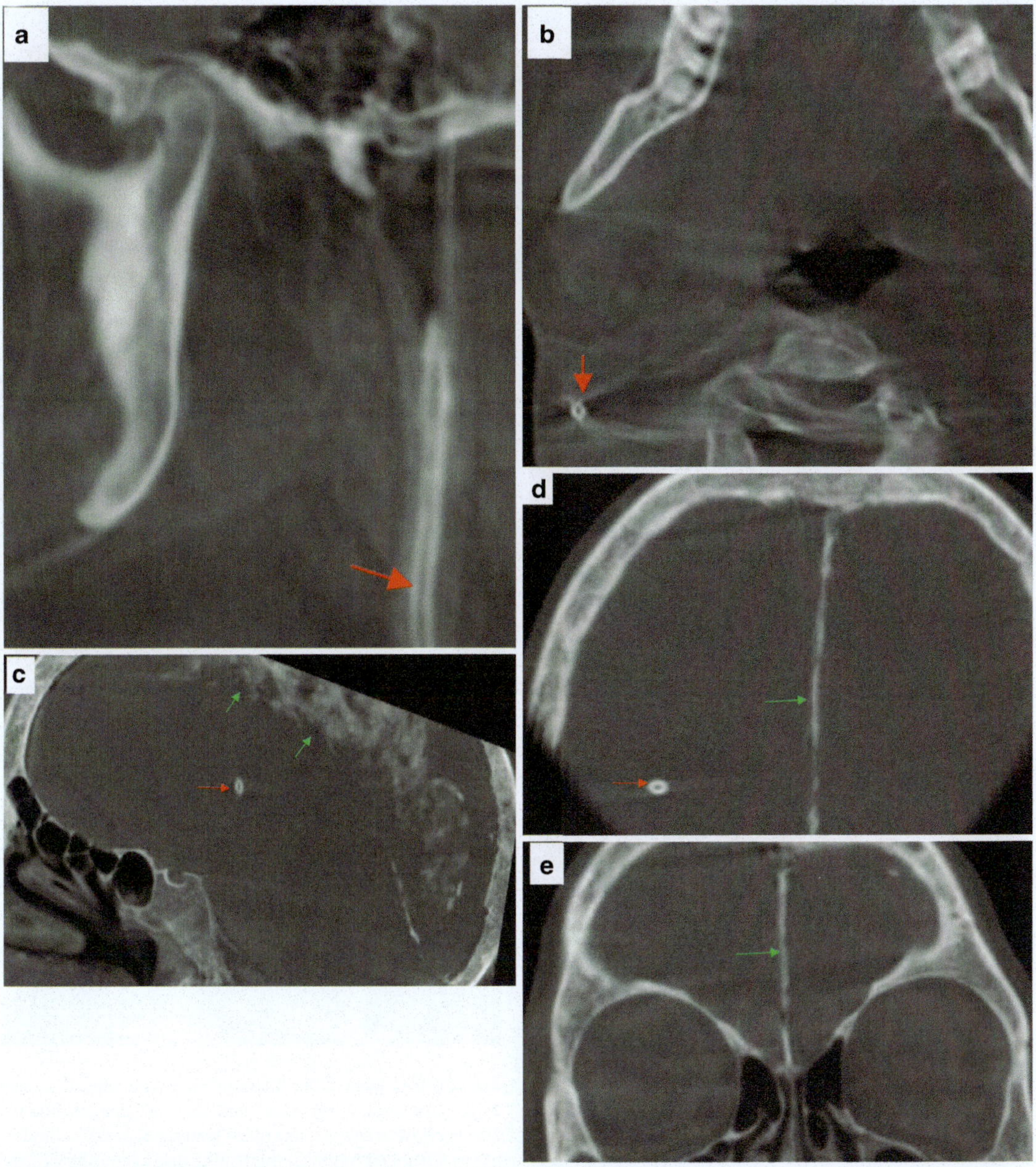

Fig. 2.26 Catheter is shown by the red arrows on sagittal (**a**) and axial (**b**) views (sagittal view shows the length and axial view shows the diameter of the catheter tunneled under the skin within neck tissues). Sagittal image (**c**) shows intracranial ventricular catheter (red arrow) and calcification of the falx cerebri (green arrows). Sagittal image (**c**), axial image (**d**) and coronal image (**e**) show calcification of the falex cerebri (green arrows) and intracranial catheter (red arrow). Courtesy: Dr, Kevin Smith and Dr. Steven Sullivan

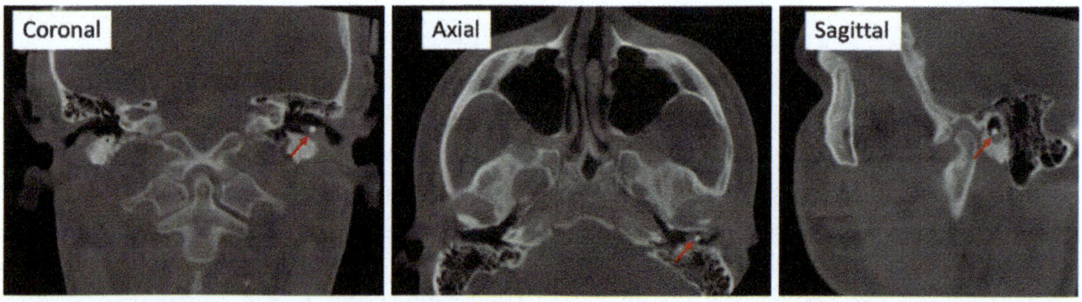

Fig. 2.27 Calcification are noted on CBCT images in ear canal with an opaque foreign object

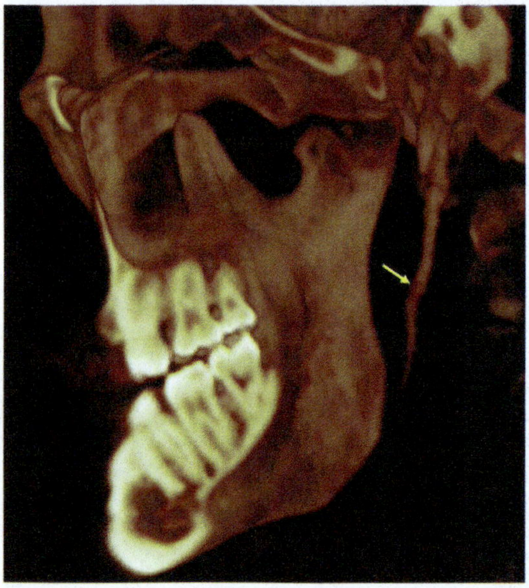

Fig. 2.28 3-D model of elongate styloid process is visible (yellow arrow)

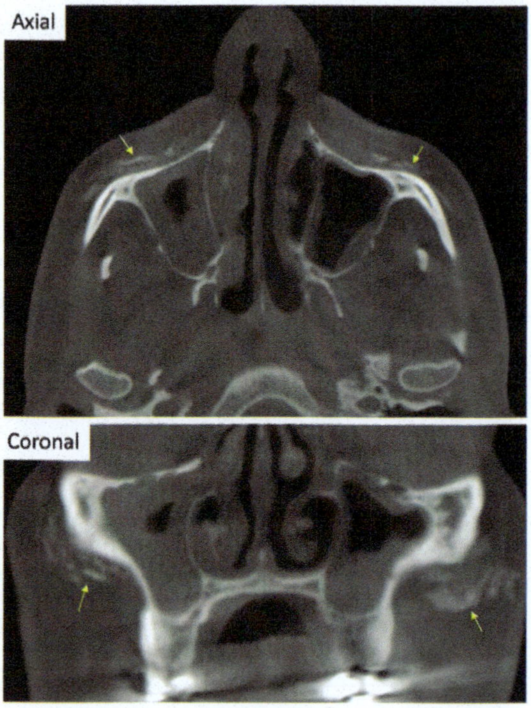

Fig. 2.29 Cheek filler material is noted buccal to the zygomatic arches on axial and coronal images bilaterally (yellow arrows). Maxillary sinuses also show moderate to advanced mucosal thickening and asymmetry of the nasal structures.Courtesy: Dr. Kevin Smith and Dr. Steven Sullivan

References

1. American Academy of Oral and Maxillofacial Radiology. Clinical recommendations regarding use of cone beam computed tomography in orthodontics. Position statement by the American Academy of Oral and Maxillofacial Radiology. Oral Surg Oral Med Oral Pathol Oral Radiol. 2013;116:238–57.
2. American Dental Association Council on Scientific Affairs. The use of cone-beam tomography in dentistry. An advisory statement from the American Dental Association Council on Scientific Affairs. J Am Dent Assoc. 2012;143:899–902.
3. Hodges RJ, Atchison KA, White SC. Impact of cone-beam computed tomography on orthodontic diagnosis and treatment planning. Am J Orthod Dentofac Orthop. 2013;143:665–74.
4. Rischen RJ, Breuning KH, Bronkhorst EM, Kuijpers-Jagtman AM. Records needed for orthodontic diagnosis and treatment planning: a systematic review. PLoS One 2013; 8: e74186. doi:https://doi.org/10.1371/journal.pone.0074186.
5. Kuijpers-Jagtman AM, Kuijpers MAR, Schols JGJH, Maal TJJ, Breuning KH, Vlijmen OJCV. Use of cone-beam computed tomography for orthodontic purposes. Seminars in Orthodontics. 2013;19(3):196–203. https://doi.org/10.1053/j.sodo.2013.03.008
6. Lai CS, Bornstein MM, Mock L, Heuberger BM, Dietrich T, Katsaros C. Impacted maxillary canines and root resorptions of neighbouring teeth: a radiographic analysis using cone-beam computed tomography. Eur J Orthod 2013; 35: 529–538. doi:https://doi.org/10.1093/ejo/cjs037.
7. Kapila S, Nervina JM. 3D Image-aided diagnosis and treatment of impacted and transposed teeth. In: Kapila S, Cone beam computed tomography in orthodontics: indications, insights and innovations. Hoboken, NJ: Wiley-Blackwell; 2014. pp. 349–381.
8. Wriedt S, Jaklin J, Al-Nawas B, Wehrbein H. Impacted upper canines: examination and treatment proposal based on 3D versus 2D diagnosis. J Orofac Orthop 2012; 73: 28–40. doi:https://doi.org/10.1007/s00056-011-0058-8.
9. Kapila S, Conley RS, Harrell WE. The current status of come beam computed tomography imaging in orthodontics. Dentomaxillofac Radiol. 2011;40(1):24–34.
10. Marquezan M, Osorio A, Sant'Anna E, Souza MM, Maia L. Does bone mineral density influence the primary stability of dental implants? A systematic review. Clin Oral Implants Res 2012; 23: 767–774. doi:https://doi.org/10.1111/j.1600-0501.2011.02228.x.
11. Hajeer MY, Millett DT, Ayoub AF, et al. Applications of 3D imaging in orthodontics: part I. J Orthod. 2004;31:62.
12. Faisal AQ, Savell TA, Palomo JM. Applications of cone beam computed tomography in the practice of oral and maxilliofacial surgery. J Oral Maxillofac Surg. 2008;66:791–6.
13. Alexiou K, Stamatakis H, Tsiklakis K. Evaluation of the severity of temporomandibular joint osteoarthritic changes related to age using cone beam computed tomography. Dentomaxillofac Radiol 2009; 38: 141–147. doi:https://doi.org/10.1259/dmfr/59263880.
14. Nielsen IL. Vertical malocclusions: etiology, development, diagnosis and some aspects of treatment. Angle Orthod. 1991;61:247–60.
15. Han S, Choi YJ, Chung CJ, Kim JY, Kim KH. Long-term pharyngeal airway changes after bionator treatment in adolescents with skeletal class II malocclusions. Korean J Orthod 2014; 44: 13–19. doi:https://doi.org/10.4041/kjod.2014.44.1.13.
16. Celikoglu M, Bayram M, Sekerci AE, Buyuk SK, Toy E. Comparison of pharyngeal airway volume among different vertical skeletal patterns: a cone-beam computed tomography study. Angle Orthod 2014; 84: 782–787. doi:https://doi.org/10.2319/101013-748.
17. Grauer D, Cevidanes LS, Styner MA, Ackerman JL, Proffit WR. Pharyngeal airway volume and shape from cone-beam computed tomography: relationship to facial morphology. Am J Orthod Dentofac Orthop 2009; 136: 805–814. doi:https://doi.org/10.1016/j.ajodo.2008.01.020.
18. Kapila S, Nervina JM. Alveolar boundary conditions in orthodontic diagnosis and treatment planning. In: Kapila S. Cone beam computed tomography in orthodontics: indications, insights and innovations. Hoboken, NJ: Wiley-Blackwell; 2014. pp. 293–316.
19. Cha JY, Mah J, Sinclair P. Incidental findings in the maxillofacial area with 3-dimensional cone-beam imaging. Am J Orthod Dentofac Orthop. 2007;132(1):7–14.
20. Drage N, Rogers S, Greenall C, Playle R. Incidental findings on cone beam computed tomography in orthodontic patients. J Orthod. 2013;40(1):29–37. https://doi.org/10.1179/1465313312Y.0000000027.
21. Avsever H, Gunduz K, Karakoc O, Akyol M, Orhan K. Incidental findings on cone-beam computed tomographic images: paranasal sinus findings and nasal septum variations. Oral Radiol. https://doi.org/10.1007/s11282-017-0283-y.
22. Edwards R, Alsufyani N, Heo G, Flores-Mir C. Agreement among orthodontists experienced with cone-beam computed tomography on the need for follow-up and the clinical impact of craniofacial findings from multiplanar and 3-dimensional reconstructed views. Am J Orthod Dentofac Orthop. 2015;148(2):264–73. https://doi.org/10.1016/j.ajodo.2015.03.024.
23. Allareddy V, Vincent SD, Hellstein JW, Qian F, Smoker WRK, Ruprecht A. Incidental findings on cone beam computed tomography images. Int J Dent. 2012;2012:871532.
24. Edwards R, Alsufyani N, Heo G, Flores-Mir C. The frequency and nature of incidental findings in large-field cone beam computed tomography scans of an orthodontic sample. Prog Orthod. 2014;15(1):37. https://doi.org/10.1186/s40510-014-0037-x.

25. Price JB, Thaw KL, Tyndall DA, Ludlow JB, Padilla RJ. Incidental findings from cone beam computed tomography of the maxillofacial region: a descriptive retrospective study. Clin Oral Implants Res. 2012;23(11):1261–8. https://doi.org/10.1111/j.1600-0501.2011.02299.x. Epub 2011 Sep 30.
26. Mutalik S, Tadinada A. Prevalence of pineal gland calcifications as an incidental finding in patients referred for implant dental therapy. Imag Sc Dent. 2017;47:175–80. https://doi.org/10.5624/isd.2017.47.3.175.
27. Rosenfeld RM, Piccirillo JF, Chandrasekhar SS, Brook I, Ashok Kumar K, Kramper M, Orlandi RR, Palmer JN, Patel ZM, Peters A, Walsh SA, Corrigan MD. Clinical practice guideline (update): adult sinusitis. Otolaryngol Head Neck Surg. 2015;152(2 Suppl):S1–S39. https://doi.org/10.1177/0194599815572097.
28. Kapila SD. Cone beam computed tomography in orthodontics: indications, insights, and innovations. Hoboken: Wiley Blackwell.
29. Turpin DL. Befriend your oral and maxillofacial radiologist. Am J Orthod Dentofac Orthop. 2007;131(6):697.
30. Çağlayan F, Tozoğlu Ü. Incidental findings in the maxillofacial region detected by cone beam CT. Diagn Interv Radiol. 2012;18:159–63. https://doi.org/10.4261/1305-3825.

Part II

Diagnosis and Treatment Planning: Analyses, Airway and TMJ

3 3D Imaging to Assess Growth and Treatment Effects

Leonardo Koerich, Eser Tufekci,
and Steven J. Lindauer

Abstract

Cone beam computed tomography (CBCT) is a three-dimensional imaging mechanism that is being used more frequently in dentistry and across many dental specialties for diagnostic and treatment planning purposes. Additionally, CBCT may be used to evaluate growth and treatment changes in individual patients over time when accurate and precise superimposition techniques are applied appropriately.

3.1 Introduction

For several years and continuing today, the superimposition of lateral cephalograms has been used to measure growth and assess treatment outcomes in orthodontics. These superimpositions can be used for the whole craniofacial complex by superimposing on the cranial base or be limited to regions such as the maxilla or mandible. These techniques are widely known and used by orthodontists and are even part of the American Board of Orthodontics' clinical examination process. However, two-dimensional (2D) evaluations have many limitations including magnification, superimposition of bilateral structures, and errors due to head rotation. Advancements in digital technology, especially in cone beam computed tomography (CBCT), provide new insights regarding how craniofacial growth and treatment outcomes can be understood in three dimensions (3D). More dramatically, questions that were answered previously in 2D studies are being asked again, and new studies are reassessing older and possibly outdated concepts with the aid of CBCT. Some of the answers remain the same. However, others have changed. This has direct implications for orthodontics and may result in changes to the protocols used to treat patients.

Because of its importance in understanding growth and treatment changes, several experts around the world have tried to establish protocols to superimpose CBCT images accurately. After almost 20 years of CBCT use, there is research showing how accurate CBCT superimposition can be using different methods or software programs. There is a significant increase in the number of publications in this area with more studies using CBCT superimposition to show changes in growth and treatment. It is important to understand how the process works to fully appreciate what CBCT can offer for longitudinal assessment.

L. Koerich (✉)
Private Practice, Charlotte, NC, USA

E. Tufekci · S. J. Lindauer
Department of Orthodontics, Virginia Commonwealth University, Richmond, VA, USA
e-mail: etufekci@vcu.edu; sjlindau@vcu.edu

O. Kadioglu, G. F. Currier (eds.), *Craniofacial 3D Imaging*,
https://doi.org/10.1007/978-3-030-00722-5_3

3.2 History of 3D Superimposition and Different Methods

A key feature that makes 2D images easy to understand is their relative simplicity. A 3D structure is summarized as one 2D image. Therefore, after tracing two sequential cephalograms, the operator can superimpose them quickly using the desired areas as references. A CBCT image, on the other hand, is a digital image that is visualized on a computer screen and is composed of several different layers or slices that depend on the voxel size and the field of view (FOV). These images can be seen in 2D by checking each one of the cross sections, axial, coronal, or sagittal, or in 3D by virtual rendering or segmentation. In other words, it is more complicated to read and understand one CBCT than one cephalometric image. The difficulty is increased when different scans are being superimposed because there are more details in the 3D images and, hence, more chances for error.

Even before CBCT was available, the research that proposed 3D superimposition used medical CT scans and landmarks as references [1, 2]. This method, using landmarks, has been used with many software applications despite its lack of accuracy. For many years, popular commercial products, such as Dolphin Imaging and InVivo, offered this method as the only technique for superimposition. Years later, research [3] showed that, despite the excellent reproducibility in identifying landmarks on a 3D image, small errors from each one of them resulted in a compounding effect that jeopardized the quality of the final superimposition. Even a small linear error in one dimension can lead to several degrees of rotational error that substantially affect distant areas.

In 1998, a method to superimpose CT scans based on visual assessment was proposed [4] in which the observers tried to match semitransparent models digitally between timepoint 1 (T1) and timepoint 2 (T2). This idea was very similar to the landmark-based method, except that the T2 image was rotated manually to fit the T1 image by matching the landmarks chosen. Although innovative for its time, this idea was not good enough to resolve the problem since minor discrepancies were still enough to reduce accuracy of the superimposition. It became more evident that significant advancements were needed as the manual methods were not good enough to address the complex problems of 3D superimpositions.

The two most recent and more accurate methods for 3D superimposition are mostly automated, requiring less human interaction and, therefore, diminishing the chance of error. First, the voxel-based superimposition method published in the journal *Dentomaxillofacial Radiology*, in 2005 [5], remains the gold standard for modern superimposition. In summary, this method compares the voxels from two different scans and tries to match them for optimal results. Given that the cranial base remains essentially unchanged once it is finished growing, adult patients were the first in which CBCT scans were superimposed successfully, because it was easier to check for accuracy. The second type of superimposition achieved was by using the cranial base for growing patients, followed by regional maxillary and mandibular superimposition methods for nongrowing patients and, finally, regional applications for growing patients.

In 2005, there was only one software product available to perform accurate cranial base superimpositions. Today, there are several software applications available that perform voxel-based superimposition with two of them commonly used by orthodontists and oral and maxillofacial surgeons. The free software available today is called "Slicer 3D" and can be downloaded quickly through the Internet at cmf.slicer.org. OnDemand 3D, Dolphin Imaging, and Maxilim are examples of commercially available software that can be used for voxel-based superimposition. The process can take a few seconds and up to 40 min [6–9], depending upon the software used and the clinician's experience using it. It is beyond the scope of this chapter to discuss in detail the advantages and disadvantages of each software product. However, the one that has had the highest number of studies testing accuracy or reproducibility is OnDemand 3D [6, 7, 9, 10].

The last and latest technique developed for superimposition of CBCT images is the surface-based superimposition method. It still has some limitations compared to voxel-based superimposition techniques. However, while voxel-based superimposition is complicated and has a steeper learning curve, surface-based superimposition is accessible through a variety of non-medical software application products. In fact, several 3D engineering software programs are great for performing this type of superimposition. Although today the voxel-based technique is much simpler than in the past, there was a time when surface-based techniques were more user-friendly, and this is most likely why it remains a popular method of superimposition used in many studies. The surface-based technique requires creation of a 3D model from a CBCT scan. The most common file extension used is .stl which stands for stereolithography. After creation of the models at T1 and T2, the software tries to match the areas to the best of its ability. Instead of using voxels as a reference, this method matches surfaces. One of the disadvantages of this method is that it requires the creation of a surface model, which could be time-consuming and usually requires more than one program since the 3D models are created in one program and the superimposition is subsequently accomplished in another program. Another major disadvantage is that multiplanar reconstruction cannot be seen, because only the 3D models are superimposed. Therefore, if inner structures are of interest for examination, their visualization is not possible.

Overall, the voxel-based technique for CBCT superimposition is simple, fast, and should be considered as the gold standard for superimposition. Surface-based techniques are very good and show results comparable to voxel-based methods [11]. However, some of the advantages that it had in the past have been surpassed by the advancements in the voxel-based technique. Although it is commonly still in use today, it is probable that surface-based methods will be used less frequently in the future. Landmark-based superimposition should not even be considered an option given its severe limitations, but it is more a method to approximate two different 3D scans in space. Despite some landmarks being suggested as reproducible to superimpose two images [12], as mentioned previously, a small error in each landmark can lead to a compounding effect that affects the final quality of the resultant superimposition.

3.3 Current Status of Voxel-Based Superimposition

Voxel-based superimposition has improved significantly since it was first developed. With the development of different techniques used to superimpose and more software products available for use today, this method can be accomplished in as fast as 10–15 s, depending on the voxel size and the software used. For this reason, the remainder of this chapter will focus on the current status of voxel-based superimposition.

Although superimposition can be accomplished using many different areas of reference, there are four main types of superimposition that interest orthodontists and oral and maxillofacial surgeons. These include (1) cranial base for nongrowing patients, (2) cranial base for growing patients, (3) regional (maxilla or mandible) for nongrowing patients, and (4) regional (maxilla or mandible) for growing patients. Each one of these will be discussed specifically.

3.4 Cranial Base Superimposition for Nongrowing Patients

Cranial base superimposition for nongrowing patients has been extensively studied by several research groups and validated using different software programs [8–11]. This type of superimposition is very easy to achieve because the whole cranial base remains stable in adult patients and, therefore, the area used to match the T1 and T2 images is large. By default, when the cranial base is superimposed in an adult, the maxilla will be superimposed as well, unless there has been some intervention applied, such as surgery, to move it. The mandible, however, because it is not rigidly connected to the rest of cranium, may have different positions between scans and is the main reason why studies that assess mandibular changes following orthognathic surgery, for example, should report changes as “displacement and/or remodeling” rather than as remodeling alone. It is very unlikely that the mandible would remain in exactly the same position relative to the cranial base before and after orthodontic or surgical treatment.

3.5 Cranial Base Superimposition for Growing Patients

Cranial base superimposition for growing patients is more challenging than in nongrowing patients, because the cranial base itself changes during growth. There are not many studies previously published that have used this technique. However, there are enough data available to support that superimposition can be achieved successfully using the voxel-based technique [9, 13]. To accomplish cranial base superimposition in growing patients, the region of interest used as a reference should only be the anterior cranial base. Therefore, it is very important in growing patients, as opposed to nongrowing patients, that the T1 and T2 scans are placed manually so that the anterior cranial base structures are close to one another to achieve optimal results (Fig. 3.1). This technique is very useful to visualize overall changes in the craniofacial complex resulting from facial growth and orthodontic, dentofacial orthopedic, and/or orthognathic surgical treatment.

3.6 Regional Superimposition for Nongrowing Patients

Regional superimposition for nongrowing patients was introduced in dentistry by Koerich et al. [6] and is relatively new compared to cranial base superimposition. Given the novelty of the technique, not many studies are available from different research groups for comparison. However, this type of superimposition is predictable and easy to achieve as reproducibility by different operators was shown to be almost perfect [6]. Similar to cranial base superimposition for growing patients, in which the area on which sequential images will be superimposed must be limited to the anterior cranial base, the key for successful regional superimposition in nongrowing patients is to approximate the T1 and T2 images of the maxilla and mandible to achieve optimal results. When the goal is to superimpose the mandible, part of the maxilla will be present in the scan as well, and vice versa. This is why it is so important to approximate the area of interest on the T1 and T2 images manually to indicate to

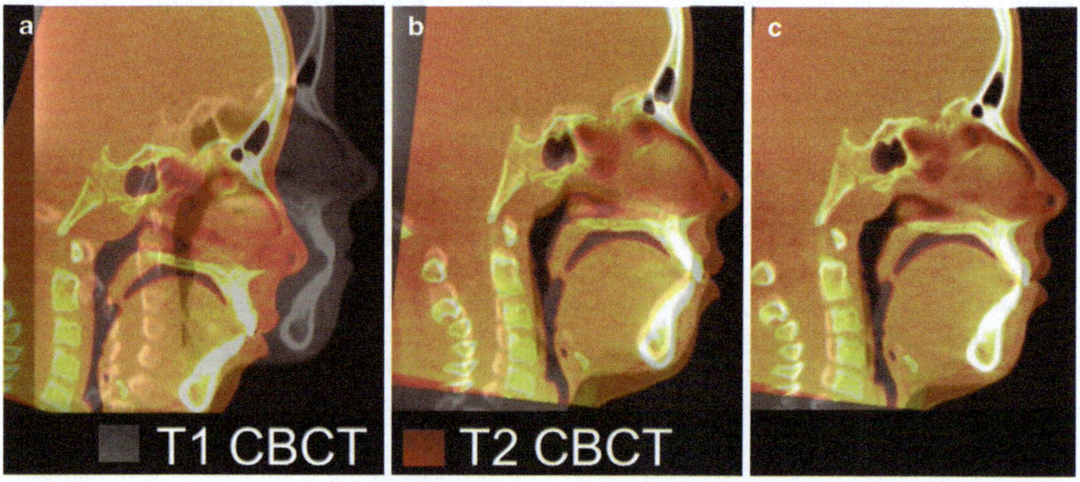

Fig. 3.1 Overlay of T1 (gray) and T2 (red) CBCT of a growing patient. (**a**) Before approximation of the cranial base, the T1 and T2 are far away. This could lead to failure during the superimposition process. (**b**) After manual approximation to optimize the superimposition. (**c**) After superimposition was done using the anterior cranial base as the reference

the program where it should look for voxels on which to superimpose. This is key to the success of this method. One of the main advantages of regional superimposition is the ability to reduce the FOV of the scan, thereby reducing the radiation dose to the patient. This technique can be applied to understand bone remodeling that has occurred in the maxilla or mandible and to track changes in tooth position over time due to orthodontic treatment, late eruption, or pathologic migration. Also, it can be used following bone grafting to assess changes in quality of bone for implant planning and placement and to track progression of condylar hyperplasia or resorption.

3.7 Regional Superimposition for Growing Patients

The latest area for development of techniques for CBCT superimposition in dentistry is in regional superimposition for growing patients. Because growth is different between the maxilla and mandible, they will be discussed separately.

The first two studies exploring regional superimposition of the mandible in growing patients, by Ruellas et al. [14] and Koerich et al. [7], both showed small differences in reliability between operators. More importantly, Ruellas et al. [14] found that some of the areas shown to be reliable for 2D mandibular superimposition in growing subjects by Bjork [15] using cephalometric films may not be stable enough for 3D superimposition. Changes in the mandibular canal and crypts of the developing third molars apparently occurred and made it difficult to superimpose sequential CBCT images reliably. These recent studies presented new findings that bring into question some of the old views that were once considered absolute truths. Today, it is reasonable to doubt whether the mandible has any areas that can be considered truly stable during periods of active growth so that they can be used reliably for superimposition purposes. Ideally, metallic implants, such as those used by Bjork [16, 17] in developing his original theories of growth, would help in developing how sequential images of growing mandibles might be accurately superimposed. Because of ethical concerns, however, this is not possible. Since miniplates and miniscrews are often used for orthodontic and orthopedic treatments, it may be possible to use these as references for superimposition. However, miniplates are malleable and bend easily as shown in Fig. 3.2. While miniscrews are believed to move sometimes in response to applied forces [18], a recent study has shown they are stable enough to be used as references for sequential 3D image superimposition at least over a short time interval [19].

The only study published on regional superimposition of the maxilla in growing patients was by Ruellas et al. [20]. They compared maxillary superimpositions in growing patients using two different areas of reference. The results showed good intra- and inter-examiner reproducibility. They did not attempt to test accuracy of the method nor to find stable structures within the maxilla to be used for superimposition. Half of the sample used were patients that had rapid maxillary expansion. Given the elaborate method used and the intense calibration of the examiners in this study, the technique used is not easily accomplished, testing by other centers using different techniques is still necessary.

In summary, the voxel-based technique is well-established for superimposition of sequential CBCT images, except for regional superimposition in growing patients. There are still improvements needed, and it is questionable whether or not there are stable reference areas that can be used to achieve reliable superimposition of the maxilla or mandible in growing patients. Rather than continuing to search for stable areas that may not exist, an alternative may be to find areas that are easier to use in order to achieve techniques with greater reproducibility of superimposition among different operators.

3.8 Evaluation of Superimpositions

Reviewing the classic longitudinal studies accomplished in orthodontics and oral surgery, there is a common pattern evident in the presentation of the methodology and analysis used. Generally, superimposition is accomplished and the results are presented by comparing the

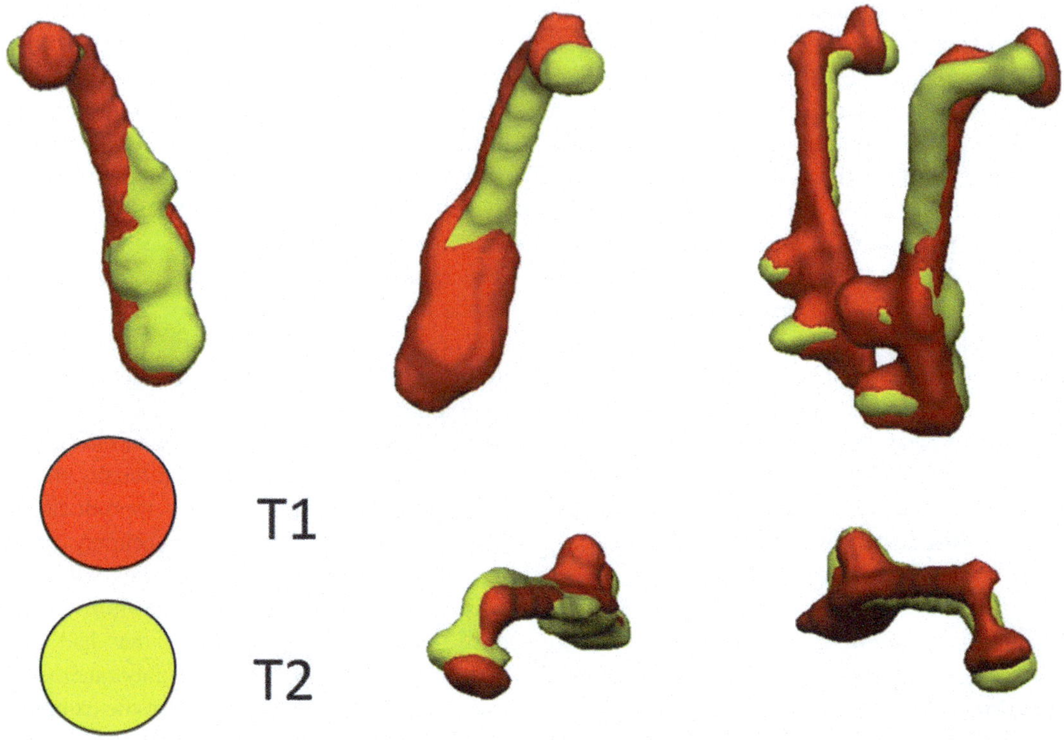

Fig. 3.2 Frontal, lateral, and superior view of segmented mandibular plates used for bone-anchored maxillary protraction. The CBCTs were taken 1 year apart, and the models were superimposed using the surface-based technique. Note that the plates were bent during the year that the patient used elastics for Class III correction. This makes it more difficult or invalid to use this type of plates as stable references to assess growth

location of landmarks, distances, and angles identified before and after treatment. From a clinical standpoint, visual assessment of a superimposition can provide a rough estimate of the changes from treatment quickly. While 3D images are superior to a 2D representation, the complexity of new technology has provided opportunities for numerous studies to be undertaken just for proposing new methods of evaluation. CBCT scans are much more detailed, extremely precise, and rich in data when compared to 2D cephalograms. However, there are two important disadvantages that should be understood. First, from a research standpoint, longitudinal studies using 3D data lack standardization in methodology, making comparison among studies more challenging. Second, for a clinician wanting to evaluate treatment outcomes, the learning curve for reading a 3D image is steeper compared to 2D. With that in mind, insights are presented on how to interpret 3D longitudinal changes from a research and clinical perspective.

3.9 Research and Clinical Evaluation

There have been several studies using different methods of 3D evaluation including volumetric changes [21], shape correspondence [22], landmark-based measurements [23], closest point technique [6, 7], and combinations [24], as well as other methods that may be found in the literature. The most commonly used ones have been measurements made directly on the multiplanar reconstruction or surface model (either landmarks, lines, or angles) or the closest point technique.

The closest point technique method has been used in the literature extensively to provide an easy way to communicate and visualize changes.

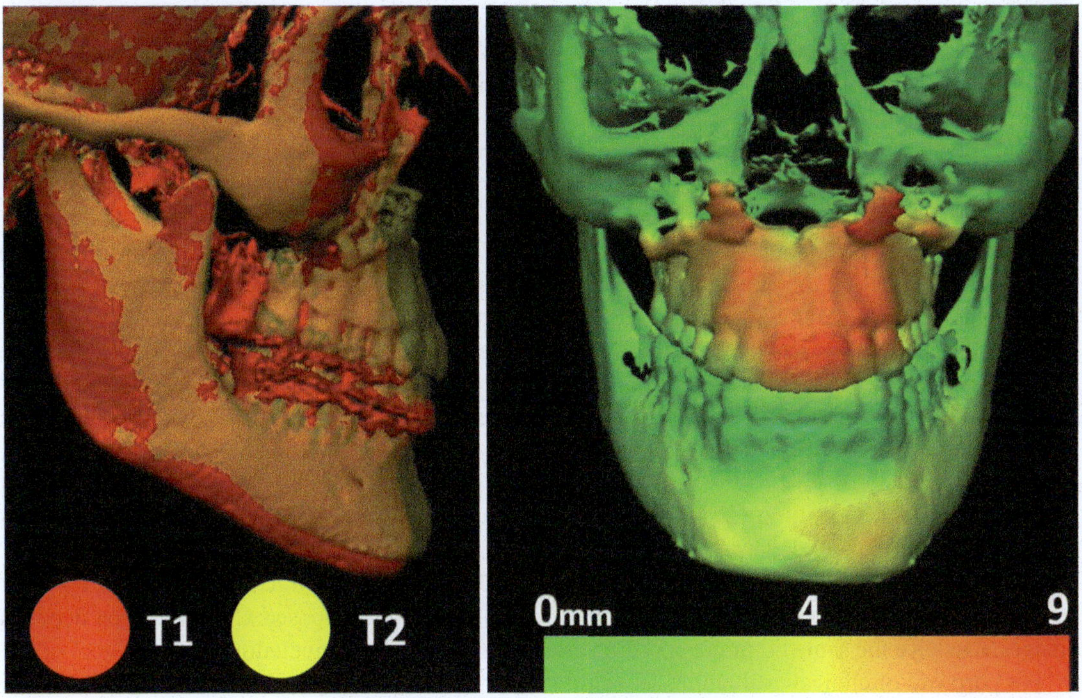

Fig. 3.3 Lateral semitransparency and frontal view of the color maps together help to communicate a more complete understanding of the treatment outcome. The color map by itself only shows inward or outward movement. Therefore, based on the color map alone, both the maxilla and mandible were advanced. When the color maps are visualized with the semitransparency, it is clear that the maxilla is not only advanced, but it was intruded posteriorly, while the incisal edges had a very small vertical change. The mandible was not advanced but rotated counterclockwise following the new maxillary position

In fact, the first study done using voxel-based superimposition also used the closest point technique for evaluation [5]. This technique is relatively easy to use and can be done using several different software applications. However, it first requires that a reliable superimposition be accomplished prior to measurements being made. It provides a simple and fast method to visualize all the changes based on color-coded maps, with each color representing a certain degree of change. On the other hand, this technique has limitations with the outcomes needed to be evaluated carefully. As mentioned by Jabar et al. [25], the closest point technique can underestimate changes, sometimes significantly. The reason for this is that the closest point technique measures the smallest distance between two surfaces rather than corresponding surfaces. Therefore, the researcher needs to fully understand the limitations of the method in order to minimize those limitations. A great aid to understanding the outcomes is to have a semitransparency in addition to the color-coded map (Fig. 3.3). Another disadvantage of this method is that only the external surfaces can be used for measurement.

The landmark-based technique has also been used extensively in the literature and is even easier to learn and perform than the closest point technique. Almost any software available can be used to provide this type of assessment, and it can be done according to the researcher's preferences. This means that landmarks, angles, and lines can be "created" according to the needs of the research being performed. The advantage is that it is a flexible technique that allows a researcher to use the best way to make measurements; however the disadvantage is that there is nonuniformity among the methods used in various studies, making it more difficult to compare results.

Results from studies using landmark-based evaluation techniques are numbers from measurements made at T1 and compared to numbers from measurements made at T2. For example, if SNA were 80° at T1 and 78° at T2, then the change from T1 to T2 could be reported as −2°. This makes changes easy to communicate and provides numbers that can be used for statistical comparisons. Since the accuracy and reliability of the measurements are not influenced by different head orientations [26] as they might be using 2D cephalometrics, prior superimposition of the T2 onto the T1 images is not necessary if the measurements do not include obtaining oriented coordinates. However, given the easy access to voxel-based superimposition and the relative standardization of the procedure, it would be recommended that superimposition of the images be accomplished so that the slices of the T1 and T2 images also have the same common coordinates.

For purposes of clinical evaluation, sequential CBCT changes can also be interpreted using the methods already presented. However, in many circumstances, the clinician is looking for a very quick and simple way of understanding changes that have occurred over time. As clinical time is valuable, techniques for clinical interpretation need to be easy to learn and able to provide the answer a clinician seeks without needing to make additional measurements, plotting landmarks, or making other complex manipulations. Clinicians would like to look at the before and after images and have an answer quickly in a few seconds or minutes. Such clinical evaluation would not be suitable for research purposes where precision and accuracy are mandated. However, a quick evaluation would suit the needs of thousands of providers who want to translate the research advancements in 3D technology to their private practice environment.

In order to make it easier to understand the changes between two longitudinally acquired CBCT scans, the first step required is the reorientation of the T2 scan to the T1 scan position. In this way, the two images will have the same coordinates and spatial orientation. The area of superimposition needs to be chosen according to the clinician's needs. As mentioned previously, a voxel-based superimposition can be accomplished in as little as 10–15 s using commercially available software such as OnDemand 3D. Once this step is done, the software enables the clinician to view the two scans side-by-side, and they are linked together. This means that, as the clinician goes through the slices in any direction on the T1 image, the slices in the T2 image will also change, so the same slice in both images can be compared easily. Figures 3.4, 3.5, and 3.6 show an example of the slices shown side-by-side. The method of evaluation is subjective, and no measurements need to be made, but measurements can be displayed if desired. This protocol can fit the needs of a dentist that does not want to spend hours and weekends learning complicated methods of performing a thorough 3D evaluation.

The decision of which method to use for evaluation of superimposed CBCT images depends on several factors including the final goal of the evaluation, the area to be measured, the knowledge of the examiner, and the availability of appropriate software. Given the underestimation of changes evident using the closest point technique, the selection of appropriate cases is the most important factor when using this type of evaluation. Landmark-based measurement always presents a great potential for successful evaluation in any case, but it is important to remember that standardization of measurements is a problem when comparing results among different studies. In addition, the three-dimensional aspect of the CBCT scan is not fully appreciated when only landmark changes are reported since landmarks themselves represent only a very small portion of the total 3D scan.

3.10 Interaction with Other 3D Modalities

Although many dentists think of 3D diagnosis as the latest development in the profession, there is evidence in the literature to suggest that attempts were made to obtain true 3D patient records more than a century ago. Dental casts themselves have been in use for more than 100 years, and Calvin Case touted the advantages of relating them to

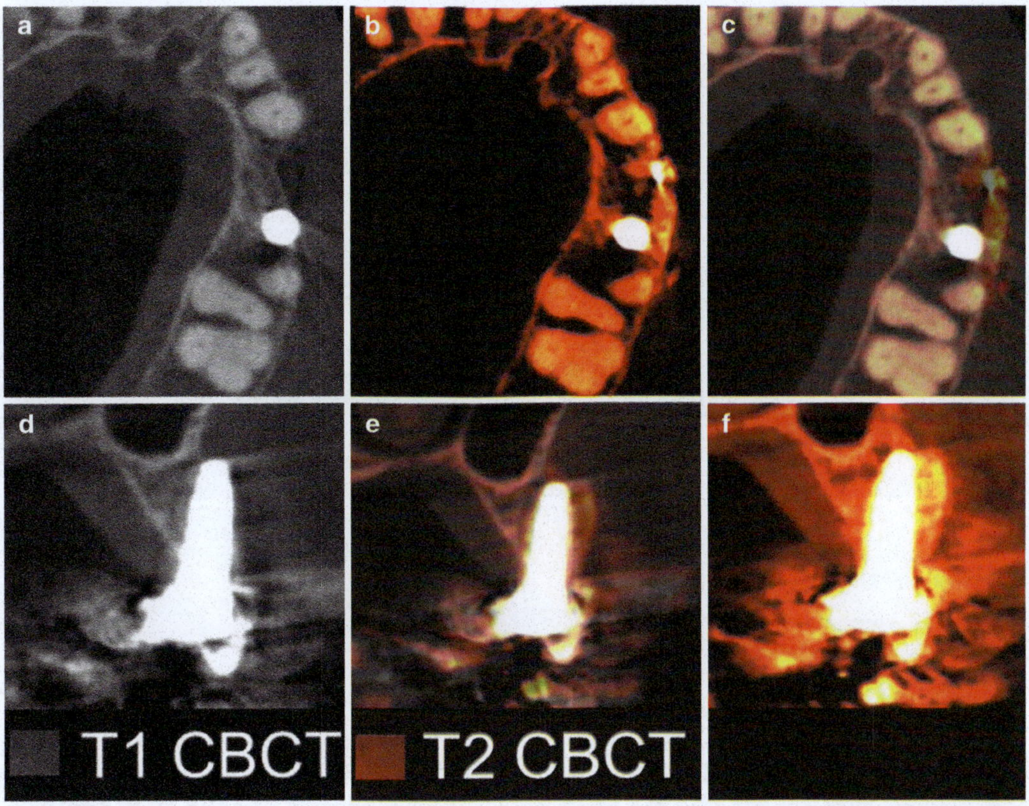

Fig. 3.4 Regional superimposition of the maxilla showing the alveolar changes after bone grafting. T1 and T2 have the same spatial coordination. Axial and coronal slices of presurgery (**a** and **d**) and 4 months after surgery (**b** and **e**). **c** and **f** show the overlay of both T1 and T2. Courtesy of Dr. Janina Golob Deeb (Richmond, VA, USA)

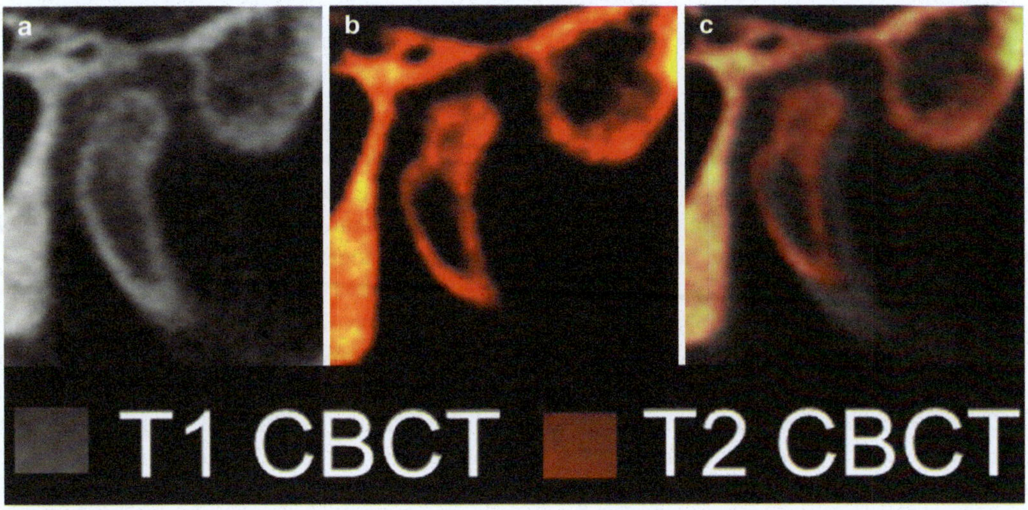

Fig. 3.5 Cranial base superimposition showing the condylar displacement after surgery. T1 and T2 have the same spatial coordination. Sagittal slices of presurgery (**a**) and 3 years after surgery (**b**). Overlay of the images is shown in (**c**). Despite posterior movement of the condyle, the posterior articular space was maintained. Surgery performed by Dr. Jonathas Claus (Florianópolis, Brazil)

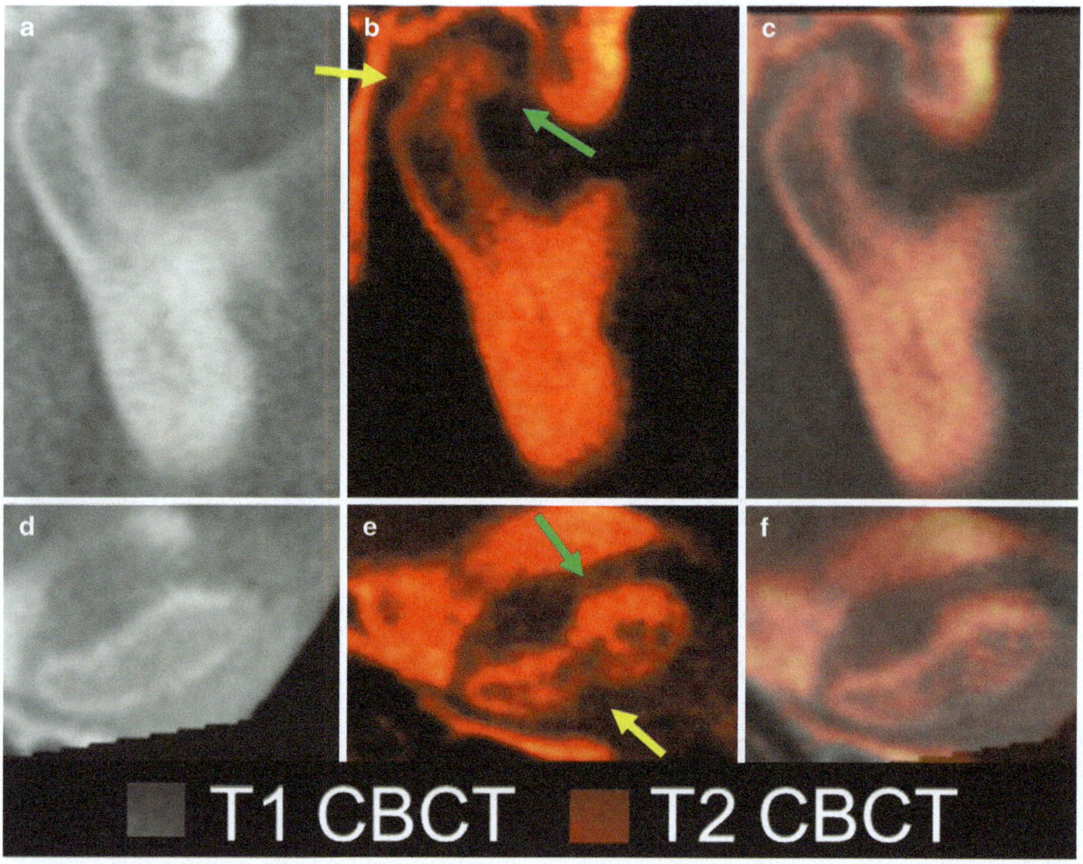

Fig. 3.6 Regional superimposition of the mandible showing the condylar changes after orthognathic surgery. T1 and T2 have the same spatial coordination. Sagittal and axial slices of presurgery (**a** and **d**) and 3 years after surgery (**b** and **e**). **c** and **f** show the overlay of both T1 and T2. Please note the areas of bone resorption (yellow arrows) and bone deposition (green arrows). The image overlay makes it easier and faster to understand the changes in the condyle. Surgery performed by Dr. Jonathas Claus (Florianópolis, Brazil)

full facial plaster casts in his book published in 1908 [27]. In 1915, Van Loon [28, 29] proposed a method for attaching facial models to dental casts and, by reproducing head orientation, creating a full 3D record. Likely, due to the laborious nature of this method, it was not used commonly. Advancements in digital technology have allowed a revival of these forgotten ideas. Obtaining an intraoral dental scan in a matter of minutes and a CBCT scan in a few seconds, using 3D stereophotogrammetry to create a facial image in a split second, and merging all that data together to create a full 3D representation of a patient are what make 3D imaging such an interesting field (Fig. 3.7). The merging of two or more types of data is commonly called image fusion and can be accomplished using CBCT and 3D stereophotogrammetry [30], CBCT and digital models [31], 3D photogrammetry and digital models [32], and all of them together [33].

The method of image fusion varies among the modalities involved. If it involves 3D stereophotogrammetry, then it will be surface- or landmark-based, since there are no voxels available for fusion in this type of image. In the literature, the fusion between CBCT and digital models can also be done using a voxel-based method since conventional impressions can be scanned by a CBCT machine and merged with the patient's CBCT scan [31]. In addition, fiducial markers are

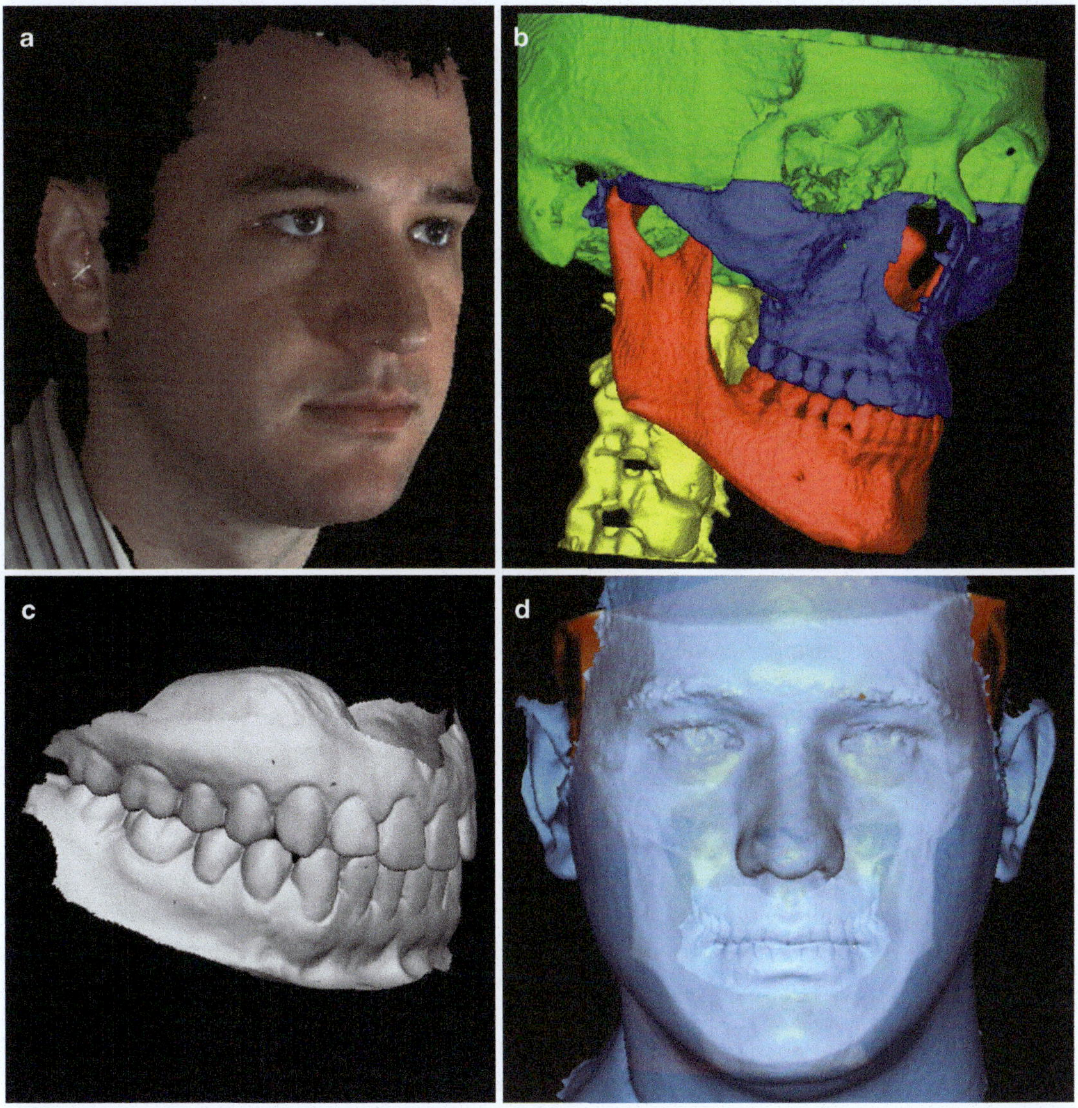

Fig. 3.7 (**a**) 3D stereolithographic photo, (**b**) CBCT scan reconstruction, (**c**) scanned model of the teeth in occlusion, and (**d**) superimposition of the three images

sometimes used in this technique to improve accuracy [34, 35]. This may be necessary since surface-based registration of the dental area may be compromised in cases where the patient has fixed orthodontic appliances in place, or if there is gutta-percha, metallic crowns, implants, or other materials present that create scatter in the CBCT image.

The application of the latest technology in private practice such as guided implant surgery [36], guided temporary anchorage device placement [37], and virtual orthognathic surgery [38] is only possible because of advancements in image fusion. Another groundbreaking advancement is the ability to track tooth root movement with one CBCT scan and multiple dental casts that can be made at various timepoints during orthodontic treatment [39, 40]. This technique allows the clinician to evaluate root position at any point during treatment without the need of an additional radiograph or CBCT, thus achieving the best clinical outcome without additional radiation.

Interaction among 3D modalities will become easier in the future as the equipment required to obtain these images becomes more accessible. As of today, despite being quite expensive, the cost-benefit of having a 3D intraoral scanner and CBCT machine justifies their purchase in the eyes of many clinicians. On the other hand, 3D stereophotogrammetry does not have the ability to improve diagnostic accuracy or optimize the workflow of the dental office. Recent studies have shown that the ability to predict soft-tissue changes following orthognathic surgery is limited even in 2D [41, 42]. In the future, if studies can determine the individual characteristics that make it possible to predict soft-tissue treatment outcomes more accurately, then 3D stereophotogrammetry will become a more helpful tool for comparing possible treatment options.

3.10.1 Case Reports

This section shows different treatment modalities from different dental specialties and how superimposition can improve the quality of diagnosis and follow-up.

3.10.2 Case 1

The patient had an implant placed at the site of the maxillary right lateral incisor by the general dentist. The implant angulation was incorrect, with almost two-thirds of the implant not covered by bone. From a functional standpoint, the crown was properly positioned, but the patient did not want to remove the implant and place a new one to correct the angulation. The patient was referred to the periodontist for a bone graft to avoid fenestration and progression of bone loss. Figure 3.4 shows the before and after surgery with the grafted bone around the implant. In this particular case, if any problem happens, a small field-of-view CBCT can be taken and superimposed to accurately assess the bone remodeling around the implant.

3.10.3 Case 2

The patient had bimaxillary advancement surgery to correct the malocclusion. A bilateral sagittal split osteotomy was done in the mandible and the patient did not have any problems after surgery. The patient was part of a study and, 3 years after two-jaw surgery, had a second CBCT taken for evaluation. Cranial base superimposition (Fig. 3.5c) showed that the condyle moved posteriorly in the glenoid fossa but the posterior articular space was maintained. This happened because the condyle was slightly resorbed 3 years after surgery as can be seen in the ramus superposition (Fig. 3.6b yellow arrow). Also, bone deposition occurred on the anterior part of the condyle (Fig. 3.6b green arrow). The different images complement each other to communicate a full understanding of the changes.

3.10.4 Case 3

This case report illustrates how emerging technology is being applied to optimize digital orthodontic treatment planning. A 28-year-old patient presented with a chief complaint of crooked teeth and deep bite (Fig. 3.8). A CBCT scan and digital models were obtained, and an initial virtual setup was generated using only the digital models (Fig. 3.9a). The teeth (including roots) were segmented from the CBCT scan and superimposed onto the crowns of the initial virtual setup using the surface-based method (Fig. 3.9b). Superimposition of the CBCT teeth and virtual setup allowed for assessment of root angulation and inclination in the virtual setup and, for this patient, several root angulation issues were found in the initial virtual setup. To obtain excellent root parallelism, a final setup was generated taking into

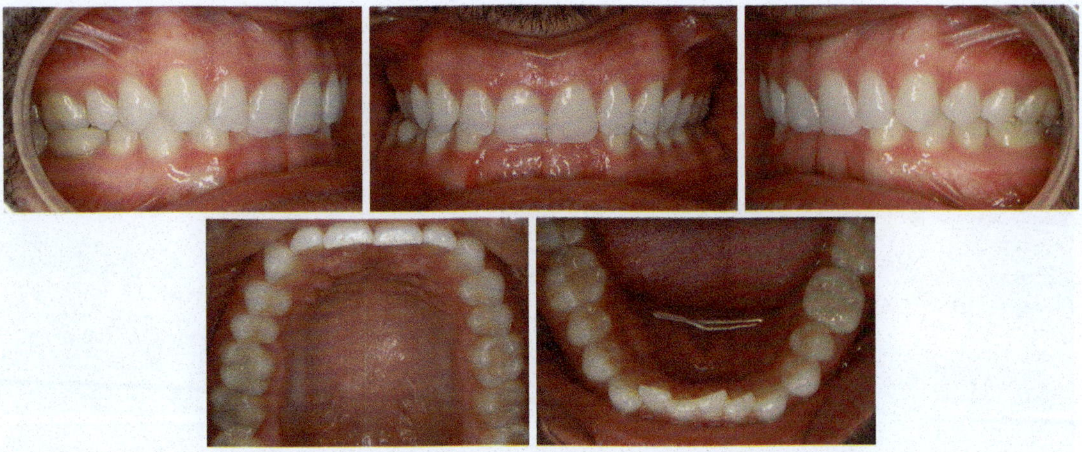

Fig. 3.8 Pretreatment intraoral pictures. Case courtesy of Dr. Hongsheng Tong (Chino, CA)

account the root position (Fig. 3.9c). INBRACE lingual braces were virtually positioned onto the final virtual setup, and the crowns with brackets of this final setup were converted back to the malocclusion maintaining the same bracket position relative to the crown (Fig. 3.10a). An indirect bonding tray was 3D printed based on the virtual bracket placement and was used to bond INBRACE lingual braces onto this patient's teeth (Fig. 3.10a). To check the accuracy of the bracket placement, an intraoral scan was obtained after bonding, and this scan was superimposed onto the malocclusion digital model with virtually positioned brackets (Fig. 3.10b). Color displacement maps found that the physical and virtual bracket placements were within 0.1 mm accuracy of each other (Fig. 3.10c). At 5 months, both maxillary and mandibular arches had custom-fabricated 0.016" NiTi archwires in place and the alignment of the teeth had improved dramatically. The patient's mandibular anterior teeth were in occlusion with maxillary anterior brackets on the lingual resulting in a bite turbo effect that caused opening of the deep bite from day one. The resulting bilateral posterior open bite was closed partially due to simultaneous extrusion of the molars. Light cross elastics were used from the buccal buttons on the mandibular right first and second molars to the lingual interdental loop between the maxillary right first and second molars to address the right-side molar crossbite problem (Fig. 3.11). To monitor root position in three dimensions during treatment for this patient, the segmented pretreatment CBCT teeth were superimposed onto a progress intraoral scan (Fig. 3.12). This method of root tracking can be done at any stage of orthodontic treatment without the need for additional CBCT scans. The treatment was finished in 18 months (Fig. 3.13).

3.11 Conclusion

Cone beam computed tomography (CBCT) for evaluating longitudinal changes is a rapidly changing and exciting area in dentistry. There are several techniques and methods of evaluation available to accomplish superimposition. As radiation dosages continue to decrease with the development of new technologies, CBCT evaluations will be more commonly performed.

There is a need for clinicians to understand how to assess the quality of the many studies being done in this area and also to learn how to evaluate their own cases in a simple and efficient manner. The learning curve in 3D technology is much steeper compared to that used when

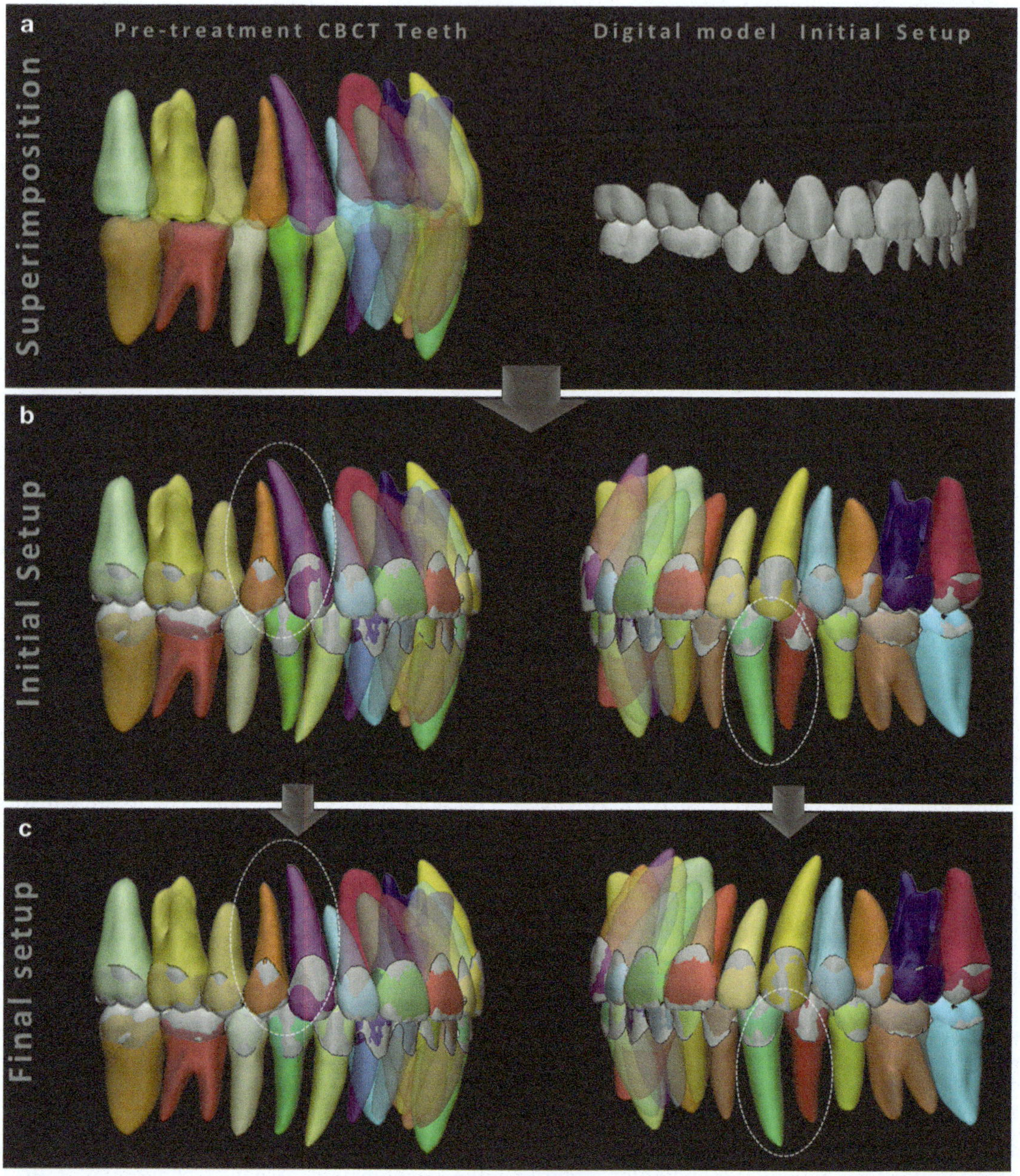

Fig. 3.9 (**a**) Pretreatment CBCT showing the tooth position and root angulation. On the right side showing a setup done using the scanned model. (**b**) Showing left and right side of the initial setup. The roots from the CBCT were merged with the teeth from the scanned model. The setup shows the improper root angulation when the digital setup takes into account the crowns only. (**c**) Final setup after considering the root position for each individual tooth

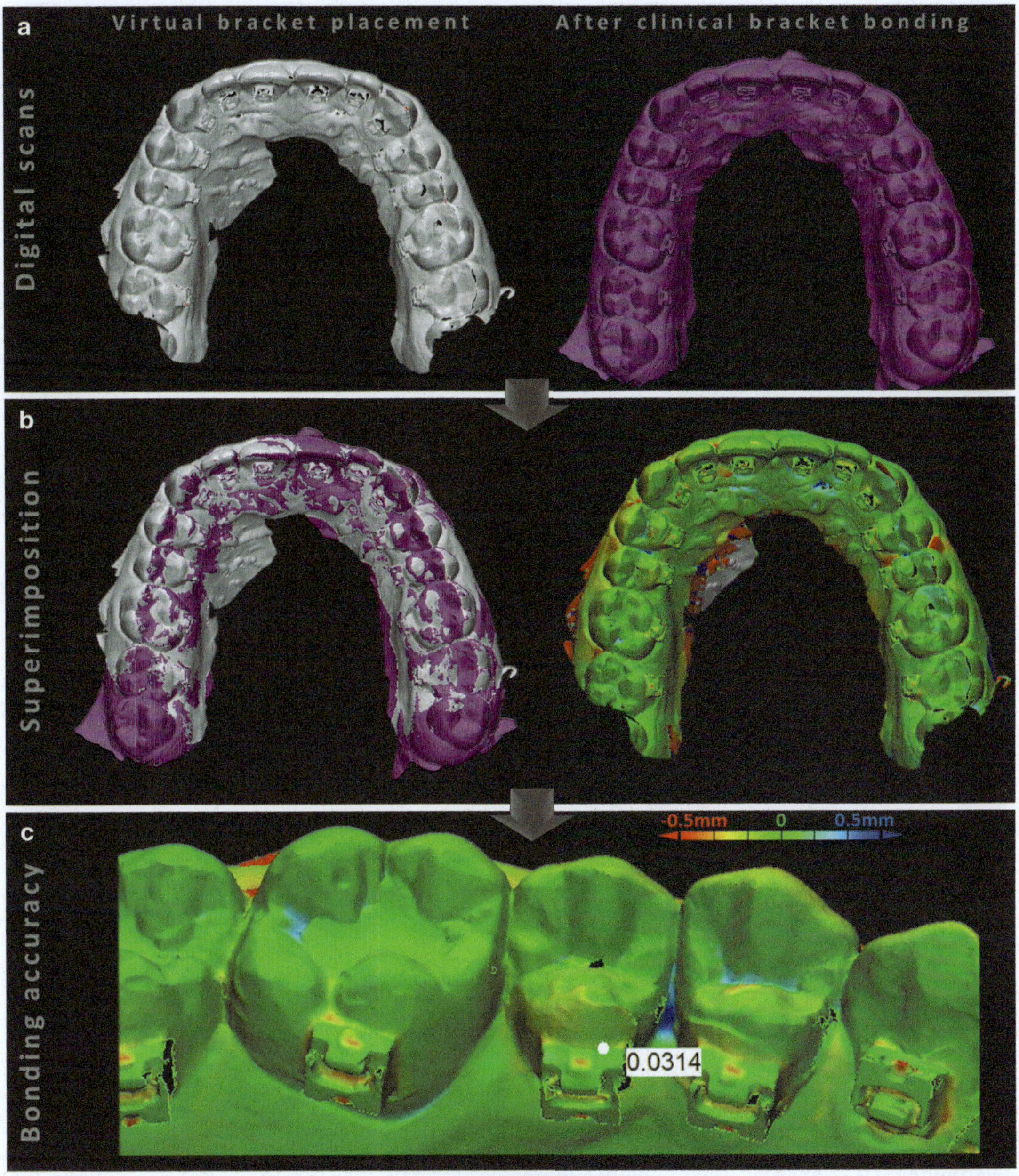

Fig. 3.10 (**a**) On the left side, the scanned teeth with the virtual brackets in place. On the right side, after bonding. (**b**) Surface-based superimposition performed using the models, and the color map shows the distance between the planned bracket position and the final bracket position. This serves to measure the accuracy of the indirect bonding technique. (**c**) Closeup view of the upper left side showing the accuracy of bracket placement (3D images courtesy from Drs. André Weissheimer, Robert Lee and John Pham)

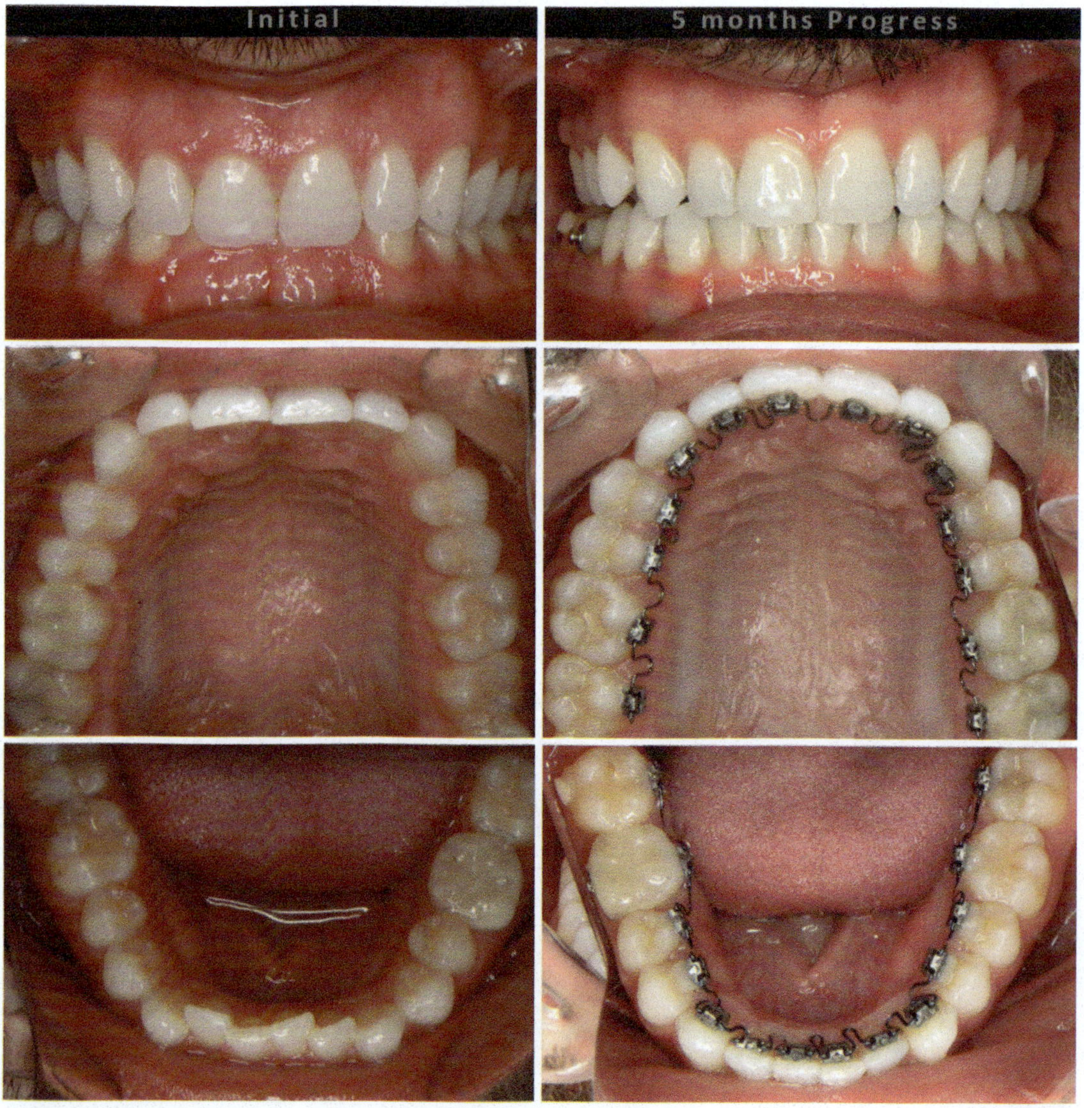

Fig. 3.11 Initial and 5 months' progress of the case treated with INBRACE

interpreting 2D images. Additionally, more changes have occurred in the 3D world in the last 10 years compared with all of the developments in 2D imaging over its lifetime. As seen in the last few years, the barrier between clinicians and the knowledge necessary to comprehend these developments will be reduced as the software programs become more automated and user-friendly.

Overall, the current literature suggests that longitudinal superimposition of CBCT images should be done using the voxel-based technique. No consensus currently exists about how the evaluation should be accomplished or interpreted, and there is room for improvement in this area. Therefore, there is a need to stay up-to-date with the latest innovations in this cutting-edge technology.

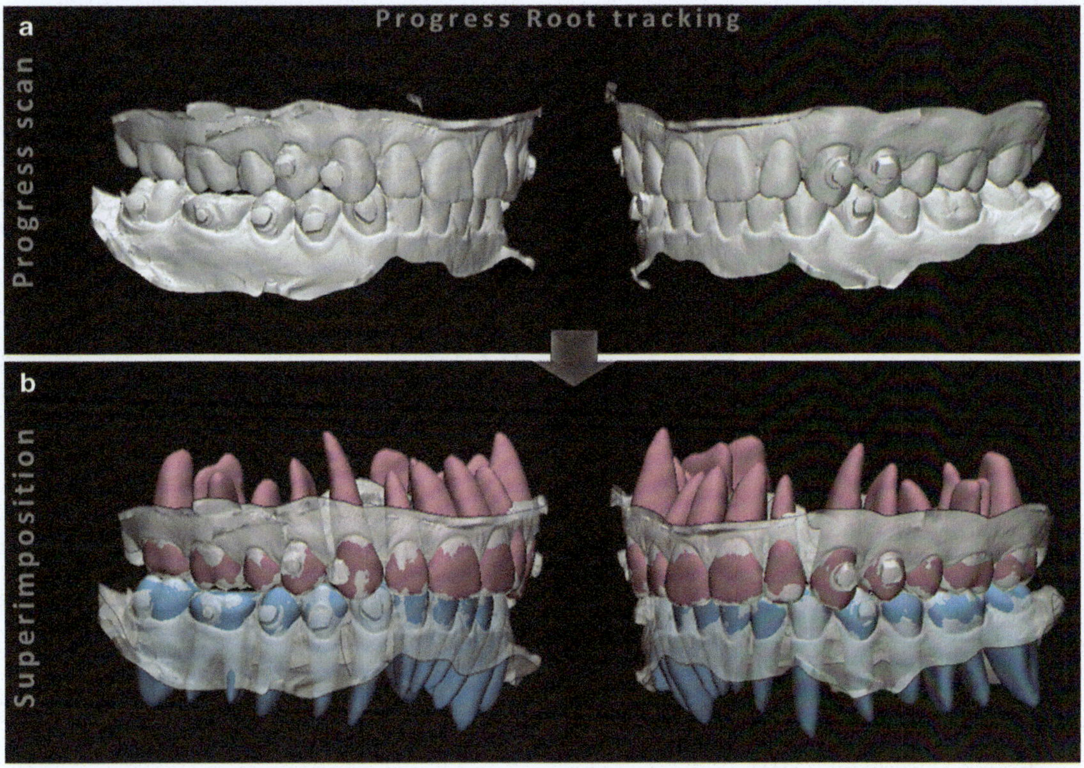

Fig. 3.12 (**a**) Scan taken during treatment. (**b**) The crown and roots from the pretreatment CBCT were merged into the new position so the doctor can track root position without additional radiation

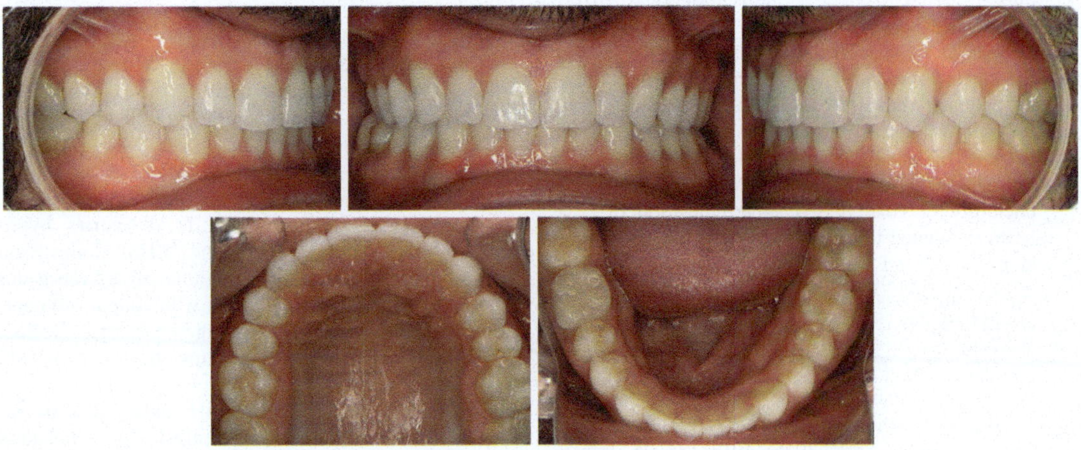

Fig. 3.13 Final pictures after 18 months of treatment

References

1. McCance AM, Moss JP, Fright WR, James DR, Linney AD. A three dimensional analysis of soft and hard tissue changes following bimaxillary orthognathic surgery in skeletal III patients. Br J Oral Maxillofac Surg. 1992;30(5):305–12.
2. McCance AM, Moss JP, Wright WR, Linney AD, James DR. A three-dimensional soft tissue analysis of 16 skeletal class III patients following bimaxillary surgery. Br J Oral Maxillofac Surg. 1992;30(4):221–32.
3. Lagravere MO, Major PW, Carey J. Sensitivity analysis for plane orientation in three-dimensional cephalometric analysis based on superimposition of serial cone beam computed tomography images. Dentomaxillofac Radiol. 2010;39(7):400–8. https://doi.org/10.1259/dmfr/17319459.
4. Kawamata A, Fujishita M, Nagahara K, Kanematu N, Niwa K, Langlais RP. Three-dimensional computed tomography evaluation of postsurgical condylar displacement after mandibular osteotomy. Oral Surg Oral Med Oral Pathol Oral Radiol Endod. 1998;85(4):371–6.
5. Cevidanes LH, Bailey LJ, Tucker GR Jr, Styner MA, Mol A, Phillips CL, et al. Superimposition of 3D cone-beam CT models of orthognathic surgery patients. Dentomaxillofac Radiol. 2005;34(6):369–75. https://doi.org/10.1259/dmfr/17102411.
6. Koerich L, Burns D, Weissheimer A, Claus JD. Three-dimensional maxillary and mandibular regional superimposition using cone beam computed tomography: a validation study. Int J Oral Maxillofac Surg. 2016;45(5):662–9. https://doi.org/10.1016/j.ijom.2015.12.006.
7. Koerich L, Weissheimer A, de Menezes LM, Lindauer SJ. Rapid 3D mandibular superimposition for growing patients. Angle Orthod. 2017;87(3):473–9. https://doi.org/10.2319/072316-574.1.
8. Nada RM, Maal TJ, Breuning KH, Berge SJ, Mostafa YA, Kuijpers-Jagtman AM. Accuracy and reproducibility of voxel based superimposition of cone beam computed tomography models on the anterior cranial base and the zygomatic arches. PLoS One. 2011;6(2):e16520. https://doi.org/10.1371/journal.pone.0016520.
9. Weissheimer A, Menezes LM, Koerich L, Pham J, Cevidanes LH. Fast three-dimensional superimposition of cone beam computed tomography for orthopaedics and orthognathic surgery evaluation. Int J Oral Maxillofac Surg. 2015;44(9):1188–96. https://doi.org/10.1016/j.ijom.2015.04.001.
10. Lee JH, Kim MJ, Kim SM, Kwon OH, Kim YK. The 3D CT superimposition method using image fusion based on the maximum mutual information algorithm for the assessment of oral and maxillofacial surgery treatment results. Oral Surg Oral Med Oral Pathol Oral Radiol. 2012;114(2):167–74. https://doi.org/10.1016/j.tripleo.2011.06.003.
11. Almukhtar A, Ju X, Khambay B, McDonald J, Ayoub A. Comparison of the accuracy of voxel based registration and surface based registration for 3D assessment of surgical change following orthognathic surgery. PLoS One. 2014;9(4):e93402. https://doi.org/10.1371/journal.pone.0093402.
12. Lemieux G, Carey JP, Flores-Mir C, Secanell M, Hart A, Lagravere MO. Precision and accuracy of suggested maxillary and mandibular landmarks with cone-beam computed tomography for regional superimpositions: An in vitro study. Am J Orthod Dentofac Orthop. 2016;149(1):67–75. https://doi.org/10.1016/j.ajodo.2015.06.025.
13. Cevidanes LH, Heymann G, Cornelis MA, DeClerck HJ, Tulloch JF. Superimposition of 3-dimensional cone-beam computed tomography models of growing patients. Am J Orthod Dentofac Orthop. 2009;136(1):94–9. https://doi.org/10.1016/j.ajodo.2009.01.018.
14. Ruellas AC, Yatabe MS, Souki BQ, Benavides E, Nguyen T, Luiz RR, et al. 3D Mandibular Superimposition: Comparison of Regions of Reference for Voxel-Based Registration. PLoS One. 2016;11(6):e0157625. https://doi.org/10.1371/journal.pone.0157625.
15. Bjork A. Prediction of mandibular growth rotation. Am J Orthod. 1969;55(6):585–99.
16. Bjork A. Facial growth in man, studied with the aid of metallic implants. Acta Odontol Scand. 1955;13(1):9–34.
17. Bjork A. The use of metallic implants in the study of facial growth in children: method and application. Am J Phys Anthropol. 1968;29(2):243–54.
18. Liou EJ, Pai BC, Lin JC. Do miniscrews remain stationary under orthodontic forces? Am J Orthod Dentofac Orthop. 2004;126(1):42–7. https://doi.org/10.1016/S0889540604002057.
19. Chen G, Chen S, Zhang XY, Jiang RP, Liu Y, Shi FH, et al. A new method to evaluate the positional stability of a self-drilling miniscrew. Orthod Craniofac Res. 2015;18(3):125–33. https://doi.org/10.1111/ocr.12065.
20. Ruellas AC, Huanca Ghislanzoni LT, Gomes MR, Danesi C, Lione R, Nguyen T, McNamara JA Jr, Cozza P, Franchi L, Cevidanes LHS. Comparison and reproducibility of 2 regions of reference for maxillary regional registration with cone-beam computed tomography. Am J Orthod Dentofac Orthop. 2016;149(4):533–42. https://doi.org/10.1016/j.ajodo.2015.09.026.
21. Grybauskas S, Locs J, Salma I, Salms G, Berzina-Cimdina L. Volumetric analysis of implanted biphasic calcium phosphate/collagen composite by three-dimensional cone beam computed tomography head model superimposition. J Craniomaxillofac Surg. 2015;43(1):167–74. https://doi.org/10.1016/j.jcms.2014.11.003.
22. de Paula LK, Ruellas ACO, Paniagua B, Styner M, Turvey T, Zhu H, Wang J, Cevidanes LHS. One-year assessment of surgical outcomes in class III patients using cone beam computed tomography. Int J Oral

Maxillofac Surg. 2013;42(6):780–9. https://doi.org/10.1016/j.ijom.2013.01.002.
23. Weissheimer A, de Menezes LM, Mezomo M, Dias DM, de Lima EM, Rizzatto SM. Immediate effects of rapid maxillary expansion with Haas-type and hyrax-type expanders: a randomized clinical trial. Am J Orthod Dentofac Orthop. 2011;140(3):366–76. https://doi.org/10.1016/j.ajodo.2010.07.025.
24. Oh KM, Seo SK, Park JE, Sim HS, Cevidanes LHS, Kim YJ, Park YH. Post-operative soft tissue changes in patients with mandibular prognathism after bimaxillary surgery. J Craniomaxillofac Surg. 2013;41(3):204–11. https://doi.org/10.1016/j.jcms.2012.09.001.
25. Jabar N, Robinson W, Goto TK, Khambay BS. The validity of using surface meshes for evaluation of three-dimensional maxillary and mandibular surgical changes. Int J Oral Maxillofac Surg. 2015;44(7):914–20. https://doi.org/10.1016/j.ijom.2015.02.005.
26. El-Beialy AR, Fayed MS, El-Bialy AM, Mostafa YA. Accuracy and reliability of cone-beam computed tomography measurements: Influence of head orientation. Am J Orthod Dentofac Orthop. 2011;140(2):157–65. https://doi.org/10.1016/j.ajodo.2010.03.030.
27. Case CS. Dental orthopedia. Chicago, IL: C. S. Case Company; 1908.
28. Van Loon JAW. A new method for indicating normal and abnormal relations of the teeth to the facial lines. Dent Cosmos. 1915;57(9):973–83.
29. Van Loon JAW. A new method for indicating normal and abnormal relations of the teeth to the facial lines (continued from page 983). Dent Cosmos. 1915;57(10):1093–101.
30. Maal TJ, Plooij JM, Rangel FA, Mollemans W, Schutyser FA, Berge SJ. The accuracy of matching three-dimensional photographs with skin surfaces derived from cone-beam computed tomography. Int J Oral Maxillofac Surg. 2008;37(7):641–6. https://doi.org/10.1016/j.ijom.2008.04.012.
31. Swennen GR, Mommaerts MY, Abeloos J, De Clercq C, Lamoral P, Neyt N, Casselman J, Schutyser F. A cone-beam CT based technique to augment the 3D virtual skull model with a detailed dental surface. Int J Oral Maxillofac Surg. 2009;38(1):48–57. https://doi.org/10.1016/j.ijom.2008.11.006.
32. Rangel FA, Maal TJ, Berge SJ, van Vlijmen OJ, Plooij JM, Schutyser F, Kuijpers-Jaqtman AM. Integration of digital dental casts in 3-dimensional facial photographs. Am J Orthod Dentofac Orthop. 2008;134(6):820–6. https://doi.org/10.1016/j.ajodo.2007.11.026.
33. Plooij JM, Maal TJ, Haers P, Borstlap WA, Kuijpers-Jagtman AM, Berge SJ. Digital three-dimensional image fusion processes for planning and evaluating orthodontics and orthognathic surgery. A systematic review. Int J Oral Maxillofac Surg. 2011;40(4):341–52. https://doi.org/10.1016/j.ijom.2010.10.013.
34. Gateno J, Xia JJ, Teichgraeber JF, Christensen AM, Lemoine JJ, Liebschner MA, Gliddon MJ, Briggs ME. Clinical feasibility of computer-aided surgical simulation (CASS) in the treatment of complex cranio-maxillofacial deformities. J Oral Maxillofac Surg. 2007;65(4):728–34. https://doi.org/10.1016/j.joms.2006.04.001.
35. Uechi J, Okayama M, Shibata T, Muguruma T, Hayashi K, Endo K, Mizoguchi I. A novel method for the 3-dimensional simulation of orthognathic surgery by using a multimodal image-fusion technique. Am J Orthod Dentofac Orthop. 2006;130(6):786–98. https://doi.org/10.1016/j.ajodo.2006.03.025.
36. Deeb GR, Soliman O, Alsaad F, Jones P, Deluke D, Laskin DM. Simultaneous virtual planning implant surgical guides and immediate laboratory-fabricated provisionals: an impressionless technique. J Oral Implantol. 2016;42(4):363–9. https://doi.org/10.1563/aaid-joi-D-15-00158.
37. Maino BG, Paoletto E, Lombardo L III, Siciliani G. A three-dimensional digital insertion guide for palatal miniscrew placement. J Clin Orthod. 2016;50(1):12–22.
38. Li Y, Jiang Y, Ye B, Hu J, Chen Q, Zhu S. Treatment of dentofacial deformities secondary to osteochondroma of the mandibular condyle using virtual surgical planning and 3-dimensional printed surgical templates. J Oral Maxillofac Surg. 2016;74(2):349–68. https://doi.org/10.1016/j.joms.2015.06.169.
39. Lee RJ, Pham J, Choy M, Weissheimer A, Dougherty HL Jr, Sameshima GT, Tong H. Monitoring of typodont root movement via crown superimposition of single cone-beam computed tomography and consecutive intraoral scans. Am J Orthod Dentofac Orthop. 2014;145(3):399–409. https://doi.org/10.1016/j.ajodo.2013.12.011.
40. Lee RJ, Weissheimer A, Pham J, Go L, de Menezes LM, Redmond WR, Loos JF, Sameshima GT, Tong H. Three-dimensional monitoring of root movement during orthodontic treatment. Am J Orthod Dentofac Orthop. 2015;147(1):132–42. https://doi.org/10.1016/j.ajodo.2014.10.010.
41. Kaipatur NR, Flores-Mir C. Accuracy of computer programs in predicting orthognathic surgery soft tissue response. J Oral Maxillofac Surg. 2009;67(4):751–9. https://doi.org/10.1016/j.joms.2008.11.006.
42. Peterman RJ, Jiang S, Johe R, Mukherjee PM. Accuracy of dolphin visual treatment objective (VTO) prediction software on class III patients treated with maxillary advancement and mandibular setback. Prog Orthod. 2016;17(1):19. https://doi.org/10.1186/s40510-016-0132-2.

Imaging and Analysis for the Orthodontic Patient

4

Jae Hyun Park and Dawn P. Pruzansky

Abstract

Cone-beam computed tomography (CBCT) has become an integral component of orthodontic diagnosis and treatment planning. The leap from 2D to 3D analysis has allowed for a more comprehensive evaluation before, during, and after orthodontic therapy. CBCT has been instrumental in localizing impacted teeth; evaluating asymmetry, airway, and temporomandibular joint anatomy; selecting sites for temporary skeletal anchorage; and assessing root length and alveolar bone dimensions. In this chapter, CBCT imaging and analysis of the orthodontic patient will be discussed.

J. H. Park (✉)
Postgraduate Orthodontic Program, Arizona School of Dentistry and Oral Health, A.T. Still University, Mesa, AZ, USA

Graduate School of Dentistry, Kyung Hee University, Seoul, South Korea
e-mail: JPark@atsu.edu

D. P. Pruzansky
Postgraduate Orthodontic Program, Arizona School of Dentistry and Oral Health, A.T. Still University, Mesa, AZ, USA
e-mail: dpruzansky@atsu.edu

O. Kadioglu, G. F. Currier (eds.), *Craniofacial 3D Imaging*,
https://doi.org/10.1007/978-3-030-00722-5_4

4.1 Introduction

Cephalometric analysis is an important component of orthodontic diagnosis and treatment planning. Together with diagnostic casts and photographs, an accurate problem list can be determined. The information in the radiographs gives practitioners important insights that cannot be seen with models or photographs alone. One is able to determine the position of the maxilla and mandible in relation to the cranial base and to one another in addition to the position of the teeth, alveolar bone support, and temporomandibular joint anatomy. Treatment planning may vary greatly with a dental versus skeletal disharmony, and the radiographs allow us to make that assessment.

4.2 History

A combination of the Broadbelt-Bolton cephalometer, developed in 1931, and long-term anthropomorphic data is the basis of the current practice of cephalometric diagnosis [1]. The cephalometer allows for reproducibility of patient positioning, which in turn has allowed for serial cephalometric studies. From these radiographic images, points and structures within craniofacial anatomy can be located and measured.

A cephalometric radiograph is obtained using a fixed frame, so that the projections are standardized and can be compared. Patients are positioned while standing, with the sagittal plane of their head 5 ft from the actual source of the x-rays and the axis of the two ear plugs aligned with the point source of the x-rays [2]. A distance of 15 cm from the midsagittal plane to the film cassette increases reproducibility, as a greater distance will magnify the image.

The limitations and problems associated with two-dimensional (2D) lateral cephalograms include magnification, superimpositions of bilateral anatomic structures, foreshortening, and elongation. There is also a "differential magnification" of structures on the right versus left side, given the difference in the distance to the film. The inability to view these bilateral structures may create an error in diagnosing certain asymmetries.

Cone-beam computed tomography (CBCT), although growing in popularity and usage, has been considered an adjunct when 2D imaging does not provide enough information for clinical diagnosis. Computed tomography (CT) has been used successfully in the medical field since 1971 [3]. In 1979, the Nobel Prize was awarded to Allan M. Cormack and Godfrey N. Hounsfield for the development of computer-assisted tomography [4, 5]. This later came to be known as computed axial tomography (CAT), and these images contributed vastly to the medical diagnostic field, since organs and systems could be viewed. Although CAT scans have increased in use for diagnostic purposes, the risks associated with radiation dosage outweighed the potential benefit to the dental field [6, 7]. In 1998, CBCT was introduced to the dental community as a lower radiation dose and lower scanning cost option to diagnostics [8, 9].

In dentistry and orthodontics, high-resolution CBCT is used to acquire a low-distortion digital image of the hard and soft tissues of the craniofacial structures. Unlike conventional CT which uses a fan-shaped beam to create multiple thin slices, CBCT machines have a cone-shaped beam. In addition, the resolution is measured in voxels instead of pixels, which results in a sharper image. These images are most commonly stored as a digital imaging and communications in medicine (DICOM) file. Panoramic and cephalometric projections that are produced by CBCT are transformed into a three-dimensional (3D) format after the data has been reformatted in a volume by computer software, the most common of which is multiplanar reconstruction (MPR) [10–12].

These images can be used to gather diagnostic information on temporomandibular joints, anatomic features of the craniofacial bones, to measure the width of alveolus and the position of teeth within the bone, to determine the position of supernumerary and impacted teeth, and to identify sites for implant placement or osteotomies [13, 14]. CBCT imaging is also used to plan for orthodontic and orthognathic surgery treatment, to assess skeletal displacements after osteotomies, to verify treatment outcomes, and to determine stability [15].

4.3 3D Cephalometric Image Preparation

Often, the 3D data is converted to a 2D image for analysis, yet the ability to perform a true 3D analysis may be a key to overcome all the traditional cephalometric disadvantages. A systematic method to digitize and analyze 3D radiographic images has not yet been well established, although much research has been devoted to this task. Kochel et al. [16, 17] developed a 3D soft tissue analysis based on the data derived from 3D stereophotogrammetric images. All the measurements were taken from the projections of the digitized points and were used to evaluate correlation of the 3D soft tissue data to variables retrieved from 2D lateral cephalometric analysis. Farronato et al. [18] proposed a 10-point 3D analysis of CBCT images directly digitized on the rendered view. They reported the reliability and the reproducibility of their method and compared it to 2D data, but the sample size prevented normative values. Bayome et al. [19] proposed a new 3D cephalometric analysis and evaluated the relationships among skeletal and dentoalveolar variables. Their study has also provided the norms of the 3D variables of a Korean normal occlusion population. As 3D analysis becomes more commonplace, normative data will be more readily available. Continued research should focus on a systematic and repeatable approach to analysis that can be integrated into diagnosis and treatment planning protocols.

4.3.1 Reorientation of Head Position

The reorientation process depends on placing the image of the head into a known, repeatable position in the coordinate system through defining the origin point and the *X*, *Y*, and *Z* planes. These definitions should be based on landmarks that are least susceptible to asymmetry and least affected by treatment procedures to strengthen the reliability and validity of the required planes. Nasion (N) and anterior nasal spines (ANS) tend to fall on or very close to the midsagittal plane in 90% of the population [20]. Therefore, Bayome et al. [19] selected N as an origin of the 3D coordinate system. The horizontal plane (*X*) was defined through the right and left orbitales (Or) and the left porion (Po), while the midsagittal plane (*Y*) was defined as the perpendicular plane passing through N and ANS. The vertical plane (*Z*) was perpendicular to both *X* and *Y* (Fig. 4.1).

Swennen et al. [21] proposed a reorientation method with the origin at Sella (S). Park et al. [22] suggested the use of the right and left zygomatic suture points or the orbitale (Or) as a stable transverse line to guide the construction of the horizontal plane in 3D coordinate systems. Kook and Kim [23] proposed a clinical method to easily reorient the head using frontal facial and intraoral photographs. Interestingly, Gupta et al. [24] studied the effect of landmark identification with and without orientation of the CBCT image and found no statistically significant differences between the two images.

4.3.2 Segmentation of CBCT Images

Volume segmentation, the allocation and separation of an anatomical structure or region of interest from the 3D volumes, should be considered prior to landmark identification. The difficulty of segmentation is attributed to the variability and complexity of the biological tissues, the large size of the data sets, and the limitations of imaging techniques, such as low contrast, motion, and noise, which may result in indistinct boundaries of the adjacent structures. Thresholds must be set to filter data depending on voxel intensities to allow for differentiation of soft tissue, bone, etc.

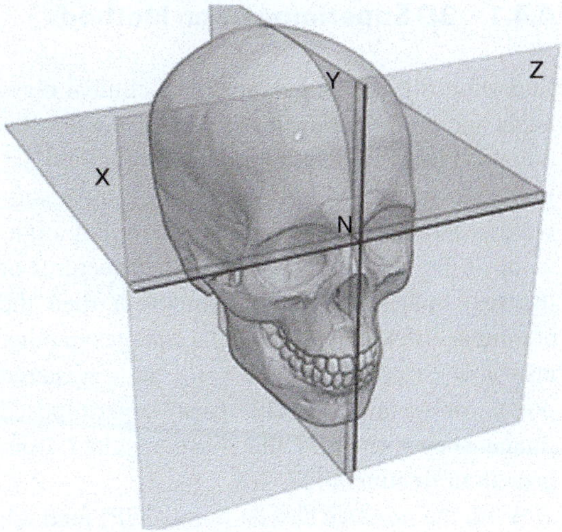

Fig. 4.1 Reorientation of the head and coordinate system. N, nasion; *X*, the horizontal plane; *Y*, the midsagittal plane; Z, the vertical plane (Reproduced with permission from Sem Orthod)

4.4 3D CBCT Superimposition

3D superimposition methods are based on either registration points or mathematical algorithms [25]. In the registration point approach, certain landmarks are registered on two volumetric images, which will coincide when the superimposition is made. With mathematical algorithms, the initial 3D CBCT scan is considered to be the volume of interest (VOI) or the reference volume [12]. Software, based on probability and information theory, then superimposes the second scan over the VOI in its best-fit position, and the fusion process of the two images occurs automatically. It is not dependent on an operator skill and is faster than manual methods [26].

4.4.1 3D Superimposition Methods

Superimposition methods used for clinical diagnosis and treatment evaluation purposes in orthodontic treatment and craniofacial surgeries vary in their benefits and limitations. With most software programs, a clinician does the initial alignment of the landmarks or anatomic structures of the two images to be superimposed; then the computer software measures the changes in other anatomic structures relative to the registered points or structures. The final superimposed image shows changes that have resulted from growth or treatment [27–30].

With the iterative closest point (ICP) method, a more accurate measurement can be made by using the same points on the same surface with fusion at different time points [31, 32]. The accuracy of linear measurements in 2D cephalograms and 3D scans are not the same because of a difference in the size and location of the objects in the two imaging systems [33]. When utilizing CBCT data, the ICP method allows for the precise fusing of two 3D images from growing patients [31, 32]. With the ICP technique, an operator manually defines a certain domain on the surface of the CBCT scans such as the outline of the anterior cranial base from the superior view (Fig. 4.2). Then the software automatically matches and registers the identical landmarks of the selected domains on the two scans and completes the superimposition process. The operator can evaluate and measure the changes relative to the registered surfaces. After mastering the use of the software, image measurements can be made with great repeatability [29]. With the ICP 3D superimposition technique, registration of the scans over the cranial base is reported to be an accurate method for superimposition [34–36]. This method can be used for a valid and reproducible assessment of treatment outcomes for growing subjects. ICP is also considered to be clinically valuable because of the manageability and 3D accuracy of data compared with MPR images (Fig. 4.3) [35].

Gianquinto et al. [37] introduced a reproducible CBCT superimposition method based on the posterior cranial base in a single software package using a step-by-step manual technique. With this method, the craniofacial volume for each of the patients is imported to 3D CBCT superimposition software. The software resamples the scans

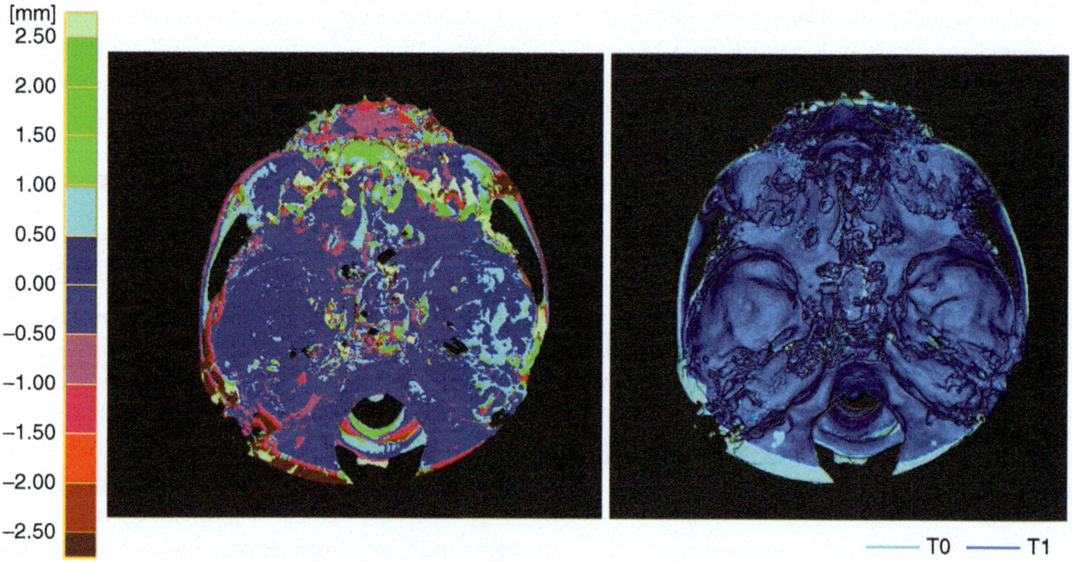

Fig. 4.2 The iterative closest point (ICP) method. (**a**) Cranial base superimposition performed on all areas of the cranial base except the peripheral growing zone. (**b**) Merged image of pre- (T0) and posttreatment (T1) CBCT scans, superimposed at the cranial base (Reproduced with permission from Sem Orthod)

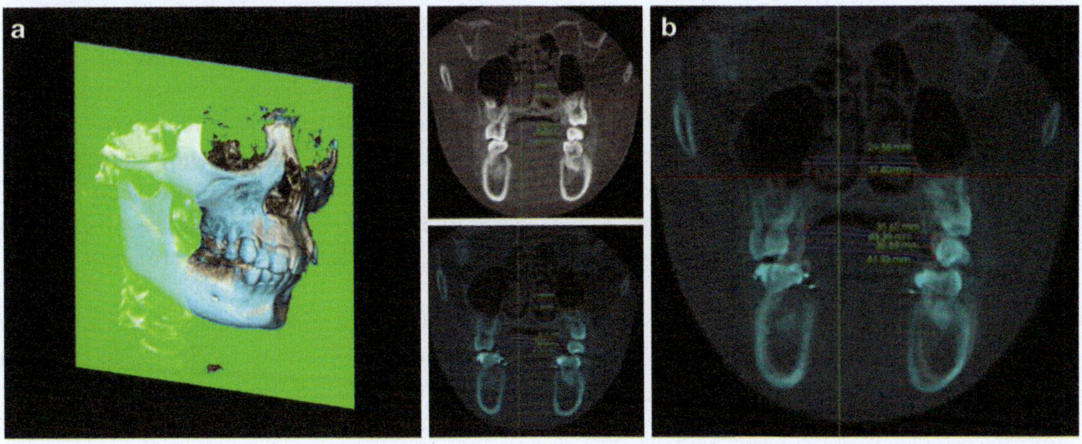

Fig. 4.3 Superimposition using the iterative closest point (ICP) method. (**a**) The combined images, pre-(gray) and (light blue) images, with illustration of an arbitrary coronal plane (green). (**b**) To facilitate measurement, the 3D sections were converted to 2D data (Reproduced with permission from Sem Orthod)

to a 0.5 mm voxel size and then superimposes the images with a mutual information algorithm. At this stage, the operator uses a semiautomatic technique to extract posterior cranial base surface data, which results in a colored map based on the distance between the two volumes relative to the cranial base [38].

The other method is voxel-based image registration, which is an accurate and reproducible semiautomated technique for 3D CBCT superimposition [39, 40]. For example, when a patient is fully grown and developed, registration of the superimposed CBCT images of the zygomatic arches can be considered as an alternative to the anterior cranial base [41]. After superimposition, the differences between the two surfaces are mapped with about 600,000 color-coded surface distances in millimeters, which helps a practitioner quantify and visually assess the hard and soft tissue changes between the two scans relative to the cranial base [27, 42].

4.4.2 Photographs and CBCT Superimposition

More recently, CBCT has been used with the registration of skin surface images [34], so clinicians can quantitatively assess 3D maxillofacial morphology and evaluate linear and angular changes in facial soft and hard tissues in clinical procedures. Standard normative 3D values for the craniofacial hard and soft tissues of normal women were calculated by Terajima et al. [43] and were then compared with 3D CT measurements before and after patients had orthognathic surgery. They reported that with this method, they were able to quantitatively assess deviations of craniofacial structures from the norm before surgery and the changes in the hard and soft tissues after surgery. Cevidanes et al. [27] also reported that because 3D surface models superimposition is currently time consuming and computing intensive, its use in routine clinical practice is not very practical at this time. Therefore, simpler analytical techniques are required for 3D superimposition techniques to be viable in routine daily practice.

Clinicians, scientists, and engineers have developed techniques for superimposing facial 2D photographs [44], 3D photographs [42, 44–46], and digital models [47–50] over CBCT scans. It is reported that the integration of 3D photographs and CBCT images has shown minimal errors in the assessment of bone and soft tissue [46]. Therefore, this process can be used as an objective tool for diagnosis and treatment planning in orthodontics and orthognathic surgery.

4.5 3D Analysis Procedures

Several software programs have been developed to view, digitize, measure, and analyze CBCT data. Ludlow et al. [51] recommended the identification of landmarks on the MPR slices due to the method's high accuracy. Another study found that digitizing landmarks on the rendered view is preferable due to its ease and shorter analysis time [19]. Several studies reported high accuracy of linear and angular measurements in 3D volume render CBCT images compared to physical measurements [52–55]. With the continued improvement of 3D cephalometric analysis, new landmarks and reference planes have been made possible [56]. The ability to section the 3D data allows practitioners to place landmarks accurately on structures that were not available on the 2D cephalograms. In turn, the 3D Cartesian system facilitated the creation of new reference planes and the evaluation of curvatures as well as their linear and angular relationships.

4.5.1 Asymmetry

Two-dimensional methods to diagnose asymmetry can often contain errors due to superimposition and magnification. CBCT allows for a more precise evaluation combining volumetric data with distance and surface area [57]. There are several techniques to identify facial asymmetry including stereophotogrammetry, 3D dynamic models, surface scanning, and CBCT [58].

Stereophotogrammetry is utilized in 3D surface imaging to assess the soft tissue morphology. Two photographs are captured to form a stereo pair which can then be reconstructed. Ras and colleagues [59] used this method to identify the best reference plane to assess facial asymmetry, which they concluded as a plane perpendicular to and bisecting the line that connects the landmarks Exocanthion.

Laser surface scanning of the right and left halves of the face simultaneously can be merged in the overlap area when the data is matched. Djordjevic and colleagues [60] compared the original facial image to a mirrored facial image and mapped the areas where symmetry deviated. They also divided the face into upper, middle, and lower thirds for a more detailed analysis.

Kook and Kim [23] suggest a stepwise process to evaluate asymmetry using CBCT data (using iCAT with Invivo5 software) (Fig. 4.4):

1. Establish the midsagittal plane in the axial section using the bottom view. Clip the axial section for better visibility, and adjust the maxillary dental midline relative to the facial midline.
2. Reorient to the frontal view, and clip the anterior part of the face to better visualize the orbital floors. Orient the image to a horizontal plane through the lower borders of the orbital floors.
3. Use the grid to identify any asymmetry in all planes. Clip images at each tooth to check for any occlusal plane canting and buccal/lingual posterior tipping in the transverse dimension.

4.5.2 Root Length and Alveolar Bone Density

Evaluation of bony housing is an important step in orthodontic diagnosis and treatment planning. Limitations to treatment may be identified if certain teeth, particularly the mandibular incisors, are unable to move due to risk of fenestration or dehiscence. The ability to visualize the alveolar process three-dimensionally is an advantage of CBCT imaging (Fig. 4.5). Evaluation of root proximity is also an important application of CBCT, since an 89% false-positive rate is seen when evaluating roots with panoramic radiographs. In a study by Wood and colleagues [61], when soft tissue is removed from the scan, alveolar bone height measurements had similar accuracy with 0.2 mm and 0.4 mm voxel sizes. In the presence of soft tissue, however, the 0.2 mm voxel-size scans were more accurate than with 0.4 mm. There is a risk of overestimating fenestrations and dehiscences on CBCT since a thin

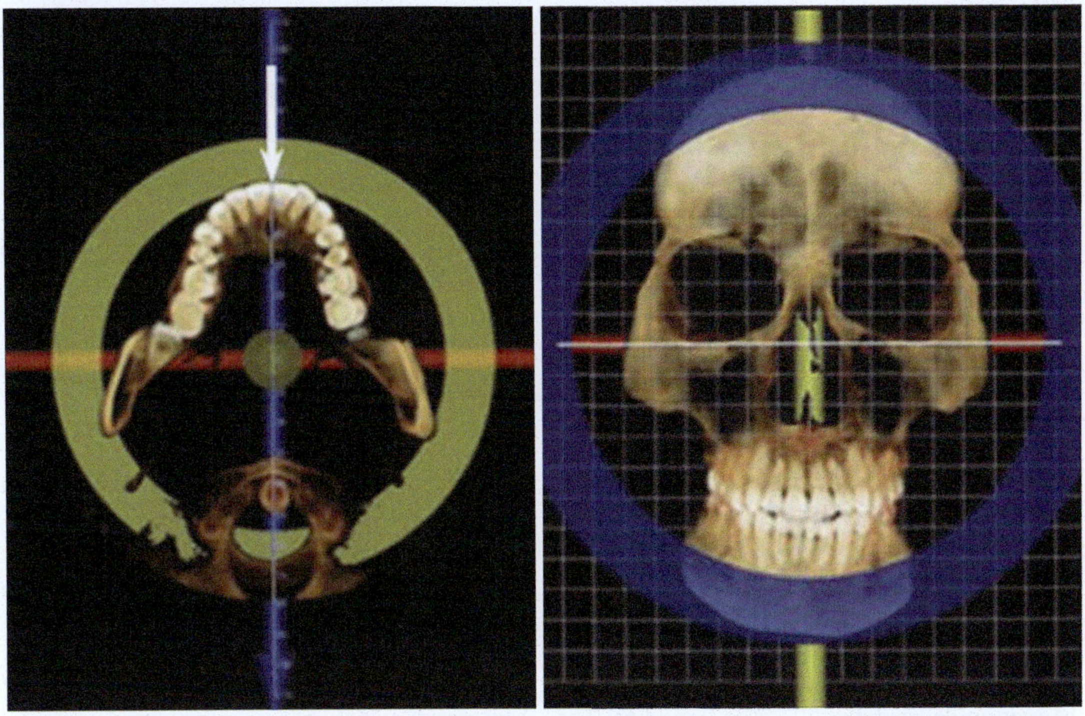

Fig. 4.4 Reorientation of CBCT for asymmetry analysis using axial and frontal views (Reproduced with permission from J Clin Orthod)

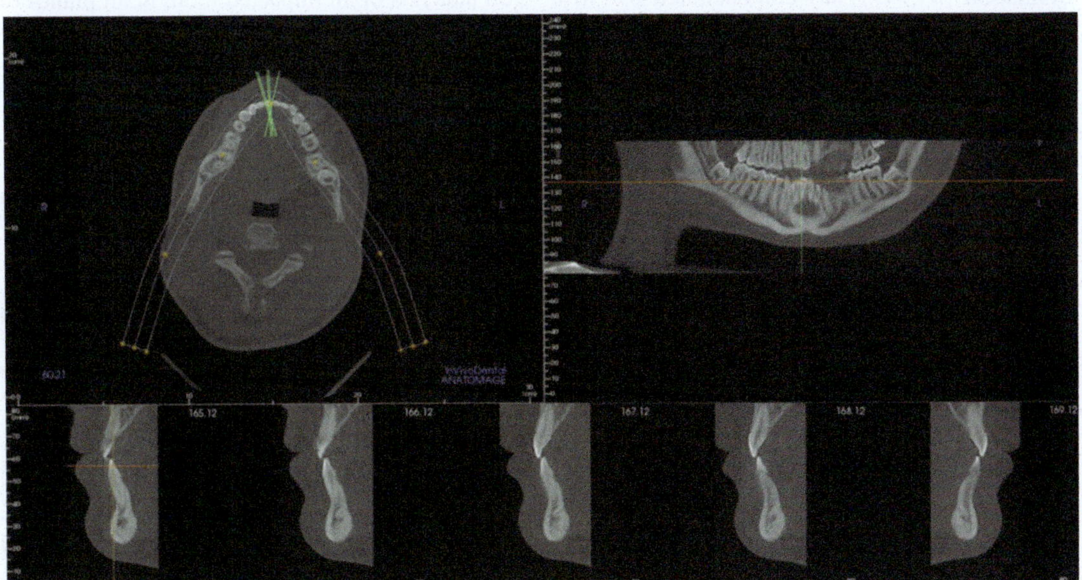

Fig. 4.5 Evaluation of thin labial alveolar bone at the mandibular incisor area

cortical labial bone layer may not be detected. Patcas et al. reported that if the mandibular anterior alveolar bone cannot be visualized on CBCT, it is less than 1 mm thick [62].

In addition, CBCT is a valuable tool in detecting early stages of root resorption and has proven to be more effective than periapical radiographs. A study by Da Silveira et al. [63] studied the influence of field of view (FOV) and voxel size in detecting resorption. They found that smaller voxel size leads to more efficient measurements, despite the FOV.

4.5.3 Temporomandibular Joint Evaluation

CBCT and newer software technologies allow for improved 3D imaging of the temporomandibular joint [11]. Bony changes such as flattening or beaking of the articular surfaces and sclerosis can be seen in degenerative diseases (Fig. 4.6). The axial slice is most useful for comparing the right and left condylar symmetries, while the sagittal slice is selected for condyle-fossa relationship evaluation [64]. CT has demonstrated 87–96% accuracy in detecting degenerative changes [65]. In addition, the joint space measurements may change to reflect the change in condylar position. Scott et al. [66] used CBCT to evaluate the condylar position in patients with temporomandibular joint dysfunction and found an increase in anterior joint space and decrease in posterior joint space when compared to the norms found by Ikeda and Kawamura [67].

4.5.4 Localization of Impacted Teeth

Before the use of CBCT, two periapical radiographs were recommended to localize the position of an impacted tooth using the buccal object rule. This method would identify whether the tooth was palatal or buccal, so that the appropriate exposure method could be identified. However, there are limitations to periapical radiographs including superimposition error, indiscernible lateral incisor root resorption in the case of impacted canines, and imprecise location of the apex of the impacted tooth. CBCT allows a full analysis of an impacted tooth in all planes of

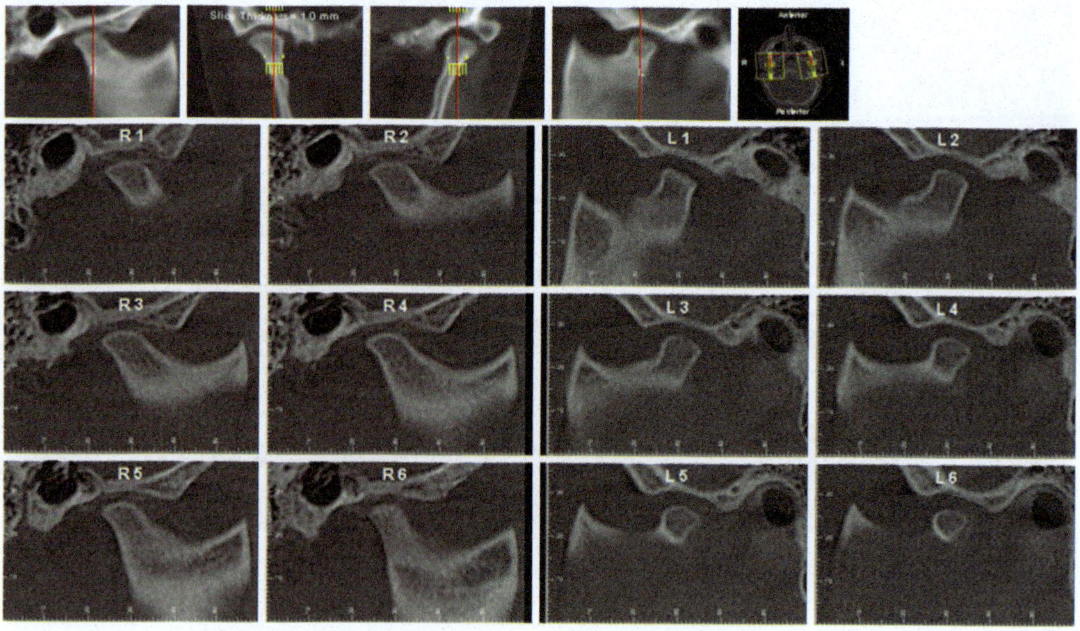

Fig. 4.6 CBCT image showing TMJ pathology. Note the flattening of the condylar heads

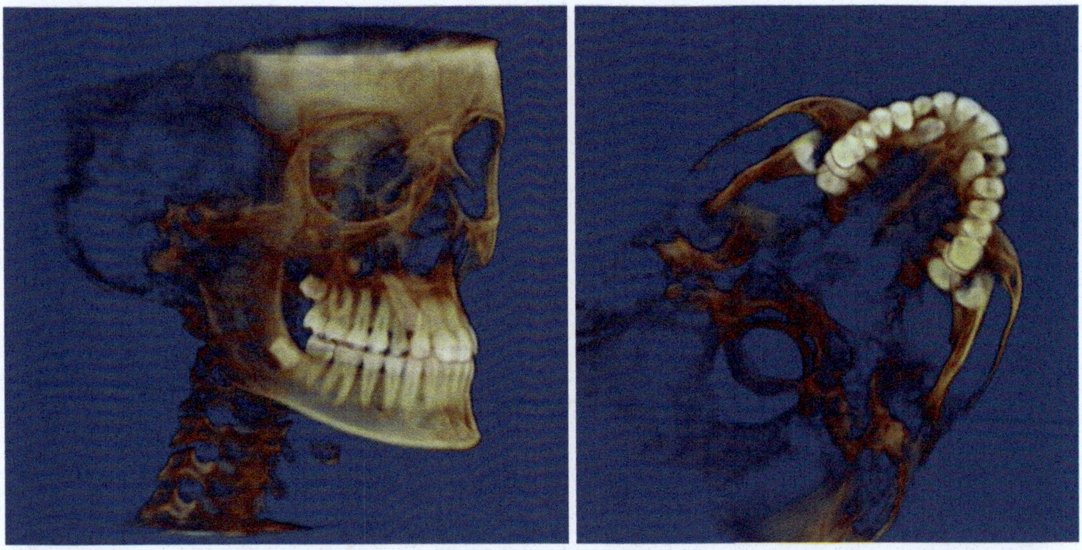

Fig. 4.7 Full visualization of an impacted maxillary canine, including proximity to adjacent lateral incisor

space, eliminating the errors encountered when using traditional radiographs, and in some cases, the treatment plan may change with the introduction of new information seen on the CBCT (Fig. 4.7). In addition to overcoming the limitations listed above, the full anatomy of the impacted tooth, surrounding teeth, and skeletal structures can be observed.

4.5.5 Evaluation of Sites for Temporary Skeletal Anchorage

CBCT can quantitatively evaluate cortical bone thickness and bone depth. The ideal length and diameter of a miniscrew for anchorage can be determined by measuring the distances between roots of premolars or molars and the distance from the intercortical bone surface to the root surface (Fig. 4.8) [11].

4.5.6 Surgical Treatment Planning

Virtual surgical planning with 3D software is becoming more popular among oral surgeons in support of surgical simulation, predicted outcomes, and surgical splint fabrication [68]. This type of software promotes inter-specialty collaboration of the final product to achieve the best surgical and orthodontic outcomes. Development of 3D imaging and models as well as virtual surgery has allowed for a more precise surgical plan (Fig. 4.9). 3D superimposition allows postsurgical evaluation to validate the predicted outcome and stability. CBCT images are a useful patient education tools to illustrate predicted surgical outcomes.

4.5.7 Airway Evaluation

Increased awareness of breathing disorders has propelled the improvement of diagnostic tools in this area. 2D cephalometric radiographs have been used in the past to evaluate the airway, but CBCT has made it possible to obtain volumetric analysis of the airway. Attempts have been made to correlate airway size and dimension to increased risk of obstructive sleep apnea (OSA) or other sleep breathing disorders (SBD) [69]. The minimum cross-sectional area (MCA) of the airway, to date, shows the highest correlation with OSA. A study by Sparks and colleagues [70] evaluated the relationship between airway and skeletal patterns. They found that Class II subjects had smaller airway volumes than Class

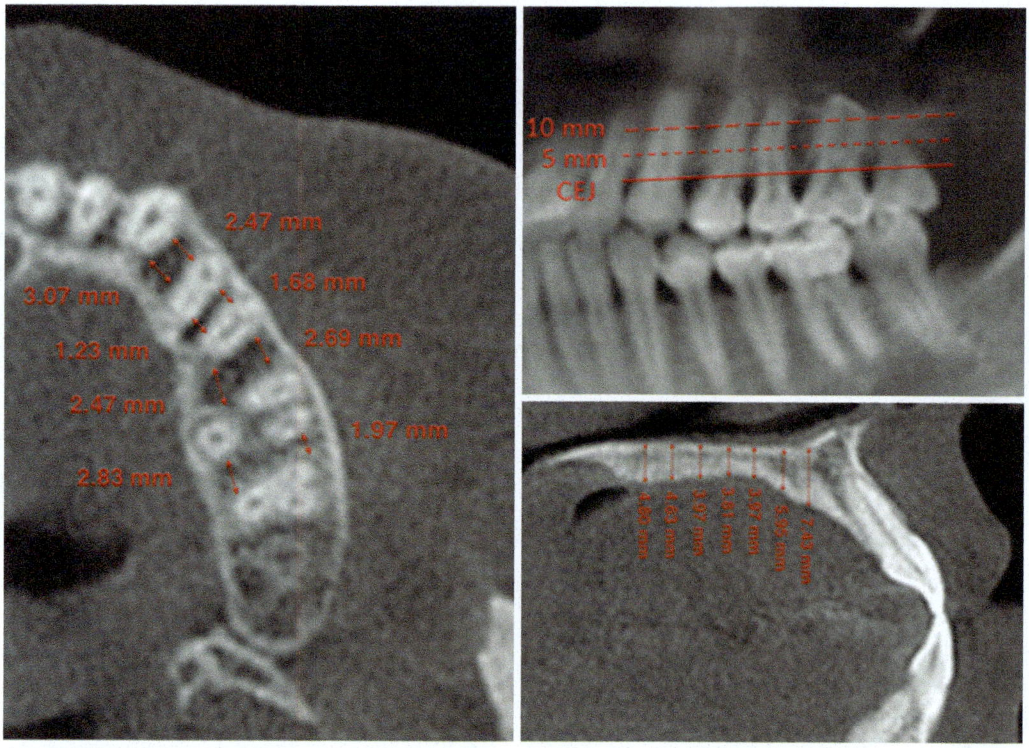

Fig. 4.8 Evaluation of cortical bone for temporary implant placement

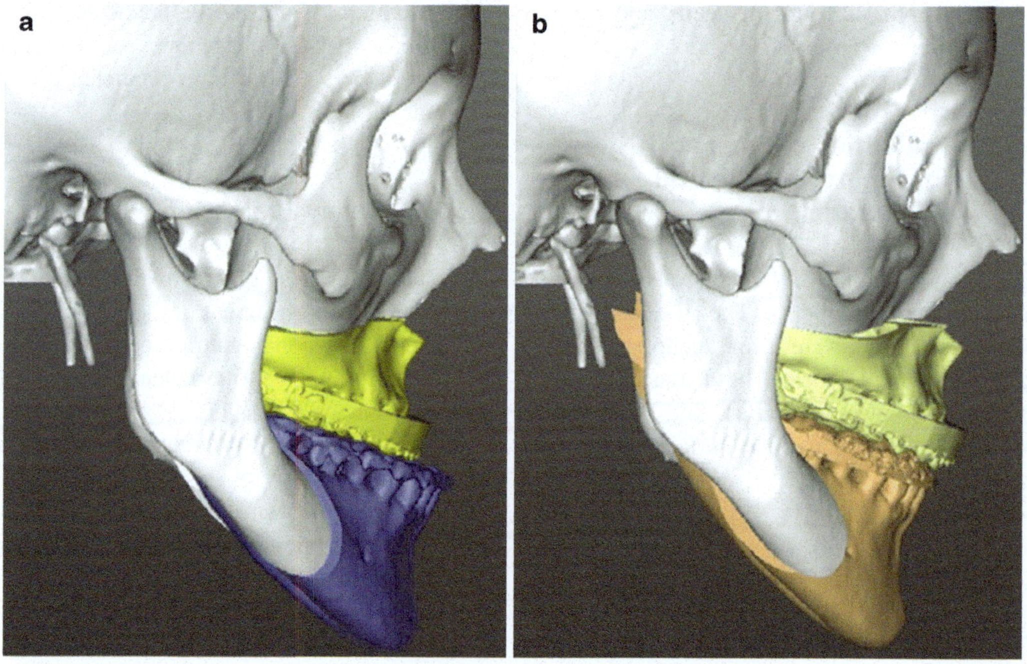

Fig. 4.9 (**a**) In this image, segmentation of CT dataset was done, and scanned images of the dental cast were transferred and merged to the CT images. (**b**) Virtual surgery was performed, and the final position of the maxilla and mandible was confirmed (Reproduced with permission from Sem Orthod)

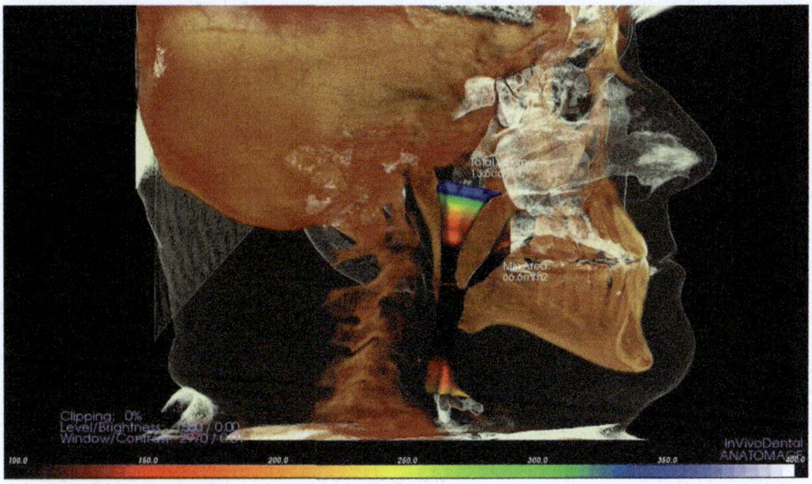

Fig. 4.10 Airway assessment before orthodontic treatment

III subjects, possibly due to tongue position. They also found Class II subjects to have the smallest MCA when compared to both Class I and Class III subjects. Airway analysis can be a useful adjunctive aid during diagnosis and may alert certain treatment plans that could potentially decrease the airway volume of a patient at high risk for OSA. In addition, posttreatment analysis can be performed to assess any effect of orthodontics on the airway dimension (Fig. 4.10).

4.6 Summary

Advanced imaging techniques, both in radiology and photography, have evolved orthodontic diagnosis and treatment planning. A thorough pretreatment evaluation allows for the most stable and predictable end product and enhances patient education and interdisciplinary communication.

References

1. Ricketts RM. The evolution of diagnosis to computerized cephalometrics. Am J Orthod. 1969;55:795–803.
2. Weems RA. Radiographic cephalometry technique. In: Jacobson A, Jacobson RL, editors. Radiographic cephalometry: from basics to 3-D imaging. Hanover Park, IL: Quintessence Publishing; 2006. p. 33–43.
3. Hounsfield GN. Computerized transverse axial scanning (tomography). Description of system. Br J Radiol. 1973;46:1016–22.
4. The Nobel Prize in physiology or medicine 1979. https://www.nobelprize.org/nobel_prizes/medicine/laureates/1979/. Accessed 6 Sep 2016.
5. Preston CB, Guan G. The relationship between conventional x-ray cephalometrics and cone-beam computed tomography. In: Park JH, editor. Computed tomography: new research. New York, NY: Nova Science Publishers, Inc.; 2013. p. 195–220.
6. Smith-Bindman R, Lipson J, Marcus R, Kim KP, Mahesh M, Gould R, et al. Radiation dose associated with common computed tomography examinations and the associated lifetime attributable risk of cancer. Arch Intern Med. 2009;169:2078–86.
7. Brenner DJ, Hall EJ. Computed tomography – an increasing source of radiation exposure. N Engl J Med. 2007;357:2277–84.
8. Mozzo P, Procacci C, Tacconi A, Martini PT, Andreis IA. A new volumetric CT machine for dental imaging based on the cone-beam technique: preliminary results. Eur Radiol. 1998;8:1558–64.
9. Cattaneo PM, Bloch CB, Calmar D, Hjortshoj M, Melsen B. Comparison between conventional and cone-beam computed tomography – generated cephalograms. Am J Orthod Dentofac Orthop. 2008;134:798–802.
10. Agrawal JM, Agrawal MS, Nanjannawar LG, Parushetti AD. CBCT in orthodontics: the wave of future. J Contemp Dent Practice. 2013;14:153–7.
11. Tai K, Yanagi Y, Park JH, Asaumi J. Clinical application of three-dimensional cone-beam computed tomography in orthodontics. In: Park JH, editor. Computed tomography: new research. New York, NY: Nova Science Publishers, Inc.; 2013. p. 255–66.
12. Park JH, Tai K, Owtad P. 3-Dimensional cone-beam computed tomography superimposition: a review. Semin Orthod. 2015;21:263–73.
13. Cevidanes LH, Styner MA, Proffit WR. Image analysis and superimposition of 3-dimensional cone-beam computed tomography models. Am J Orthod Dentofac Orthop. 2006;129:611–8.

14. da Motta AT, de Assis Ribeiro Carvalho F, Oliveira AE, Cevidanes LH, de Oliveira Almeida MA. Superimposition of 3D cone-beam CT models in orthognathic surgery. Dent Press J Orthod. 2010;15:39–41.
15. Becker A, Chaushu S, Casap-Caspi N. Cone-beam computed tomography and the orthosurgical management of impacted teeth. JADA. 2010;141(Suppl 3):14S–8S.
16. Kochel J, Meyer-Marcotty P, Strnad F, et al. 3D soft tissue analysis—part 1: sagittal parameters. J Orofac Orthop. 2010;71:40–52.
17. Kochel J, Meyer-Marcotty P, Kochel M, et al. 3D soft tissue analysis—part 2: vertical parameters. J Orofac Orthop. 2010;71:207–20.
18. Farronato G, Garagiola U, Dominici A, et al. "Ten-point" 3D cephalometric analysis using low-dosage cone beam computed tomography. Prog Orthod. 2010;11:2–12.
19. Bayome M, Park JH, Kook YA. New three-dimensional cephalometric analyses among adults with a skeletal class I pattern and normal occlusion. Korean J Orthod. 2013;43:62–73.
20. Harvold E. Cleft lip and palate: morphologic studies of the facial skeleton. Am J Orthod. 1954;40:493–506.
21. Swennen GR, Schutyser F, Barth EL, De Groeve P, De Mey A. A new method of 3-D cephalometry part I: the anatomic Cartesian 3-D reference system. J Craniofac Surg. 2006;17:314–25.
22. Park JU, Kook YA, Kim Y. Assessment of asymmetry in a normal occlusion sample and asymmetric patients with three-dimensional conebeam computed tomography: a study for a transverse reference plane. Angle Orthod. 2012;82:860–7.
23. Kook YA, Kim Y. Evaluation of facial asymmetry with three-dimensional cone-beam computed tomography. J Clin Orthod. 2011;45:112–5.
24. Gupta A, Kharbanda OP, Balachandran R, Sardana V, Kalra S, Chaurasia S, et al. Precision of manual landmark identification between as-received and oriented volume-rendered cone-beam computed tomography images. Am J Orthod Dentofac Orthop. 2017;151:118–31.
25. Heon JC. Three-dimensional superimposition. PCSO Bull. 2010;82:23–6.
26. Kapila S, Conley RS, Harrell WE Jr. The current status of cone beam computed tomography imaging in orthodontics. Dentomaxillofac Radiol. 2011;40:24–34.
27. Cevidanes LH, Heymann G, Cornelis MA, DeClerck HJ, Tulloch JF. Superimposition of 3-dimensional cone-beam computed tomography models of growing patients. Am J Orthod Dentofac Orthop. 2009;136:94–9.
28. Mah JK, Yi L, Huang RC, Choo H. Advanced applications of cone beam computed tomography in orthodontics. Semin Orthod. 2011;17:57–71.
29. Cevidanes LH, Oliveira AE, Grauer D, Styner M, Proffit WR. Clinical application of 3D imaging for assessment of treatment outcomes. Semin Orthod. 2011;17:72–80.
30. Nguyen T, Cevidanes L, Paniagua B, Zhu H, Koerich L, De Clerck H. Use of shape correspondence analysis to quantify skeletal changes associated with bone-anchored class III correction. Angle Orthod. 2014;84:329–36.
31. Tai K, Park JH, Mishima K, Shin JW. 3-Dimensional cone beam computed tomography analysis of transverse changes with Schwarz appliances on both jaws. Angle Orthod. 2011;81:670–7.
32. Tai K, Park JH. Superimposition of 3-dimensional conebeam computed tomography for 2-dimensional image analysis. In: Park JH, editor. Computed tomography: new research. New York, NY: Nova Science Publishers, Inc.; 2013. p. 457–75.
33. Rayapudi N, Padmalatha C, Gandikopta CS, Yudhistar PV, Tircoveluri S. A comparative study of linear measurements of facial skeleton using computed tomography and traditional cephalometry. APOS Trends Orthod. 2013;3:7.
34. Naudi KB, Benramadan R, Brocklebank L, Ju X, Khambay B, Ayoub A. The virtual human face: superimposing the simultaneously captured 3D photo realistic skin surface of the face on the untextured skin image of the CBCT scan. Int J Oral Maxillofac Surg. 2013;42:393–400.
35. Tai K, Park JH, Mishima K, Hotokezaka H. Using superimposition of 3-dimensional cone-beam computed tomography images with surface-based registration in growing patients. J Clin Pediatr Dent. 2010;34:361–7.
36. Tai K, Hotokezaka H, Park JH, Tai H, Miyajima K, Choi M, et al. Preliminary cone-beam computed tomography study evaluating dental and skeletal changes after treatment with a mandibular Schwarz appliance. Am J Orthod Dentofacial Orthop. 2010;138:262.e1–e11.
37. Gianquinto JR, Tuncay OC, Sciote JJ, Yang J. A method of superimposition of CBCT volumes in the posterior cranial base. Philadelphia, PA: The Temple University Digital Library. The Temple University; 2011.
38. Kau CH, Olim S, Nguyen JT. The future of orthodontic diagnostic records. Semin Orthod. 2011;17:6.
39. Heymann GC, Cevidanes L, Cornelis M, DeClerck HJ, Tulloch JF. Three-dimensional analysis of maxillary protraction with intermaxillary elastics to miniplates. Am J Orthod Dentofac Orthop. 2010;137:274–84.
40. Swennen GR, Mollemans W, DeClercq C, Abeloos J, Lamoral P, Lippens F, et al. A cone-beam computed tomography triple scan procedure to obtain a three-dimensional augmented virtual skull model appropriate for orthognathic surgery planning. J Craniofac Surg. 2009;20:297–307.
41. Nada RM, Maal TJ, Breuning KH, Berge SJ, Mostafa YA, Kuijpers-Jagtman AM. Accuracy and reproducibility of voxel based superimposition of conebeam computed tomography models on the anterior cranial base and the zygomatic arches. PLoS One. 2011;6:e16520.

42. Cevidanes LH, Motta A, Proffit WR, Ackerman JL, Styner M. Cranial base superimposition for 3-dimensional evaluation of soft-tissue changes. Am J Orthod Dentofac Orthop. 2010;137(Suppl 4):S120–9.
43. Terajima M, Yanagita N, Ozeki K, Hoshino Y, Mori N, Goto TK, et al. Three-dimensional analysis system for orthognathic surgery patients with jaw deformities. Am J Orthod Dentofac Orthop. 2008;134:100–11.
44. Grauer D, Cevidanes LS, Proffit WR. Working with DICOM craniofacial images. Am J Orthod Dentofac Orthop. 2009;136:460–70.
45. Kau CH. Creation of the virtual patient for the study of facial morphology. Facial Plast Surg Clin North Am. 2011;19:615–22.
46. Jayaratne YS, McGrath CP, Zwahlen RA. How accurate are the fusion of cone-beam CT and 3-D stereophotographic images? PLoS One. 2012;7:e49585.
47. Park TJ, Lee SH, Lee KS. A method for mandibular dental arch superimposition using 3D conebeam CT and orthodontic 3D digital model. Korean J Orthod. 2012;42:169–81.
48. Chenin DL, Chenin DA, Chenin ST, Choi J. Dynamic cone-beam computed tomography in orthodontic treatment. J Clin Orthod. 2009;43:507–12.
49. Lin HH, Chiang WC, Lo LJ, Sheng-Pin Hsu S, Wang CH, Wan SY. Artifact-resistant superimposition of digital dental models and cone-beam computed tomography images. J Oral Maxillofac Surg. 2013;71:1933–47.
50. Cevidanes LH, Tucker S, Styner M, Kim H, Chapuis J, Reyes M, et al. Three-dimensional surgical simulation. Am J Orthod Dentofac Orthop. 2010;138:361–71.
51. Ludlow JB, Gubler M, Cevidanes L, Mol A. Precision of cephalometric landmark identification: cone-beam computed tomography vs conventional cephalometric views. Am J Orthod Dentofac Orthop. 2009;136:e1–10.
52. Nguyen E, Boychuk D, Orellana M. Accuracy of cone beam computed tomography in predicting the diameter of unerupted teeth. Am J Orthod Dentofac Orthop. 2011;140:e59–66.
53. Lagravere MO, Carey J, Toogood RW, Major PW. Three-dimensional accuracy of measurements made with software on cone-beam computed tomography images. Am J Orthod Dentofac Orthop. 2008;134:112–6.
54. Stratemann SA, Huang JC, Maki K, Miller AJ, Hatcher DC. Comparison of cone beam computed tomography imaging with physical measures. Dentomaxillofac Radiol. 2008;37:80–93.
55. Periago DR, Scarfe WC, Moshiri M, Scheetz JP, Silveira AM, Farman AG. Linear accuracy and reliability of cone beam CT derived 3-dimensional images constructed using an orthodontic volumetric rendering program. Angle Orthod. 2008;78:387–95.
56. Bayome M, Park JH, Kim Y, Kook YA. 3D analysis and clinical applications of CBCT images. Semin Orthod. 2015;21:254–62.
57. Nur RB, Cakan DG, Arun T. Evaluation of facial hard and soft tissue asymmetry using cone-beam computed tomography. Am J Orthod Dentofac Orthop. 2016;149:225–37.
58. Akhil G, Senthil Kumar KP, Raja S, Janardhanan K. Three-dimensional assessment of facial asymmetry: a systematic review. Pharm Bioall Sci. 2015;7(Suppl 2):S433–7.
59. Ras F, Habets LLMH, van Ginkel FC, Prahl-Andersen B. Method for quantifying facial asymmetry in three dimensions using stereophotogrammetry. Angle Orthod. 1995;65:233–9.
60. Djordjevic J, Toma AM, Zhurov AI, Richmond S. Three-dimensional quantification of facial symmetry in adolescents using laser surface scanning. Eur J Orthod. 2014;36:125–32.
61. Wood R, Sun Z, Chaudhry J, Tee BC, Kim DG, Leblebicioglu B, et al. Factors affecting the accuracy of buccal alveolar bone height measurements from cone-beam computed tomography images. Am J Orthod Dentofac Orthop. 2013;143:353–63.
62. Patcas R, Muller L, Ullrich O, Peltomaki T. Accuracy of cone-beam computed tomography at different resolutions assessed on the bony covering of the mandibular anterior teeth. Am J Orthod Dentofac Orthop. 2012;141:41–50.
63. Da Silveira PF, Fontana MP, Oliveira HW, Vizzotto MB, Montagner F, Silveira HL, et al. CBCT-based volume of simulated root resorption – influence of FOV and voxel size. Int Endontic J. 2015;48:959–65.
64. Rodrigues AF, Fraga MR, Vitral RWF. Computed tomography evaluation of the temporomandibular joint in class I malocclusion patients: condylar symmetry and condyle-fossa relationship. Am J Orthod Dentofac Orthop. 2009;136:192–8.
65. Ahmad M, Hollender L, Anderson Q, Kartha K, Ohrbach R, Truelove EL. Research diagnostic criteria for temporomandibular disorders (RDC/TMD): development of image analysis criteria and examiner reliability for image analysis. Oral Surg Oral Med Oral Pathol Oral Radiol Endod. 2009;107:844–60.
66. Scott B, Kulbersh R, Kaczynski R. An evaluation of condylar position in patients with temporomandibular dysfunction using cone-beam computed tomography. In: Park JH, editor. Computed tomography: new research. New York, NY: Nova Science Publishers, Inc.; 2013. p. 379–92.
67. Ikeda K, Kawamura A. Assessment of optimal condylar position with limited cone-beam computed tomography. Am J Orthod Dentofac Orthop. 2009;135:495–501.
68. Park JH, Papademetriou M, Kwon YD. Orthodontic considerations in orthognathic surgery: who does what, when, where and how? Semin Orthod. 2016;22:2–11.
69. Kim KB. How has our interest in the airway changed over 100 years? Am J Orthod Dentofac Orthop. 2015;148:740–7.
70. Sparks R, Ngan P, Martin C, Razmus T, Mah J, Gunel E. A comparison of airway dimensions among different skeletal craniofacial patterns. In: Park JH, editor. Computed tomography: new research. New York, NY: Nova Science Publishers, Inc.; 2013. p. 401–26.

5 Dose Adjustments for Accuracy: Ultralow Dose Radiation 3D CBCT for Dental and Orthodontic Application

Maria Therese S. Galang-Boquiren, Budi Kusnoto, Zhang Zheng, and Xiaochuan Pan

Abstract

Cone beam computed tomography (CBCT) is increasingly popular when gathering initial patient imaging records for diagnosis and treatment planning. Although traditional two-dimensional panoramic or cephalometric radiographs can provide sufficient information to perform treatment in most cases, clinicians have become aware of the distortion inherent with these radiographs that can affect angular and linear measurements and, more importantly, tooth location and tooth-bone-jaw relationships.

A problem with CBCT technology is that its routine use poses a health risk as a source of ionizing radiation, especially in orthodontic patients who are mostly growing patients, preadolescent, and adolescent.

But what if there was a way to reduce the radiation dose and still reap the benefits of this technology to better serve our patients? This chapter will discuss dose adjustment methods used in the medical arena and their applications in the dental profession, with special focus on orthodontics.

There are many benefits of CBCT as compared to traditional two-dimensional (2D) imaging. In orthodontics, the additional accuracy of three-dimensional (3D) images eliminates common discrepancies, such as image magnification and the distortion that are typically encountered in 2D radiographs. This becomes particularly useful when localizing supernumerary or impacted teeth, estimating unerupted tooth sizes, measuring tooth roots, airway assessment, and treatment planning for skeletal asymmetries.

To address concerns about radiation dose, several dental organizations, including the American Association of Orthodontists (AAO), the American Dental Association (ADA), and the American Academy of Oral and Maxillofacial Radiology (AAOMR), have formulated clinical guidelines, essentially enforcing the universal "as low as reasonably achievable" ALARA tenet. In other words, the use of the CBCT should be performed on an as-needed basis [1–3].

The CBCT manufacturers have also implemented various technological advancements over

M. T. S. Galang-Boquiren · B. Kusnoto (✉)
Department of Orthodontics, University of Illinois at Chicago College of Dentistry, Chicago, IL, USA
e-mail: mgalang@uic.edu; bkusno1@uic.edu

Z. Zheng
Department of Radiology, The University of Chicago, Chicago, IL, USA
e-mail: zhangzh@uchicago.edu

X. Pan
Department of Radiology, The University of Chicago, Chicago, IL, USA

Department of Radiation and Cellular Oncology, The University of Chicago, Chicago, IL, USA
e-mail: xpan@uchicago.edu

O. Kadioglu, G. F. Currier (eds.), *Craniofacial 3D Imaging*,
https://doi.org/10.1007/978-3-030-00722-5_5

the years, improving image quality and, more importantly, addressing concerns about radiation dose [4].

In medical radiology, much research has been conducted on reconstruction algorithms (RA) and their application in CBCT scan analysis [5]. For example, utilization of an ultralow-dose CBCT RA integrating the Barzilai-Borwein gradient method limits the radiation dose while preserving image quality [6]. These algorithms have recently been applied specifically in orthodontics, showing that it is possible to decrease patient exposure while ensuring the diagnostic value of the orthodontic CBCT.

5.1 Clinical Significance and Applications

One of the most common orthodontic problems to benefit immensely from CBCT imaging is impacted/ectopic eruption of teeth, particularly the maxillary permanent canine. Several problems are linked to this diagnosis, such as the resulting adjacent root resorption and even subsequent loss of the maxillary permanent incisors. Further complications arise with the fact that an impacted canine does not follow a predictable pattern of impaction. The inclination can be horizontal, vertical, or a combination of the two [7]. Thus, the orthodontic treatment is often extended due to various maneuvers undertaken to minimize hazardous outcomes.

Although traditional orthodontic imaging, such as 2D panoramic or cephalometric radiographs, can provide sufficient information to perform treatment in most cases, clinicians should be aware of the distortion inherent in those radiographs that can affect angular and linear measurements [7, 8]. CBCT scanning circumvents this issue and aids in creating a treatment plan as it provides a more accurate, precise, and detailed view of both the bone and teeth [9, 10]. The information provided from a CBCT scan can be used to determine the exact position and relationships between teeth, which is particularly useful in the case of impacted teeth. The position of the impacted tooth in relation to other teeth can be clearly visualized with the use of CBCT technology. Figure 5.1 illustrates a patient with multiple missing teeth. This same patient's CBCT (Fig. 5.2) exhibits the exact location and position of those missing/impacted teeth. In cases when two teeth are in close proximity, as is often the case with canine impactions, CBCT technology is crucial in order to view the exact anatomical position of the impacted tooth. This helps both the orthodontist in planning the most efficient biomechanics to be utilized in bringing in the impacted tooth/teeth without damaging adjacent teeth and bone and the oral surgeon in locating the best access in cases that need surgical exposure. Figure 5.3 illustrates various examples of impacted canines, and how they can be visualized in CBCT images in terms of their exact location and even the extent of damage, when it exists, on adjacent teeth pre-treatment. In such cases, CBCT scans are utilized so that the best and most efficient treatment plan can be determined [11].

One may argue that traditional, full-mouth series radiographs and occlusal radiographs, when properly read, will lead to determining

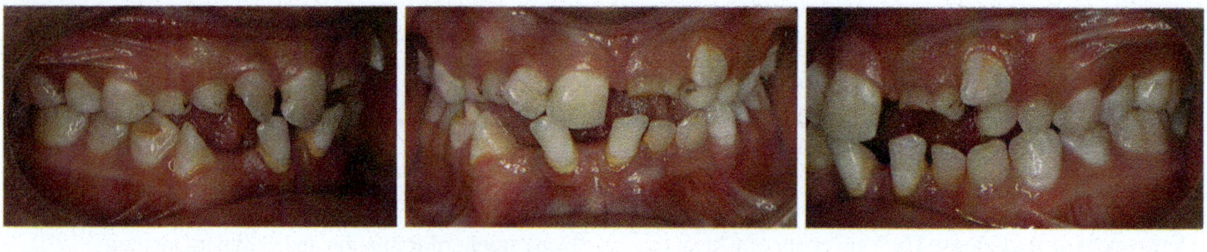

Fig. 5.1 Intraoral photos of a patient with multiple impacted/missing teeth

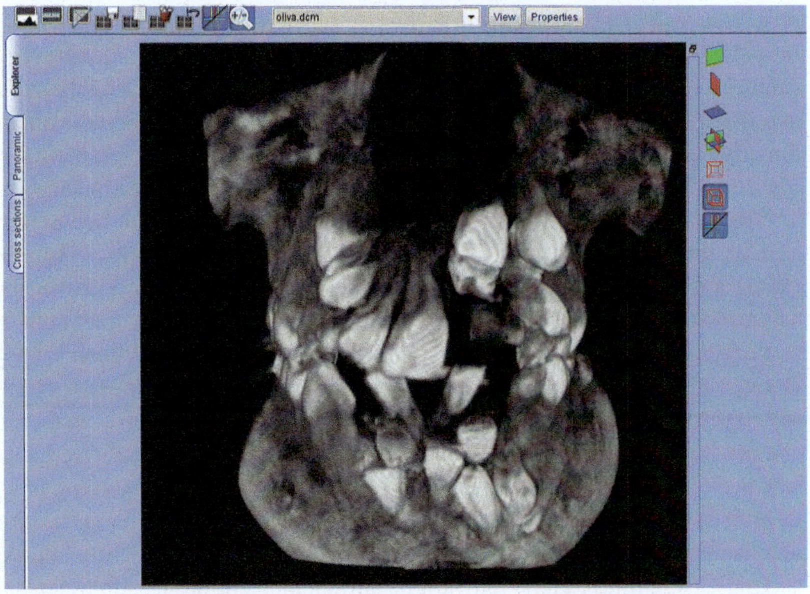

Fig. 5.2 CBCT image capture of patient's anterior dentition showing the number and location of the impacted teeth

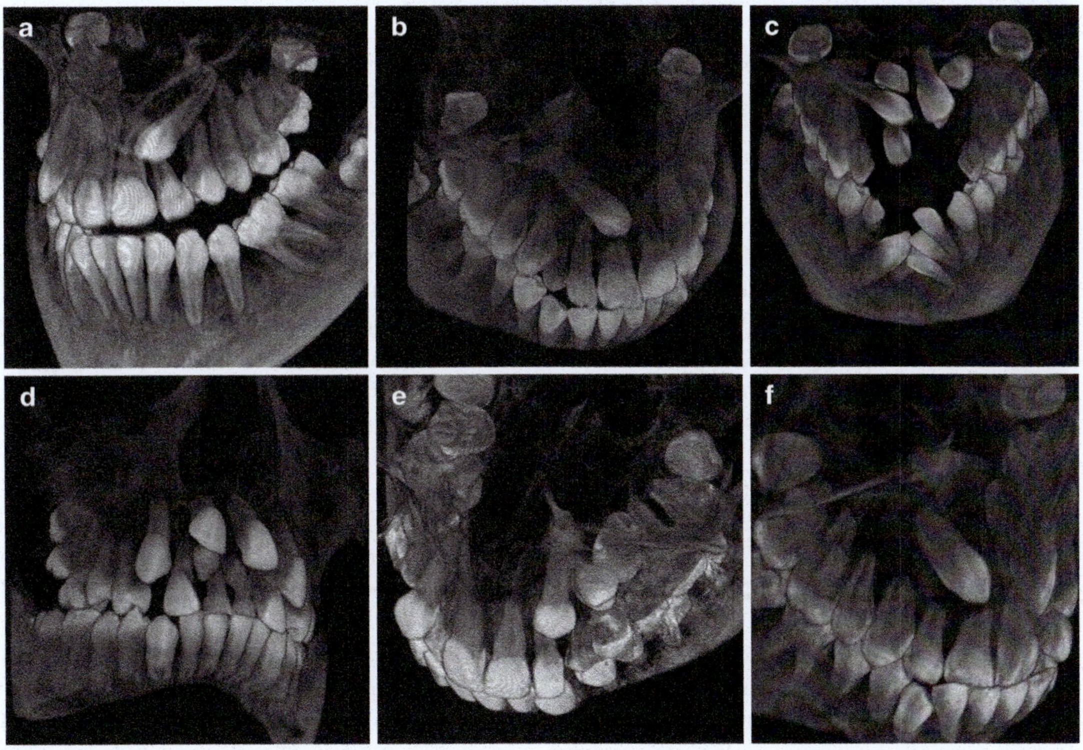

Fig. 5.3 Examples of CBCT image captures of various impacted teeth. (**a**) and (**b**) Show different views of impacted canines from various angles captured from a CBCT image. (**c**) Illustrates severe crowding sometimes typically associated with impacted teeth. (**d**) Shows a supernumerary tooth in the patient's midline. (**e**) Shows impacted posterior teeth. (**f**) Illustrates impacted teeth with associated incisor root resorption

proper access of impacted teeth (whether facial or palatal in the case of impacted maxillary canines), but troubles arise when the crown to be exposed lies directly above the alveolar crest or directly above a tooth root, as illustrated in Figs. 5.1–5.3.

The range of potential problems associated with an impacted canine suggests that the use of radiography at an early age may be beneficial to assist in interceptive orthodontic treatment, thus preventing the occurrence of actual damage. Because CBCT scanning is useful in determining the exact position of the impacted canine with respect to the surrounding teeth, interceptive treatment such as expansion, or even extraction of primary teeth, may prevent the development of any damage to adjacent structures.

One CBCT image alone may not be sufficient when considering the above scenario in a patient with an impacted tooth. During treatment, periodic assessments are necessary to assess the outcomes of the treatment mechanics implemented and to verify if the planned treatment is being executed correctly. Progress records, i.e., radiographs, are taken to evaluate treatment in its various stages in order to modify or amplify techniques accordingly. The use of radiographic imaging at multiple time points before, during, and/or after any planned dental or orthodontic treatment, needless to say, results in increased radiation exposure for the patient.

The increase in ionizing radiation exposure with the use of CBCT scanners has resulted in an argument against CBCT as the imaging technique of choice for comprehensive orthodontic assessment. This has led to the development of general guidelines to deal with justification, optimization, and referral criteria for users of dental CBCT [1–3, 12, 13].

Canine impaction, or impacted teeth in general, is just one of the various uses of dental CBCT imaging, yet it illustrates the undeniable benefits of this technology at multiple time points during treatment of these types of cases. Not to mention the other uses of CBCT, such as precise localization of unerupted teeth and better assessment of unerupted tooth sizes, assessment of root resorption, identifying and quantifying asymmetry, visualizing airway abnormalities, assessment of periodontal structures, identifying specific endodontic problems, viewing condylar positions and temporomandibular joint bony structures, planning for dental implant placement, assessing bone density, visualizing root proximity and resorption, and even providing the imaging data to support treatment simulation, surgical guidance, and dental appliance construction [12]. Surely the need for CBCT will increase for dental practitioners and specialists. Though many enhancements to reduce radiation exposure have been incorporated in more recent generations of CBCT machines, much improvement is still needed for early intervention methods in growing children, adolescents, and young adults who require repetitive imaging and follow-up.

5.2 Algorithm-Enabled Imaging Dose Reduction

Because of the potential risk to patients with frequent use of CBCT dental imaging, the development of innovative scanning configurations for dose reduction is an interesting research topic. In the past decade, dose reduction has been an active area of research in many medical imaging applications of CBCT, including diagnostic CBCT, image-guided surgery, and image-guided radiation therapy [14–25]. In general, there are four approaches to reducing the CBCT imaging dose, (1) sparse-view imaging, in which fewer projections are taken over the same angular range; (2) low-exposure imaging, in which the radiation exposure per projection view is reduced; (3) short-scan imaging, in which the scan is taken over a smaller angular range; and (4) region-of-interest (ROI) imaging, in which the illuminated volume is reduced. The discussion below will address each of the four imaging configurations, the current challenges, and the potential impact of advanced RA in dental CBCT applications.

5.2.1 Approaches for Low-Dose Dental CBCT Imaging and Current Challenges

Sparse-View Imaging

One can reduce imaging dose by reducing the total number of projection views over one rotation while maintaining the same imaging dose per view. However, when the projection views are sparsely sampled, the clinically used, analytic-based RAs, which require densely sampled projection views, can result in prominent streak artifacts [26–28]. Efforts have been made to interpolate additional projection views from measured data. Such an approach may be useful for certain scanning configurations and particular subjects; however, it often results in additional artifacts that blur the reconstructed images along the angular direction.

Low-Exposure Imaging

Another way to reduce the X-ray dose is to lower the exposure, or mAs, in dental CBCT data acquisition protocols; however, this approach often results in an insufficient number of detected X-ray photons and hence elevates the noise level in the projection data. Accordingly, the quality of the CBCT images reconstructed with the clinically used, analytic-based algorithm will be degraded by the low-dose, high-noise data [29].

Short-Scan Imaging

Short-scan imaging, in which the patient is scanned with a short-scan angular range (180° plus the fan angle), is an existing protocol in most commercial dental CBCT scanners. By employing the short-scan configuration, the imaging dose can be significantly reduced by maintaining the same angular sampling density as in a full scan. Clinically used RAs offer exact reconstruction from short-scan data for the middle plane but approximate reconstructions for off-middle planes. In particular, clinical reconstruction from short-scan data often leads to artifacts in off-middle planes depending upon the data quality, subject structure, and distance of the off-middle plane from the middle plane [30]. To date, there is no effective approach for solving this problem for clinically used, analytic-based algorithms.

ROI Imaging

In dental clinical practice, it is common to be interested in detailed information only within an ROI. For example, the clinician may only require an image of a single root canal. In this scenario, the X-ray beam's field of view (FOV) can be confined to illuminate only the ROI in an attempt to reduce the radiation dose. Regions outside the ROI are thus scanned partially, resulting in incomplete, truncated projection data. Direct application of clinically used analytic-based algorithms produces bright shading artifacts near the edge of the FOV. Extrapolation is often applied to the data to help reduce such truncation artifacts. While this method may help reduce artifacts in some cases, it is known to have limited utility for subject sizes considerably larger than the FOV [31].

5.2.2 Optimization-Based Image Reconstruction

In the past decade, a great body of work has been dedicated to the investigation of optimization-based, iterative algorithms [26–39] for image reconstruction from medical CBCT data. Optimization-based algorithms generally possess a higher degree of flexibility than clinically used, analytic-based algorithms and accommodate image reconstruction for a wide variety of imaging conditions of practical significance. Among all the algorithms, those seeking solutions to optimization problems with image total-variation (TV) constraints have attracted considerable attention because of their potential to exploit image sparsity and improve image quality.

The adaptive-steepest-descent (ASD)-projection-onto-convex-set (POCS) algorithm [26–28, 40–43] and a primal-dual algorithm developed by Chambolle and Pock (CP) [30, 31, 37, 44–46] are two such algorithms that have

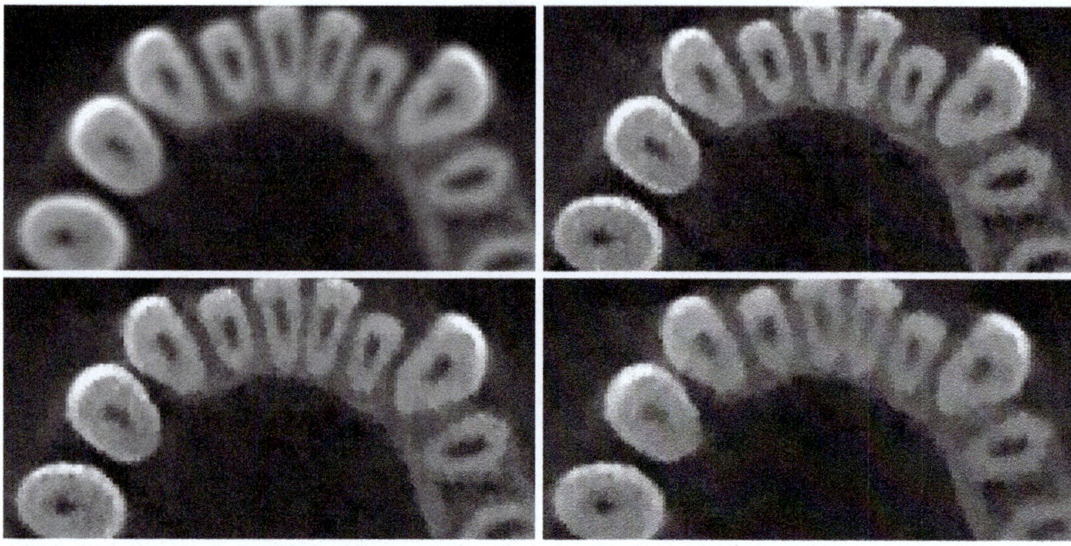

Fig. 5.4 A transverse image of a patient's roots. Top left, clinical FDK image; top right, ASD-POCS image reconstructed from 300-view data; bottom left, ASD-POCS image reconstructed from 151-view data; bottom right, ASD-POCS image reconstructed from 76-view data. Data were acquired by use of an iCAT dental CBCT scanner over a 2π angular range. Display window: [0, 1800] HU

been studied extensively in simulation and real-data studies. Use of the ASD-POCS or CP algorithms has shown that reconstructions (Fig. 5.4) from sparse-view data using half of the projections collected in a clinical protocol can have comparable quality to clinical images reconstructed using all of the collected data; those from one quarter of projections, although slightly degraded in image quality, can also be useful in certain practical applications [42].

Optimization-based reconstruction has shown the capability to suppress the noise in images reconstructed from low-dose CBCT data due to low X-ray exposure (or low mAs). A recent study (Fig. 5.5) found that the ASD-POCS reconstruction with appropriately selected reconstruction parameters appears to preserve better contrast while suppressing noise, yielding a relative high utility for the task of low-contrast visualization [29].

Another study indicates that, in a scenario where data contain severe noise, a CP algorithm that solves an optimization problem with an image-TV constraint and Gaussian blur operator may further suppress the image noise while preserving the low-contrast region and fine structures [47].

Optimization-based algorithms also provide a way to reduce or eliminate the artifacts in off-middle planes of clinical images reconstructed from short-scan data. A recent study (Fig. 5.6) showed that the CP and ASD-POCS algorithms can effectively reduce FDK-reconstruction artifacts in CBCT with a short-scan configuration [30]. A direct benefit of such an artifact reduction is improved contrast of low-contrast anatomic structures.

Optimization-based algorithms can handle reconstruction from truncated data without data extrapolation since they do not typically show significant bright shading artifacts near the edge of the FOV. In addition, recent studies (Fig. 5.7) using the CP algorithm show that reconstruction of the solution of an image-TV constrained optimization program with a weighted data fidelity,

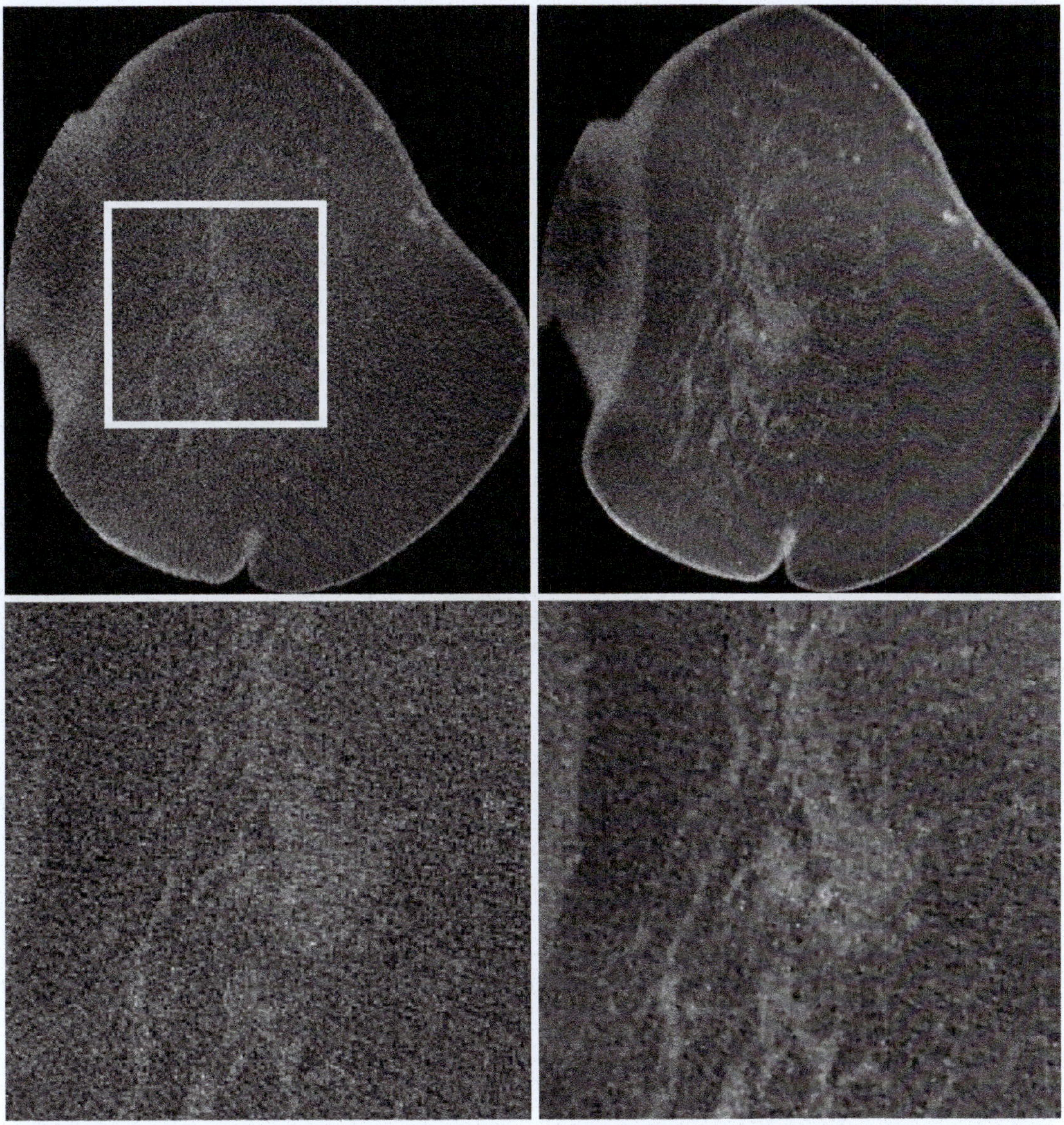

Fig. 5.5 Clinical FDK (left) and ASD-POCS (right) reconstructions for a large-size breast. Data were acquired by use of a dedicated breast CT scanner at low-radiation exposure with 500 projection views over 2π angular range. Beneath each of the reconstructions, we display the corresponding zoomed-in view of the ROI within the box depicted in the clinical FDK reconstruction. Display window: [0.15, 0.25] cm^{-1}

including a data-derivative term, can further suppress truncation artifacts [31]. Moreover, such reconstructions can reveal more details of the anatomic structures outside the FOV in both the transverse and axial directions by comparison to clinical images. This may provide more useful information to dentists/orthodontists and increase their confidence in treatment planning.

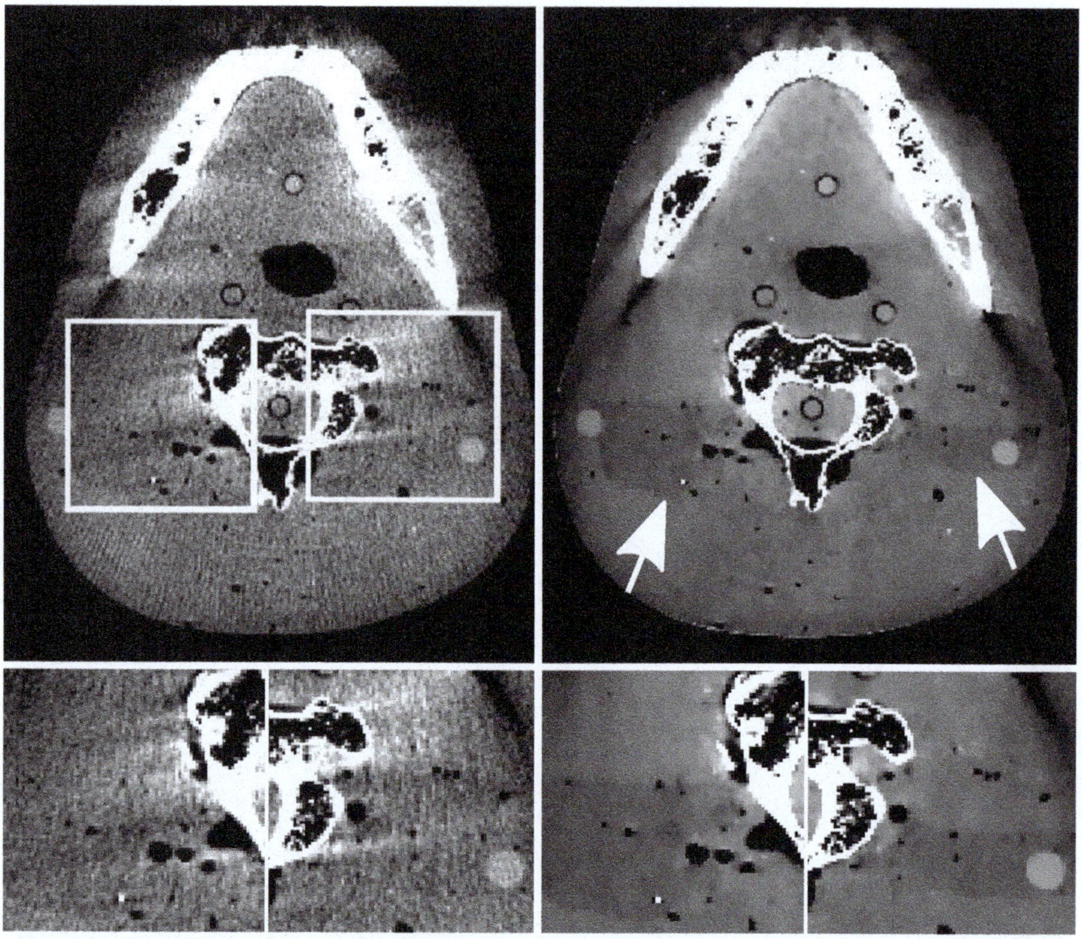

Fig. 5.6 Clinical FDK (left) and CP (right) reconstructions of a RANDO phantom within a transverse slice at 7.32 cm from the middle plane. Data were acquired by use of Varian On-Board Imager (OBI) with 347 projection views over short-scan angular range (196°). Two ROIs images enclosed in the FDK image are displayed in zoomed-in views below. Arrows in the CP reconstruction highlight the low-contrast structures in the RANDO phantom. Display window: [0.22, 0.30] cm^{-1}

5.3 Conclusion

CBCT imaging is truly beneficial in a subset of patients. The main issue is determining when these benefits outweigh the health risks. As discussed above, these concerns have led to vast research and development of clinical techniques to lower radiation exposure. The next logical step is for the 3D CBCT manufacturers to implement these research findings into the development of their instruments, which might lead to significant reduction of radiation level while obtaining full 3D diagnostic capability, thus improving patient care with considerably less radiation risk to the patient.

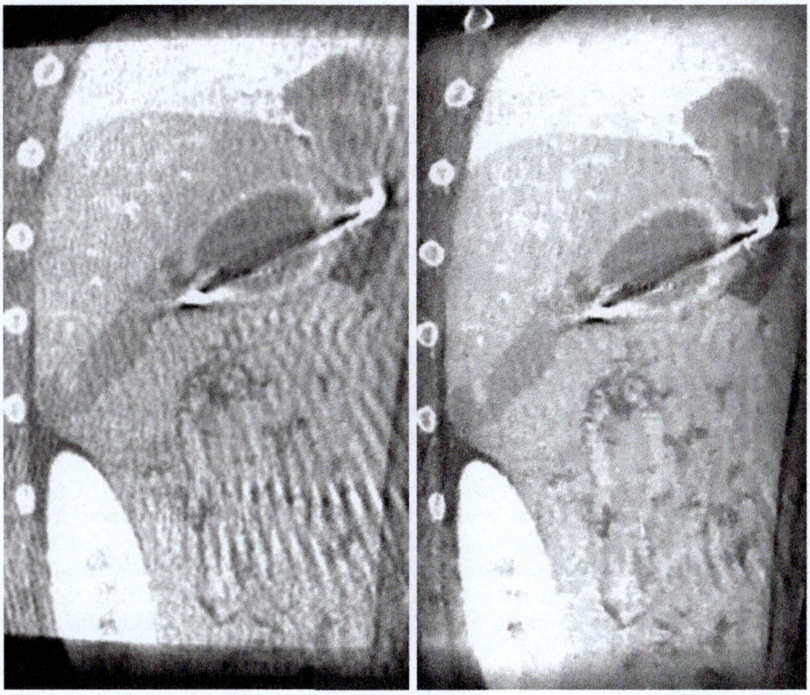

Fig. 5.7 Clinical FDK (left) and CP (right) reconstructions from truncated swine data within a coronal slice. Data were acquired by use of a GE C-arm scanner with 148 projection views over short-scan angular range (194°). Display window: [0.11, 0.23] cm^{-1}

References

1. American Association of Orthodontists. Statement on the role of CBCT in orthodontics. Creve Coeur, MO: American Association of Orthodontists; 2010. p. 26-10H.
2. American Dental Association Council on Scientific Affairs. The use of cone beam computed tomography in dentistry: an advisory statement from the American Dental Association Council on Scientific Affairs. J Am Dent Assoc. 2012;143:899–902.
3. American Academy of Oral and Maxillofacial Radiology. Clinical recommendations regarding use of cone beam computed tomography in orthodontics. Position statement by the American Academy of Oral and Maxillofacial Radiology. Oral Surg Oral Med Oral Pathol Oral Radiol. 2013;116:238–57. https://doi.org/10.1016/j.OOOO.2013.06.002.
4. Ludlow JB. A manufacturer's role in reducing the dose of cone beam computed tomography examinations: effect of beam filtration. Dentomaxillofac Radiol. 2011;40:115–22. https://doi.org/10.1259/dmfr/31708191.
5. Qiu W, Pengpan T, Smith ND, Soleimani M. Evaluating iterative algebraic algorithms in terms of convergence and image quality for cone beam CT. Comput Methods Prog Biomed. 2013;109:313–22. https://doi.org/10.1016/j.cmpb.2012.09.006.
6. Park JC, Song B, Kim JS, Park SH, Kim HK, Liu Z, et al. Fast compressed sensing-based CBCT reconstruction using Barzilai–Borwein formulation for application to on-line IGRT. Med Phys. 2012;39:1207–17. https://doi.org/10.1118/1.3679865.
7. Walker L, Enciso R, Mah J. Three-dimensional localization of maxillary canines with cone-beam computed tomography. Am J Orthod Dentofac Orthop. 2005;128:418–23.
8. Hatcher DC, Miller A, Peck JL, Sameshima GT, Worth P. Mesiodistal root angulation using panoramic and cone beam CT. Angle Orthod. 2007;77:206–13.
9. Monnerat C, Restle L, Mucha JN. Tomographic mapping of mandibular interradicular spaces for placement of orthodontic mini-implants. Am J Orthod Dentofac Orthop. 2009;135:428–9.
10. Baik HS, Kim KD, Lee KJ, Park SH, Yu HS. A proposal for a new analysis of craniofacial morphology by 3-dimensional computed tomography. Am J Orthod Dentofac Orthop. 2006;129:600. e23–600. e34.
11. Nakajima A, Arai Y, Dougherty H Sr, Homme Y, Sameshima GT, Shimizu N. Two- and three-

dimensional orthodontic imaging using limited cone beam–computed tomography. Angle Orthod. 2005;75:895–903.
12. Larson BE. Cone-beam computed tomography is the imaging technique of choice for comprehensive orthodontic assessment. Am J Orthod Dentofac Orthop. 2012;141:402–4., 406 passim. https://doi.org/10.1016/j.ajodo.2012.02.009.
13. Halazonetis DJ. Cone-beam computed tomography is not the imaging technique of choice for comprehensive orthodontic assessment. Am J Orthod Dentofac Orthop. 2012;141:403–5., 407 passim. https://doi.org/10.1016/j.ajodo.2012.02.010.
14. Wiesent K, Barth K, Navab N, Durlak P, Brunner T, Schuetz T, et al. Enhanced 3-D-reconstruction algorithm for C-arm systems suitable for interventional procedures. IEEE Trans Med Imaging. 2000;19:391–403.
15. Lauritsch G, Boese J, Wigström L, Kemeth H, Fahrig R. Towards cardiac C-arm computed tomography. IEEE Trans Med Imaging. 2006;25:922–34.
16. Wallace MJ, Kuo MD, Glaiberman C, Binkert CA, Orth RC, Soulez G. Three-dimensional C-arm cone-beam CT: applications in the interventional suite. J Vasc Interv Radiol. 2008;19:799–813. https://doi.org/10.1016/j.jvir.2008.02.018.
17. Orth RC, Wallace MJ, Kuo MD. C-arm cone-beam CT: general principles and technical considerations for use in interventional radiology. J Vasc Interv Radiol. 2008;19:814–20. https://doi.org/10.1016/j.jvir.2008.02.002.
18. Grass M, Koppe R, Klotz E, Proksa R, Kuhn M, Aerts H, et al. Three-dimensional reconstruction of high contrast objects using C-arm image intensifier projection data. Comput Med Imaging Graph. 1999;23:311–21.
19. Siewerdsen JH, Moseley DJ, Burch S, Bisland SK, Bogaards A, Wilson BC, et al. Volume CT with a flat-panel detector on a mobile, isocentric C-arm: preclinical investigation in guidance of minimally invasive surgery. Med Phys. 2005;32:241–54.
20. Hott JS, Deshmukh VR, Klopfenstein JD, Sonntag VK, Dickman CA, Spetzler RF, et al. Intraoperative Iso-C C-arm navigation in craniospinal surgery: the first 60 cases. Neurosurgery. 2004;54:1131–7.
21. De Vos W, Casselman J, Swennen G. Cone-beam computerized tomography (CBCT) imaging of the oral and maxillofacial region: a systematic review of the literature. Int J Oral Maxillofac Surg. 2009;38:609–25. https://doi.org/10.1016/j.ijom.2009.02.028.
22. Jaffray DA, Siewerdsen JH, Wong JW, Martinez AA. Flat-panel cone-beam computed tomography for image-guided radiation therapy. Int J Radiat Oncol Biol Phys. 2002;53:1337–49.
23. Oldham M, Létourneau D, Watt L, Hugo G, Yan D, Lockman D, et al. Cone-beam-CT guided radiation therapy: a model for on-line application. Radiother Oncol. 2005;75:271–8.
24. Smitsmans MH, De Bois J, Sonke J-J, Betgen A, Zijp LJ, Jaffray DA, et al. Automatic prostate localization on cone-beam CT scans for high precision image-guided radiotherapy. Int J Radiat Oncol Biol Phys. 2005;63:975–84.
25. Grills IS, Hugo G, Kestin LL, Galerani AP, Chao KK, Wloch J, et al. Image-guided radiotherapy via daily online cone-beam CT substantially reduces margin requirements for stereotactic lung radiotherapy. Int J Radiat Oncol Biol Phys. 2008;70:1045–56.
26. Sidky EY, Kao C-M, Pan X. Accurate image reconstruction from few-views and limited-angle data in divergent-beam CT. J Xray Sci Technol. 2006;14:119–39.
27. Sidky EY, Pan X. Image reconstruction in circular cone-beam computed tomography by constrained, total-variation minimization. Phys Med Biol. 2008;53:4777–807. https://doi.org/10.1088/0031-9155/53/17/021.
28. Bian J, Siewerdsen JH, Han X, Sidky EY, Prince JL, Pelizzari CA, et al. Evaluation of sparse-view reconstruction from flat-panel-detector cone-beam CT. Phys Med Biol. 2010;55:6575–99. https://doi.org/10.1088/0031-9155/55/22/001.
29. Bian J, Yang K, Boone JM, Han X, Sidky EY, Pan X. Investigation of iterative image reconstruction in low-dose breast CT. Phys Med Biol. 2014;59:2659–85. https://doi.org/10.1088/0031-9155/59/11/2659.
30. Zhang Z, Han X, Pearson E, Pelizzari C, Sidky EY, Pan X. Artifact reduction in short-scan CBCT by use of optimization-based reconstruction. Phys Med Biol. 2016;61:3387–406. https://doi.org/10.1088/0031-9155/61/9/3387.
31. Xia D, Langan DA, Solomon SB, Zhang Z, Chen B, Lai H, et al. Optimization-based image reconstruction with artifact reduction in C-arm CBCT. Phys Med Biol. 2016;61:7300–33.
32. Delaney A, Bresler Y, Sunnyvale C. Globally convergent edge-preserving regularized reconstruction: an application to limited-angle tomography. IEEE Trans Image Process. 1998;7:204–21. https://doi.org/10.1109/83.660997.
33. Elbakri I, Fessler J. Statistical image reconstruction for polyenergetic X-ray computed tomography. IEEE Trans Med Imaging. 2002;21:89–99.
34. Pan X, Sidky EY, Vannier M. Why do commercial CT scanners still employ traditional, filtered back-projection for image reconstruction? Inverse Probl. 2009;25:123009. https://doi.org/10.1088/0266-5611/25/12/123009.
35. Tang J, Nett BE, Chen G-H. Performance comparison between total variation (TV)-based compressed sensing and statistical iterative reconstruction algorithms. Phys Med Biol. 2009;54:5781–804. https://doi.org/10.1088/0031-9155/54/19/008.
36. Stsepankou D, Arns A, Ng S, Zygmanski P, Hesser J. Evaluation of robustness of maximum likelihood cone-beam CT reconstruction with total variation regularization. Phys Med Biol. 2012;57:5955–70. https://doi.org/10.1088/0031-9155/57/19/5955.
37. Sidky EY, Jørgensen JH, Pan X. Convex optimization problem prototyping for image reconstruction

in computed tomography with the Chambolle–Pock algorithm. Phys Med Biol. 2012;57:3065–91. https://doi.org/10.1088/0031-9155/57/10/3065.

38. Wang AS, Stayman JW, Otake Y, Kleinszig G, Vogt S, Gallia GL, et al. Soft-tissue imaging with C-arm cone-beam CT using statistical reconstruction. Phys Med Biol. 2014;59:1005–26. https://doi.org/10.1088/0031-9155/59/4/1005.
39. Nien H, Fessler JA. Fast splitting-based ordered-subsets X-ray CT Image reconstruction. Proceedings of the third international conference on image form Xray computed tomography. 2014. pp. 291–294. https://web.eecs.umich.edu/~fessler/papers/files/proc/14/web/nien-14-FSBpdf. Accessed 16 Jul 2017.
40. Han X, Bian J, Eaker DR, Kline TL, Sidky EY, Ritman EL, et al. Algorithm-enabled low-dose micro-CT imaging. IEEE Trans Med Imaging. 2011;30:606–20. https://doi.org/10.1109/TMI.2010.2089695.
41. Bian J, Wang J, Han X, Sidky EY, Shao L, Pan X. Optimization-based image reconstruction from sparse-view data in offset-detector CBCT. Phys Med Biol. 2012;58:205–30. https://doi.org/10.1088/0031-9155/58/2/205.
42. Zhang Z, Han X, Kusnoto B, Sidky EY, Pan X. Preliminary evaluation of dental cone-beam CT image reconstruction from reduced projection data by constrained TV-minimization. Proceedings of the third international conference on image form Xray computed tomography. 2014. pp. 299–302. https://www.researchgate.net/publications/312068633_preliminary_Evaluation_of_Dental_Cone-beam_CT_Image_from_Reduced_Projection_Data_by_Constrained-TV_minimization. Accessed 16 Jul 2017.
43. Kusnoto B, Kaur P, Salem A, Zhang Z, Galang-Boquiren MT, Viana G, et al. Implementation of ultra-low-dose CBCT for routine 2D orthodontic diagnostic radiographs: cephalometric landmark identification and image quality assessment. Semin Orthod. 2015;21:233–47. https://doi.org/10.1053/j.sodo.2015.07.001.
44. Chambolle A, Pock T. A first-order primal-dual algorithm for convex problems with applications to imaging. J Math Imaging Vis. 2011;40:120–45. https://doi.org/10.1007/s10851-010-0251-1.
45. Sidky EY, Kraemer DN, Roth EG, Ullberg C, Reiser IS, Pan X. Analysis of iterative region-of interest image reconstruction for x-ray computed tomography. J Med Imaging (Bellingham). 2014;1:031007. https://doi.org/10.1117/1.JMI.1.3.031007.
46. Sidky EY, Chartrand R, Boone JM, Pan X. Constrained TpV minimization for enhanced exploitation of gradient sparsity: application to CT image reconstruction. IEEE J Trans Eng Health Med. 2014;2:1–18.
47. Zhang Z, Ye J, Chen B, Perkins AE, Rose S, Sidky EY, et al. Investigation of optimization-based reconstruction with an image-total-variation constraint in PET. Phys Med Biol. 2016;61:6055–84. https://doi.org/10.1088/0031-9155/61/16/6055.

The Upper Airway

6

Aaron M. Laird, Enver Yetkiner, Onur Kadioglu, and G. Fräns Currier

Abstract

The upper airway is one of the essentials for healthy respiration. Anatomically, it can be evaluated in three sections; nasopharynx, oropharynx and laryngopharynx. These structures are affected by several factors such as growth and development, gender variability and body-mass index. In this chapter; normal dimensions of the upper airway, methods for its evaluation along with factors affecting its anatomy, volume and minimal correctional area will be discussed.

The upper airway is a structure responsible for one of the main vital functions in the human organism—breathing.

6.1 Airway and Associated Anatomy

The normal pharynx is a 12- to 14-cm-long musculomembranous tube extending from the cranial base to the lower border of the cricoid cartilage or the level of the sixth cervical vertebra where it transitions into the esophagus. The width of the pharynx varies constantly, because it is dependent on muscle tone, especially the constrictors. Volume and dimension may also fluctuate throughout the day and night as a result of head position, sleep cycles, or the activity of the autonomic nervous system. The pharynx lies posterior and is continuous with the regions of the nasopharynx, oropharynx, and laryngopharynx.

6.1.1 Nasopharynx

The nasopharynx lies above the soft palate but behind the posterior nares, which allows for respiratory passage through the nasal cavity into the nasopharynx. The nasal septum separates the two posterior nares and within these air spaces exist the inferior and middle nasal conchae. The walls of the nasopharynx are rigid, with the exception of the soft palate, with its cavity much less fluctuant in size and shape unlike the oro- and laryngopharynx. The nasal and oral portions of the pharynx communicate through the pharyngeal isthmus, which lies between the posterior border

A. M. Laird
Private Practice, Katy, TX, USA

E. Yetkiner
Department of Orthodontics, School of Dentistry, Ege University, Izmir, Turkey

O. Kadioglu (✉) · G. F. Currier
Division of Orthodontics, Department of Developmental Sciences, University of Oklahoma Health Sciences Center College of Dentistry, Oklahoma City, OK, USA
e-mail: onur-kadioglu@ouhsc.edu; frans-currier@ouhsc.edu

O. Kadioglu, G. F. Currier (eds.), *Craniofacial 3D Imaging*,
https://doi.org/10.1007/978-3-030-00722-5_6

of the soft palate and the posterior pharyngeal wall. The soft palate is a mobile flap suspended from the posterior border of the hard palate, sloping downward and backward between the oral and nasal portions of the pharynx. Their sides blend with the pharyngeal wall. Elevation of the soft palate and constriction of the palatopharyngeal sphincter close the isthmus during swallowing.

6.1.2 Oropharynx

The roof and posterior wall form a continuous concave slope from the nasal septum to the oropharynx. It is bounded above by mucosa overlying the posterior part of the sphenoid and further posterior by the basilar part of the occipital bone. A mass of lymphatic, adenoid tonsillar tissue lies in the mucosa of the upper portion of the roof and posterior wall near the midline. It protects the tissues of the upper respiratory tract. The size of the tonsil is largest at 5 years of age relative to the volume of the nasopharynx. This can account for the frequent problems in nasal breathing of preschool children besides a higher incidence of adenoidectomy in this age group. The oropharynx extends from the soft palate to the upper border of the epiglottis. It opens into the mouth through the oropharyngeal isthmus and faces the pharyngeal aspect of the tongue.

6.1.3 Laryngopharynx

The laryngopharynx, also known clinically as the hypopharynx, extends from the superior border of the epiglottis to the inferior border of the cricoid cartilage where it becomes continuous with the esophagus. At rest, the laryngopharynx extends posteriorly from the lower portion of the third cervical vertebral body to the upper part of the sixth. During deglutition, the hyoid elevators may elevate it considerably. The laryngeal inlet lies in its incomplete anterior wall, while the posterior surfaces of the arytenoid and cricoid cartilages, which encase the extending airway, lie just below this opening further constituting the anterior wall of the laryngopharynx.

The main muscles of the pharyngeal tube are the constrictors. The superior ones possess attachment points on the mandible, the tongue, the pterygomandibular raphe, the medial pterygoid, and the medial pharyngeal raphe on the posterior aspects of the pharynx. The middle constrictors possess attachment points to the greater and lesser cornus of the hyoid bone, the pharynx, and the pharyngeal raphe posteriorly. The inferior constrictors possess minimal direct or indirect attachment to structures in the oral cavity. However, they can be readily observed with changes in the skeletal position of the maxilla and mandible during orthognathic surgery that will either directly or indirectly influence the musculature that attach to, support, and direct the shape of all regions of the pharynx [1].

6.2 Upper Airway Evaluation Methods

There are two general methods to evaluate the upper airway: the clinical examination and the radiological evaluations.

6.2.1 Clinical Examination

The physical examination begins with observations of craniofacial morphology, skin color, use of accessory muscles of respiration, nasal flaring, chest wall retractions, and level of consciousness. The profile of the mandible should be evaluated for the presence of retrognathia or micrognathia, both of which may lead to airway obstruction, especially during sleep.

(a) *Posterior rhinoscopy*: This procedure is used to examine the posterior part of the nasal cavity and nasopharynx.
(b) *Nasopharyngoscopy*: It is the procedure which enables the examination of the internal surfaces of the nose and nasopharynx. It provides a direct view of every part of the upper airway from the nasal passages down to the throat to the larynx.
(c) *Esophageal manometry*: It is a complex technique, which may be affected by the place-

ment of the probe and the position of the catheter. Moreover, it is time-consuming, may affect the patient's sleep, and is not widely available. The correct positioning of the esophageal catheter requires experience and clinical practice. The use of a small catheter has improved tolerance to the procedure.

(d) *Plethysmography*: It measures changes in volume in different areas of your body with blood pressure cuffs or other sensors. These are attached to a machine called a plethysmograph. Plethysmography is especially effective in detecting changes caused by blood flow. It helps the clinician to calculate the air volume that the lungs can hold.

(e) *The simultaneous nasal and oral respirometric technique (SNORT)*: A custom-fitted face mask with separate valves to the nose and mouth that is attached to a flowmeter, air pressure transducer, recorder, and computer. It can give the nasal versus the oral inspirations, expirations, and their ratios.

(f) *Acoustic rhinometry (AR)*: This technique is based on the principle that a sound pulse propagating in the nasal cavity is reflected by local changes in acoustic impedance. Acoustic rhinometry is a quick, painless, noninvasive method that can be used to estimate the dimensions of nasal obstructions, evaluate nasal cavity geometry, monitor nasal disorders, and assess surgical results and response to medical treatment. However, lack of standardization is one of the main problems with this method [2].

(g) *Fluoroscopy*: It can be used to clarify uncertain radiographic findings or to study functional aspects. Furthermore, fluoroscopic examination can be used to evaluate children suspected of foreign-body aspiration. The amount of radiation delivered, even in a short fluoroscopy, is considerable and thus limits this technique except in exceptional circumstances. Somnofluoroscopy is a lateral fluoroscopic examination of the upper airway with synchronous polysomnography that provides information about the dynamic function of the airway and the level of stenosis or occlusion during sleep.

6.2.2 Radiographic Examination

Traditional radiographic cephalometry, magnetic resonance imaging (MRI), and computed tomography (CT) are widely used static imaging modalities for the assessment of airway anatomy and volume.

Cephalometric Airway Analysis

For the last century, the gold standard method for analysis of craniofacial development has been cephalometry with linear and angular measurements made from lateral headfilms. Serial cephalographs were used to assess both growth and outcomes of orthodontic and surgical interventions. Airway studies predate the use of radiographic images; digital examination and interpretation of facial phenotype led Meyer in the 1870s to remove adenoid vegetations (tonsils) from the nasopharynx of patients who were mouth breathers with reduced pulmonary function [3]. In the years that followed, lateral cephalographs were incorporated into airway studies to determine anteroposterior measurements at defined landmarks according to the author's specific objectives. Advantages of cephalometric analyses include its wide availability, simplicity, low expense, and ease of comparison to extensive normative data and other studies [4]. Studies by Winnberg et al. [5], Muto et al. [6], Pae et al. [7], Saitoh [8], and many others have utilized the cephalographs to evaluate the changes in airway-related structures as a result of head position, breathing pattern, and orthognathic surgery.

Studies by Mehra et al. [9] and Saitoh [8] utilized defined, reproducible reference points and planes to examine anteroposterior changes in airway structures. By orienting head films to Frankfort horizontal (FH—an imaginary line connecting porion to orbitale) and establishing a reference plane that ran perpendicular to Frankfort through porion, anteroposterior measurements on the cephalograph could be made. The soft palate, posterior nasal spine, base of the tongue, and epiglottis were all utilized to determine the changes in linear measurements occurring within these regions.

A thorough examination of the anatomy of pharynx and its regions within the head and neck is necessary to understanding the potential effects that may occur as a result of skeletal and dental manipulation of the maxilla and mandible. The use of landmarks on the lateral cephalometric images has allowed clinicians to define regions of the airway within the head and the neck and examine the effects that occur within these spaces as a result of treatment. However, as a two-dimensional representation of three-dimensional (3D) structures, lateral cephalographs offer limited information about the airways [10]. Information regarding axial cross-sectional areas and overall volumes can only be determined by 3D imaging modalities [11].

Magnetic Resonance Imaging (MRI)

MRI is accurate and reliable; and in comparison to radiographs that are two-dimensional projections, MRI provides an intrinsically scaled, three-dimensional image of all tissues composing the structure of the upper airway. Moreover, MRI provides superior resolution for soft tissues as compared with other techniques commonly used to assess the upper airway structures both in normal children and in children evaluated for obstructive sleep apnea (OSA) [12].

Three-Dimensional Analysis

Early two-dimensional claims of the effects on airway have been challenged with the introduction of three-dimensional tomographic evaluations. The ability to perform precise measurements of various cross-sectional areas, three-dimensional (3D) reconstructions, and volumetric measurements of the upper airway are some of the advantages of CT technology when compared with cephalometric techniques.

Medical computed tomography is a 3D imaging modality used in medicine but not as a routine method for airway analysis because of its high cost both financially and in terms of radiation. These drawbacks have been overcome with the introduction of cone-beam computed tomography (CBCT). Following the introduction of CBCT in 1998, this technology has improved with lower costs, less radiation exposure to patients, and better accuracy in identifying the boundaries of soft tissues and air spaces [13]. The CBCT allows for the segmentation and visualization of hollow structures, such as the airway in three dimensions, permitting the transition from lengths and angles to volumes and cross-sectional areas [14]. Although MRIs operate without the need for ionizing radiation, an MRI requires significantly longer operating times that result in decreased airway image quality due to motion artifacts [15]. CBCTs have led to a better understanding of upper airway anatomy and physiology. The CBCT machine completes a 360° rotation around the patient's head, acquiring the digital images which provide the raw data for the reconstruction of the examined volume.

The upper airway has been an area of interest, because the oropharyngeal and nasopharyngeal structures play important roles in the growth and development of the craniofacial and orodental complex [16]. The upper airway is an irregular lumen. As a result of individual differences, the volume of the upper airway cannot reflect the narrowest position of the airway. Thus, cross-sectional area is a better indicator than volume with which to evaluate changes in the size of the upper airway [17].

Studies have evaluated the accuracy of CBCT images in performing linear measurements of landmarks on dry skulls and have found them to be accurate to the submillimeter level with error less than 1% [18, 19]. The accuracy of CBCTs for measuring the airway has also been evaluated. Aboudara et al. [11] utilized plastic tubes of known volume to test the accuracy of volumetric measurements and found that after six repeated measurements, the volume assessment by CBCT was accurate and repeatable. Similar findings utilizing soft tissue phantoms for volumetric measurements by Yamashina et al. [20] found that the volume acquired from CBCT is nearly a 1:1 representation of the real volume. Stratemann et al. [21] concluded that the CBCT is suited to improving understanding of the upper airway by evaluation of the cross-sectional area, volume, three-dimensional form, and a more accurate review of the anatomy than the two-dimensional lateral view, defining key characteristics that modify airway flow.

To visualize a CBCT scan, Digital Imaging and Communications in Medicine (DICOM) viewer software is necessary and is the accepted file format (see Fig. 6.1). The evaluation of the size, shape, and volume of the upper airway starts by defining the volume corresponding to the airway passages, a process called segmentation. The software allows for viewing, measuring, segment-

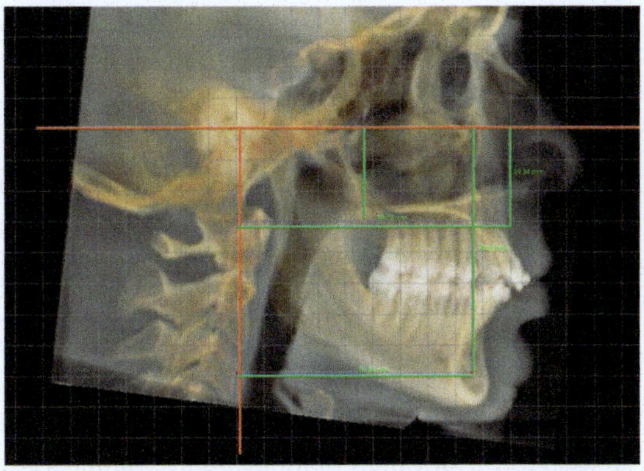

Fig. 6.1 Lateral view of a CBCT scan with reference lines for various measurements

ing, and completing analysis of the CBCT scan. To segment and structure the airway means to delineate and remove all other surrounding structures for a clearer analysis and visualization [14]. This segmentation can be performed manually by the user or automatically by the software program. The manual approach, a slice-by-slice analysis by the user, is time-consuming and impractical for clinical application [22]. With the semiautomatic approach, the computer differentiates the air space and surrounding soft tissues by using the differences in density values of these structures. In some programs, the semiautomatic segmentation includes two user-guided interactive steps: placement of initial seed regions in the axial, coronal, and sagittal slices and selection of an initial threshold. Density values are called Hounsfield units. Because the airway is radiolucent, the Hounsfield units for the airway are lower than the values for the surrounding soft tissues, allowing easy and automatic differentiation.

The studies by El and Palomo [14] and Weissheimer et al. [22] examined the reliability and accuracy of imaging software programs in airway volumetric analysis. They concluded that all major software programs were reliable in repeated measurements of a given volume but showed poor accuracy. Serial measurements of the same patient utilizing popular software programs with defined, repeatable landmarks delineating the boundaries of the airway should give a reliable assessment of the changes occurring in airway volumes.

The effectiveness of using CBCTs to analyze the pharyngeal airway has been investigated in several studies. Souza et al. [23] and Guijarro-Martinez et al. [10] found 3D analysis using CBCT to be accurate and reliable when measuring volumes and areas, even the narrowest areas, within the pharyngeal airway. Between examiners with different backgrounds, Mattos et al. [24] observed that airway assessments were reliable in linear and volumetric measurements in CBCT. Aboudara et al. [11] compared nasopharyngeal airway size between lateral cephalograms and CBCT and stated that the CBCT is a simple and effective method to accurately analyze the airway with only 0–5% error.

The reliability of three-dimensional CBCT scans has been compared to two-dimensional lateral cephalometric radiographs as a tool for measuring airway size and shape. The results of studies have shown that 3D analysis provided multiple measures (airway length, cross-sectional area, and volumetric measurements) that are not available from two-dimensional lateral cephalometric radiographs. It has also been shown that volumetric airway estimates from two-dimensional lateral cephalometric radiographs are questionable and that three-dimensional CBCT scans offer a simple and effective method to accurately analyze the airway [11].

6.3 The Factors Affecting the Dimensions of the Upper Airway

The upper airway is a complex structure that is outlined by the soft tissues forming the nasopharynx and oropharynx within the skeletal boundaries of the mid- and lower face. The effect of growth should play a role when evaluating dimensions of the pharyngeal airway. It undergoes morphologic changes as the result of growth of the skeleton and surrounding tissues as well as the functional changes throughout childhood. It has been found that greater rate of changes in the soft tissue measurements of the posterior pharyngeal wall occurred between 6 and 9 years of age as well as between 12 and 15 years [25, 26]. Vogler et al. [27] found that the adenoid pad continued to grow linearly throughout the first decade of life and was at maximum size between 7 and 10 years of age and then progressively diminishing until 60 years of age. Mislik et al. [28] reported that the smallest distance from the soft palate to the posterior pharyngeal wall and the retroglossal dimension showed high inter-individual variations. The retroglossal dimension decreases slightly between 6 and 12 years of age and then slightly increases up to 17 years of age. The smallest distance from the soft palate to the posterior pharyngeal wall displayed a slight continuous increase of about 1.03 mm between 6 and 17 years of age. Laird [29] also found, when evaluated cross-sectionally, that the chronologic age demonstrated a statistically significant effect on all aspects of the airway. In each aspect of the airway, except for nasopharynx volume, the mean value increases from the ≥7 to <10 age group to the ≥16 to <18 age group and then decreases in the ≥18 age group. With nasopharynx volume, the mean continues to increase throughout all age groups. The skeletal age displayed a statistically significant effect on all aspects of the airway as well. This finding signifies an increase in the airway as one matures skeletally (see Fig. 6.2).

Sexual dimorphism in craniofacial dimensions is a fact that has been established in various analyses. In general, women are smaller in stature than men (with less muscle mass and smaller heads) and subsequently requiring less oxygen. If airways in women were similar dimensions to those in men, it follows that their airways must be larger in relative terms, and this may be one of the reasons that women would be less prone to obstructive sleep apnea (OSA) than men. It may be important that females have smaller cross-sectional areas of the tongue than males as measured from lateral cephalograms and that females reach adult values earlier [30]. Laird [29] also found that gender had a statistically significant effect on the airway in selected areas. The mean values of total airway volume, oropharynx volume, MCA (most constricted area), and MCA-AP (most constricted area-anteroposterior) were all found to be greater in males than in females (see Tables 6.1 and 6.2). However, in the studies of Değerliyurt et al. [31] and Kim et al. [32], no differences were observed between genders.

Laird [29] also evaluated the body mass index (BMI) and showed a statistically significant effect on four aspects of the airway: total airway volume, oropharynx volume, nasopharynx volume, and FH to MCA (see Tables 6.3 and 6.4). The mean total airway volume was highest in the normal category and lowest in the underweight category with overweight and obesity categories displaying similar mean values. The mean oropharynx volume was

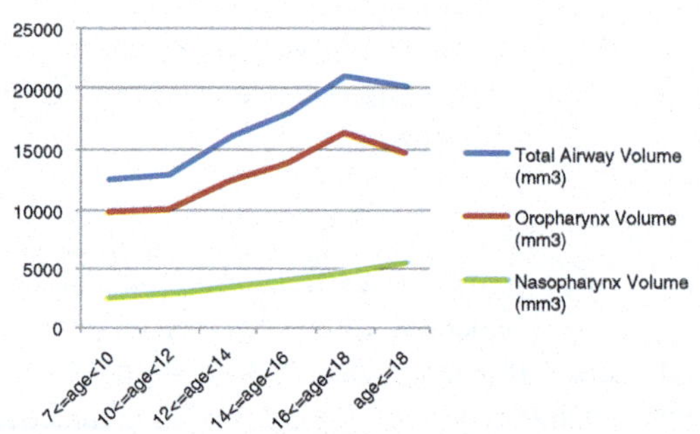

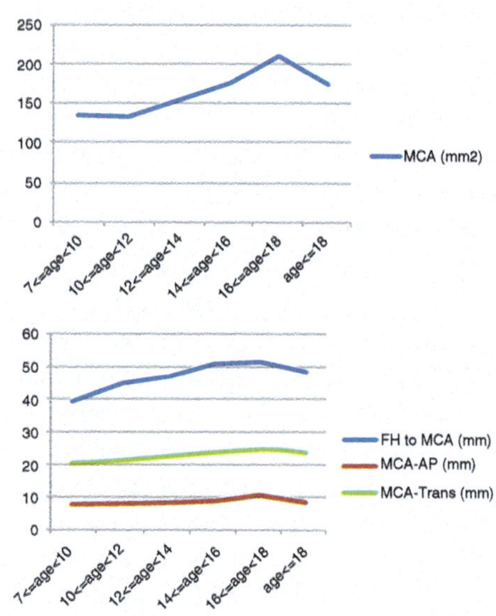

	7<=age<10	10<=age<12	12<=age<14	14<=age<16	16<=age<18	age>=18
Total Airway Volume (mm3)	12593.44	12937.67	16001.76	18021.55	21121.36	20380.55
Oropharynx Volume (mm3)	9825.37	10019.04	12368.12	13872.57	16430.23	14804.05
Nasopharynx Volume (mm3)	2768.07	2918.63	3633.64	4148.98	4691.13	5576.5
MCA (mm2)	137.14	134.78	156.59	175.57	212.63	177.08
FH to MCA (mm)	39.72	44.89	46.78	50.99	51.55	48.55
MCA-AP (mm)	8.25	8.52	8.83	9.32	10.58	8.99
MCA-Trans (mm)	21.04	21.58	23.11	24.22	25.03	24.16

Fig. 6.2 Airway norms; total, nasopharynx, oropharynx volumes, minimum cross sectional area dimensions and its location

Table 6.1 Gender univariate ANOVA on all aspects of the airway

Dependent	Levels	*N*	Mean	Std	Min	Max	Pr > *F*
Total airway volume (mm³)	F	488	17,094.09	6506.31	4485.00	47,386.00	0.0233
	M	371	18,268.43	8639.02	5332.00	59,599.00	
Oropharynx volume (mm³)	F	488	12,872.50	5453.23	2333.00	39,004.00	0.0052
	M	371	14,100.14	7388.99	3686.00	51,955.00	
Nasopharynx volume (mm³)	F	488	4221.59	1900.41	236.00	12,775.00	0.6865
	M	371	4168.29	1936.65	248.00	12,232.00	

Table 6.2 Gender univariate ANOVA on all aspects of the airway

Dependent	Levels	*N*	Mean	Std	Min	Max	Pr > *F*
MCA (mm²)	F	488	164.21	85.07	12.30	488.20	0.0674
	M	371	175.77	99.59	13.60	574.80	
FH to MCA (mm)	F	488	47.10	17.97	0.00	78.88	0.0172
	M	371	50.12	18.87	2.39	85.06	
MCA-AP (mm)	F	488	8.84	3.53	1.28	27.49	0.0058
	M	371	9.52	3.66	1.36	23.48	
MCA-Trans (mm)	F	488	23.58	6.11	5.06	39.87	0.8665
	M	371	23.66	6.89	5.87	45.64	

Table 6.3 BMI univariate ANOVA on all aspects of the airway

Dependent	Levels	*N*	Mean	Std	Min	Max	Pr > *F*
Total airway volume (mm³)	Underweight	178	16,484.06	7421.19	4529.00	45,174.00	0.0399
	Normal	424	18,301.81	7804.79	4485.00	59,599.00	
	Overweight	155	17,282.08	7286.28	5283.00	51,337.00	
	Obesity	102	17,124.04	6570.50	5739.00	44,433.00	
Oropharynx volume (mm³)	Underweight	178	12,693.22	6301.78	2333.00	37,992.00	0.0228
	Normal	424	14,083.46	6723.90	3359.00	51,955.00	
	Overweight	155	12,781.44	5987.41	4076.00	44,730.00	
	Obesity	102	12,755.18	5393.56	4089.00	32,201.00	
Nasopharynx volume (mm³)	Underweight	178	3790.83	1704.91	415.00	10,467.00	0.0050
	Normal	424	4218.35	1910.67	236.00	12,775.00	
	Overweight	155	4500.64	2074.47	1039.00	11,263.00	
	Obesity	102	4368.86	1936.75	838.00	12,232.00	

Table 6.4 BMI univariate ANOVA on all aspects of the airway

Dependent	Levels	*N*	Mean	Std	Min	Max	Pr > *F*
MCA (mm²)	Underweight	178	163.29	89.77	13.60	574.80	0.0652
	Normal	424	177.61	94.30	12.30	491.70	
	Overweight	155	159.50	89.82	29.80	457.70	
	Obesity	102	159.33	85.09	31.80	485.10	
FH to MCA (mm)	Underweight	178	48.28	18.36	0.00	80.47	0.0213
	Normal	424	48.92	19.24	0.00	83.22	
	Overweight	155	50.42	17.21	8.00	85.06	
	Obesity	102	43.42	15.99	6.04	80.81	
MCA-AP (mm)	Underweight	178	8.85	3.43	1.74	23.48	0.4723
	Normal	424	9.29	3.58	1.59	21.99	
	Overweight	155	8.94	3.98	1.28	27.49	
	Obesity	102	9.25	3.37	2.64	18.87	
MCA-Trans (mm)	Underweight	178	23.29	6.38	7.78	40.26	0.1019
	Normal	424	24.16	6.57	5.87	45.64	
	Overweight	155	22.91	6.43	5.06	39.85	
	Obesity	102	23.00	6.03	6.80	37.52	

Table 6.5 Angle's classification univariate ANOVA on all aspects of the airway

Dependent	Levels	*N*	Mean	Std	Min	Max	Pr > *F*
Total airway volume (mm^3)	Class I	360	17876.54	7808.77	4485.00	51337.00	0.1189
	Class II	352	16992.47	6567.87	5283.00	47386.00	
	Class II Div. 2	27	19968.07	8889.66	5615.00	43311.00	
	Class III	120	18028.84	8746.23	4682.00	59599.00	
Oropharynx volume (mm^3)	Class I	360	13690.13	6651.15	3359.00	44730.00	0.1275
	Class II	352	12830.27	5545.28	2333.00	39004.00	
	Class II Div. 2	27	14880.15	6929.59	5379.00	33521.00	
	Class III	120	13887.23	7584.53	3508.00	51955.00	
Nasopharynx volume (mm^3)	Class I	360	4186.41	1983.56	248.00	12775.00	0.1081
	Class II	352	4162.20	1818.55	417.00	10467.00	
	Class II Div. 2	27	5087.93	2421.97	236.00	9875.00	
	Class III	120	4141.62	1831.31	549.00	10104.00	

greatest in the normal category and lowest in the underweight category. The means for overweight and obesity followed closely behind that of underweight. The mean nasopharynx volume was highest in the overweight and lowest in the underweight categories. Lastly, the mean FH to MCA value was greatest in the overweight group and lowest in the obesity group.

Obstructive processes of morphologic, physiologic, or pathologic nature, such as hypertrophy of adenoids and tonsils, chronic and allergic rhinitis, irritant environmental factors, infections, congenital nasal deformities, nasal traumas, polyps, and tumors, are predisposing factors to a blocked upper airway. When that happens, a functional imbalance results in an oral breathing pattern that can alter facial morphology and dental arch forms, generating a malocclusion. Laird [29] reported that the variable, adenoidectomy/tonsillectomy, appeared to have a statistically significant effect on the nasopharyngeal volume and MCA ($p < 0.05$). The mean for nasopharynx volume in those patients who did not state having received an adenoidectomy or tonsillectomy procedure was revealed to be higher than in those patients who confirmed having undergone the procedure. The same finding was noted with MCA. The mean for MCA was higher in those who did not receive the adenoidectomy or tonsillectomy procedure than in those who did.

Considering the functional matrix theory proposed by Moss, the association of respiratory and masticatory functions and swallowing might act on craniofacial development [33].

Since Angle [34] showed that Class II Division 1 malocclusion was associated with obstruction of the pharyngeal airway space (PAS) and mouth-breathing subjects, multiple studies have examined the pharyngeal airway and its dimensions among the three dentofacial skeletal classes.

According Laird [29], no statistically significant differences were noted among the mean values of the four different Angle's classifications of malocclusion (Table 6.5). The mean value for MCA-Trans was noted to be the highest in the Class II Division 2 group, whereas those for the other groups were fairly equivalent (Table 6.6).

Considering skeletal classification, Laird reported that ANB was found to have a statistically significant effect on the total airway volume and oropharynx volume. The mean values for all three aspects of the airway were highest in the ANB Class III patient subjects and lowest in the Class II subjects. In the remaining aspects of the airway, a significant difference was not illustrated among the mean values for the three different skeletal classification groups, according to the ANB variable. As SNB and facial angle increased by 1°, the mean total airway increased by 225.27 mm^3 and 252.39 mm^3, respectively. On the other hand, the mean total airway volume

Table 6.6 Angle's classification univariate ANOVA on all aspects of the airway

Dependent	Levels	*N*	Mean	Std	Min	Max	Pr > *F*
MCA (mm²)	Class I	360	173.16	96.17	13.60	574.80	0.1252
	Class II	352	161.50	83.53	12.30	488.20	
	Class II Div. 2	27	195.77	111.34	27.90	434.60	
	Class III	120	173.94	95.36	29.90	441.80	
FH to MCA (mm)	Class I	360	49.24	18.40	0.00	84.72	0.1370
	Class II	352	48.56	18.10	0.00	83.22	
	Class II Div. 2	27	50.75	17.24	9.89	79.79	
	Class III	120	44.91	19.39	3.24	85.06	
MCA-AP (mm)	Class I	360	9.15	3.49	1.28	23.48	0.4897
	Class II	352	9.04	3.67	1.59	27.49	
	Class II Div. 2	27	10.16	4.48	3.72	21.99	
	Class III	120	9.12	3.51	2.33	17.65	
MCA-Trans (mm)	Class I	360	23.88	6.53	5.87	45.64	0.0473
	Class II	352	23.23	6.42	5.06	39.87	
	Class II Div. 2	27	26.60	5.83	15.92	41.02	
	Class III	120	23.29	6.36	6.80	39.85	

Table 6.7 ANB univariate ANOVA on all aspects of the airway

Dependent	Levels	*N*	Mean	Std	Min	Max	Pr > *F*
Total airway volume (mm³)	0 < ANB < 5	509	17,443.02	7281.11	4529.00	51,337.00	0.0261
	ANB >= 5	216	17,010.63	6682.48	4485.00	39,552.00	
	ANB <= 0	134	19,154.56	9329.69	5425.00	59,599.00	
Oropharynx volume (mm³)	0 < ANB < 5	509	13,316.71	6189.62	2333.00	44,730.00	0.0047
	ANB >= 5	216	12,658.20	5455.81	3359.00	30,259.00	
	ANB <= 0	134	14,929.53	8089.31	3490.00	51,955.00	
Nasopharynx volume (mm³)	0 < ANB < 5	509	4126.32	1844.81	236.00	12,232.00	0.3427
	ANB >= 5	216	4352.42	2068.40	248.00	12,775.00	
	ANB <= 0	134	4225.03	1920.55	549.00	10,807.00	

Table 6.8 ANB univariate ANOVA on all aspects of the airway

Dependent	Levels	*N*	Mean	Std	Min	Max	Pr > *F*
MCA (mm²)	0 < ANB < 5	509	168.46	90.58	13.60	574.80	0.0128
	ANB >= 5	216	158.96	88.35	12.30	454.00	
	ANB <= 0	134	188.56	98.98	21.60	488.20	
FH to MCA (mm)	0 < ANB < 5	509	47.89	18.34	0.00	84.72	0.5966
	ANB >= 5	216	49.33	17.35	2.39	83.22	
	ANB <= 0	134	48.87	20.35	0.00	85.06	
MCA-AP (mm)	0 < ANB < 5	509	8.98	3.47	1.28	21.99	0.1381
	ANB >= 5	216	9.15	3.68	2.57	22.69	
	ANB <= 0	134	9.68	3.90	1.36	27.49	
MCA-Trans (mm)	0 < ANB < 5	509	23.46	6.38	6.80	45.64	0.5124
	ANB >= 5	216	23.62	6.64	5.87	39.87	
	ANB <= 0	134	24.19	6.48	5.06	39.85	

decreased by 224.23, 122.57, and 130.76 mm^3 for each 1° increase in ANB, FMA, and angle of convexity, respectively (see Tables 6.7 and 6.8). In the study, regarding oropharynx volume, Class III subjects were shown to have a larger oropharynx volume compared to both Class II and Class I subjects. For each 1 mm increase in transverse, A pt. horizontal, A pt. vertical, D pt. horizontal, D pt. vertical, and PNS vertical, the mean oropharynx volume increased in a positive direction [29].

Concerning nasopharynx volume, a statistically significant association was indicated with SNA, SNB, Witts, FMA, facial angle, IMPA, U1 to NA (mm), L1 to NB (degrees and mm), and all six skeletal linear measurements (A pt. horizontal and vertical, D pt. horizontal and vertical, PNS vertical, and a transverse dimension). For each 1° increase in SNA, SNB, facial angle, IMPA, and L1 to NB, the mean nasopharynx volume increased by 48.80 mm^3, 46.09 mm^3, 40.47 mm^3, 30.66 mm^3, and 24.85 mm^3, respectively. For each 1 mm increase in Witts and L1 to NB, the mean nasopharynx volume increased by 42.51 mm^3 and 95.62 mm^3, respectively. The mean nasopharynx volume decreased by 25.56 mm^3 for each 1° increase in FMA. As with total airway and oropharynx volumes, the same observation was noticed in nasopharynx volume in relation to the six skeletal linear measurements. The mean oropharynx volume increased in a positive direction for each 1 mm increase in transverse, A pt. horizontal, A pt. vertical, D pt. horizontal, D pt. vertical, and PNS vertical [29].

Considering MCA, a statistically significant relationship was detected with SNA, SNB, ANB, FMA, facial angle, angle of convexity, and all six skeletal linear measurements. The mean MCA increased by 2.06, 3.68, and 3.41 mm^2 for each 1° increase in SNA, SNB, and facial angle, respectively. Conversely, the mean MCA decreased by 3.56, 1.73, and 1.75 mm^2 for each 1° increase in ANB, FMA, and angle of convexity, respectively.

Looking at FH to MCA, a statistically significant correlation was revealed with Witts, FMA, facial angle, and all six skeletal linear measurements. For each 1 mm increase in Witts, the mean FH to MCA increased by 0.45 mm. For each 1° increase in facial angle, the mean FH to MCA increased by 0.40 mm. In contrast, the mean FH to MCA decreased by 0.39 for each 1° increase in FMA. For each 1 mm increase in the transverse, A pt. horizontal, A pt. vertical, D pt. horizontal, D pt. vertical, and PNS vertical skeletal linear measurements, the mean FH to MCA increased in a positive direction. Thus, the location of MCA became more inferior from FH as each skeletal linear measurement augmented [29].

Grauer et al. [35] assessed differences in airway shape and volume using CBCTs and found that both varied between Class II and III facial patterns. For Class II, the airway was inclined forward, and for Class III it was oriented more vertically. Iwasaki et al. [36] further observed that Class III malocclusions were associated with a large and flat oropharyngeal airway as compared with Class Is. They also reported that changes in facial growth resulted from respiratory obstruction caused by enlarged adenoids or tonsils, and obstructions of different parts of the upper airway caused by adenoids and enlarged tonsils were associated with different forms of malocclusion. However, nasal airway resistance can result from not only adenoids and enlarged tonsils but also nasal airway shape and tongue position. The same authors, in 2017, evaluated the influences of other factors of nasal airway ventilation state, adenoid size, tonsil size, tongue posture (inferior and anterior), and airway form from CBCT data. They investigated the relationships among upper airway factors (i.e., nasal obstruction, adenoids, enlarged tonsils, and inferior and anterior tongue posture) and maxillofacial form difference between Class II and Class III children. Nasal obstruction and adenoids were confirmed as upper airway features of Class II children. Relative constriction of the maxillary dentition correlated with nasal obstruction, enlarged tonsils, and an inferior and anterior tongue posture. The upper airways of Class III children were characterized by no nasal obstruction and a large pharyngeal airway diameter. Protrusion of the mandibular incisors was associated with enlarged tonsils and an anterior tongue posture [37]. Primozic et al. [38] reported that Class III subjects had a significantly inferior tongue posture when compared with Class I subjects, and an inferior tongue posture was also associated with increased mandibular arch width.

Subjects with a Class II skeletal pattern have a narrower anteroposterior pharyngeal dimension, and this narrowing was specifically noted in the nasopharynx area at the hard palate level and in the oropharynx at the level of the tip of the soft palate and the mandible. Ceylan et al. [39] investigated the pharyngeal size on lateral cephalograms in adolescents and stated that as the oropharynx area became smaller, the ANB increased. Similarly, Kim et al. [40] perceived the mean total airway volume in subjects with a larger ANB to be significantly smaller than a normal AP skeletal relationship. The oropharyngeal airway volumes in Class II patients were significantly smaller than Class I as well as Class III patients, according to two separate studies conducted by El et al. [14, 41]. The uvulo-glosso-pharyngeal dimensions in subjects with different anteroposterior (AP) jaw relationships were investigated by Abu Alhaija et al. [42]. They found that the AP skeletal pattern showed a weak but significant correlation with the inferior pharyngeal airway space. They also observed that the vertical airway length (VAL) was reduced in Class II males. Claudino et al. [43] further discovered that in Class II subjects, the lower pharyngeal portion, velopharyngeal, and oropharyngeal areas all displayed a decreased airway volume as compared airways in Class I and III subjects.

Other studies have found no differences in pharyngeal airway dimensions between Class I and II types. Comparing upper and lower pharyngeal widths, de Freitas et al. [44] determined that Class I and II patients with vertical growth patterns had significantly narrower upper pharyngeal widths. Shigeta et al. [45] examined the influence of aging and BMI on oropharynx configuration and noticed that although larger in length and volume in males, the airway lengthens and collapses with age. Regarding AP as well as gender variables, Mislik et al. [28] concluded that there were no significant differences between genders and significant correlations with AP variables.

An initial airway assessment and understanding changes of the desired volume should be relayed to the treating orthodontist so that a proper orthodontic plan can be devised that will complement desired objectives. Once a thorough treatment plan defining skeletal and orthodontic movements has been prepared, predictions can be made about the effects on the airway and its management. However, controversial findings on the relationship between vertical craniofacial patterns and pharyngeal airway were also demonstrated [35, 46].

6.3.1 Orthodontic Treatment Effects on Airway

The upper airway has an important role in respiration, swallowing, and pronunciation. The size of the tongue, soft palate, and parapharyngeal fat pads and the position of the lateral pharyngeal walls in the mandible and maxilla are all important determinants of upper airway morphology [47, 48]. Narrowing in one or more segments of the upper airway may induce breathing problems. Therefore, many studies have shown a possible relationship between pharyngeal airway and skeletal structures, soft tissues, and the musculature after orthodontic treatment mechanotherapy. Orthodontists and oral surgeons have long been in the forefront of airway evaluation and related concerns.

Rapid maxillary expansion (RME) and facemask (FM) were further introduced as a treatment option for increasing the pharyngeal airway dimension and diversifying the paradigms of clinical orthodontic treatment [49]. In addition, changes in the size of nasopharyngeal airway have been reported following RME [50]. Many studies have reported that rapid maxillary expansion improves nasal airway ventilation [51–53]. Christie et al. [54] concluded in their cone-beam computed tomography study that nasal cavity increases significantly (2.73 mm) following maxillary expansion. Baratieri et al. [55] performed a meta-analysis evaluating at the effects of RME on airway volume of the nasopharynx and found that RME therapy during the growth period caused increases in the width of the nasal cavity and in the posterior nasal airway with an associated reduced nasal airway resistance and increased total nasal flow with the stability lasting at least 1 year. The nasopharyngeal airway was increased with maxillary protraction in skeletal Class III children [56].

In growing children, intraoral appliances such as mandibular advancement devices (i.e., activators) have been used for many decades to modify mandibular growth in skeletal Class II patients. In contrast to these devices used in adults, activators are considered to enhance skeletal growth of the mandible. There are numerous studies that have examined skeletal changes with functional orthopedic treatment in Class II children. There appears to be a few studies that have investigated pharyngeal airway dimension changes caused by these devices. Hanggi et al. [57] reported that activator-headgear therapy had the potential to increase pharyngeal airway dimensions, such as the smallest distance between the tongue base and the posterior pharyngeal wall, or the pharyngeal area.

Extractions of permanent teeth have long been a part of the orthodontic treatment. Most common indications for extractions in orthodontics are due to excessive crowding or anteroposterior changes, as in Class II or Class III dental camouflages. Depending on the diagnosis and treatment planning, two or four premolars are usually extracted. Dental extractions have been a topic of discussion and a cause of clinical disagreement ever since they were introduced to orthodontics. The debate on permanent tooth extractions is ongoing. Only now, it is not only the esthetics [58] and stability [59] that are discussed but also temporomandibular joint considerations and upper airway volumes. One of the main issues of the current dispute is the dilemma on whether extracting teeth, therefore reducing the length of the dental arch, would deprive the tongue of its essential space and affect the upper airway. Thus, the mechanobiological response of the upper airway should be taken into consideration during large incisor retraction. Germec-Cakan et al. [60] reported a decrease of airway space behind the tip of soft palate and tongue in subjects with extraction orthodontic treatment. Wang et al. [61] also evaluated the effects of four premolars' extraction in bimaxillary protrusive patients; they found that the dimension of the velopharynx, glossopharynx, and hypopharynx was decreased after maximal retraction of anterior teeth. Chen et al. [4] also observed the greatest changes in the hypopharynx. This can be explained by the supportive bone and cartilage of the nasopharynx and hypopharynx, while the antetheca of the palatopharynx and glossopharynx was made up from the soft palate and tongue, which were easily affected by the change in the surrounding tissues. These studies confirmed that maximal retraction of the anterior teeth did influence pharyngeal airway dimension in adults. Retraction of incisors with extraction of four premolars and use of miniscrews decreased the oral volume, which in turn reduced the tongue's space in terms of the sagittal plane, and then the tongue retracted to press the soft palate. This movement resulted in an adaptation leading to the diminution of the upper airway [62]. In addition, they also found a backward and downward movement of the hyoid bone. Other studies by Valiathan et al. [62] and Maaitah et al. [63] indicated that orthodontic treatment with extraction of four premolars did not influence oropharyngeal airway volume in adolescents. The authors attributed the negative finding mainly to the pharynx growth [62].

Combined orthodontic and orthognathic surgical treatment is a common treatment modality for the correction of facial deformities. An important aspect of orthognathic surgery is the effect of skeletal movements and changes in the position of the hyoid bone, tongue, soft palate, and posture (thus, pharyngeal airway). The pharyngeal upper airway has attracted much attention because snoring and sleep apnea are known to be closely associated with its volume. Many authors have reported a decrease in airway dimensions after Class III orthognathic surgery and, accordingly, extension of the head posture [6, 64]. In some studies, posterior and inferior movements of the hyoid bone were detected after surgery in the short term. However, this movement returned to its original position in the long term [65].

With isolated mandibular setback surgery or bimaxillary surgery, it is generally accepted that the position of the hyoid bone and the tongue is changed and the pharyngeal airway space is narrowed [8, 65]. Most reports have stated that the pharyngeal airway space is narrowed immediately after orthognathic surgery and then changes continually over time [66, 67]. However, bimaxillary surgery rather than isolated mandibular setback surgery enables better airflow, because clockwise rotation of the maxilla and maxillary advancement may make the pharyngeal airway space wider, whereas mandibular setback causes narrowing of the pharyngeal airway space [4, 68]. When the maxilla and mandible are protruded, widening occurs in the velopharyngeal airway with the elevation of the tissues attached to the maxilla, mandible, and hyoid bone [69]. Kim et al. [32] found that the volume of pharyngeal airway sections decreased significantly after surgery for both genders. Hart et al. [70] showed that the negative effects of posterior skeletal movements could be minimized when both jaws were simultaniously operated to compansate for possible diminishing effects of single jaw surgeries.

Snoring and OSA are described as two aspects of the same basic disorder, namely, sleep-related narrowing of the upper airways, which differ only in severity [71]. The patency of the upper airway depends on the balance between the negative intrapharyngeal pressure developed during inspiration and its counteraction by dilating muscles [72]. It is clear that upper airway collapse most often results from a combination of anatomical factors that predispose the airway to collapse during inspiration, plus neuromuscular compensation that is insufficient during sleep to maintain airway patency [73]. Therefore, it is possible that small pharyngeal dimensions established early in life may predispose to OSA and snoring later when subsequent soft tissue changes [74] caused by age, obesity, or genetic background further reduce the available oropharyngeal airway. Consequently, it can only be regarded as beneficial if functional orthopedic treatment in children [75] or surgical mandibular advancement [64] results in a permanent increase in pharyngeal airway dimensions. The mechanobiological response of the upper airway should be taken into consideration during orthodontic treatment.

References

1. Standring S. Gray's anatomy: the anatomical basis of clinical practice, vol. 39. Philadelphia, PA: Elsevier, Inc; 2005. p. 619–31.
2. Hilberg O, Jackson AC, Swift DL, Pedersen OF. Acoustic rhinometry: evaluation of nasal cavity geometry by acoustic reflection. J Appl Physiol. 1989;66(1):295–303.
3. Meyer W. On adenoid vegetations in the nasopharyngeal cavity: their pathology, diagnosis, and treatment. Med Chir Trans. 1870;53:191–216.
4. Chen F, Terada K, Hua Y, Saito I. Effects of bimaxillary surgery and mandibular setback surgery on pharyngeal airway measurements in patients with class III skeletal deformities. Am J Orthod Dentofac Orthop. 2007;131(3):373–7.
5. Winnberg A, Pancherz H, Westesson PL. Head posture and hyo-mandibular function in man. A synchronized electromyographic and videofluorographic study of the open-close-clench cycle. Am J Orthod Dentofacial Orthop. 1988;94(5):393–404.
6. Muto T, Yamazaki A, Takeda S, Kawakami J, Tsuji Y, Shibata T, Mizoguchi I. Relationship between the pharyngeal airway space and craniofacial morphology, taking into account head posture. Int J Oral Maxillfac Surg. 2006;35(2):132–6.
7. Pae EK, Lowe AA, Sasaki K, Price C, Tsuchiya M, Fleetham JA. A cephalometric and electromyographic study of upper airway structures in the upright and supine positions. Am J Orthod Dentofac Orthop. 1994;106(1):52–9.
8. Saitoh K. Long-term changes in pharyngeal airway morphology after mandibular setback surgery. Am J Orthod Dentofac Orthop. 2004;125(5):556–61.
9. Mehra P, Downie M, Pita MC, Wolford LM. Pharyngeal airway space changes after counterclockwise rotation of the maxillomandibular complex. Am J Orthod Dentofac Orthop. 2001;120(2):154–9.
10. Guijarro-Martinez R, Swennen GR. Cone-beam computerized tomography imaging and analysis of the upper airway: a systematic review of the literature. Int J Oral Maxillofac Surg. 2011;40(11):1227–37.
11. Aboudara C, Nielsen I, Huang JC, Maki K, Miller AJ, Hatcher D. Comparison of airway space with conventional lateral headfilms and 3-dimensional reconstruction from cone-beam computed tomography. Am J Orthod Dentofac Orthop. 2009;135(4):468–79.
12. Fujioka M, Young LW, Girdany BR. Radiographic evaluation of adenoidal size in children: adenoidal-nasopharyngeal ratio. AJR Am J Roentgenol. 1979;133(3):401–4.

13. Lenza MG, Lenza MM, Dalstra M, Melsen B, Cattaneo PM. An analysis of different approaches to the assessment of upper airway morphology: a CBCT study. Orthod Craniofac Res. 2010;13(2):96–105.
14. El H, Palomo JM. Airway volume for different dentofacial skeletal patterns. Am J Orthod Dentofac Orthop. 2011;139(6):511–21.
15. De Backer JW, Vos WG, Verhulst SL, De Backer W. Novel imaging techniques using computer methods for the evaluation of the upper airway in patients with sleep-disordered breathing: a comprehensive review. Sleep Med Rev. 2008;12(6):437–47.
16. McNamara JA. Influence of respiratory pattern on craniofacial growth. Angle Orthod. 1981;51(4):269–300.
17. Houston WJ. The analysis of errors in orthodontic measurements. Am J Orthod. 1983;83(5):382–90.
18. Kobayashi K, Shimoda S, Nakagawa Y, Yamamoto A. Accuracy in measurement of distance using limited cone-beam computerized tomography. Int J Oral Maxillofac Implants. 2004;19(2):228–31.
19. Loubele M, Guerrero ME, Jacobs R, Suetens P, van Steenberghe D. A comparison of jaw dimensional and quality assessments of bone characteristics with cone-beam CT, spiral tomography, and multi-slice spiral CT. Int J Oral Maxillofac Implants. 2007;22(3):446–54.
20. Yamashina A, Tanimoto K, Sutthiprapaporn P, Hayakawa Y. The reliability of computed tomography values and dimensional measurements of the oropharyngeal region using cone-beam CT: comparison with multidector CT. Dentomaxillofac Radiol. 2008;37(5):245–51.
21. Stratemann S, Huang JC, Make K, Hatcher D, Miller AJ. Three-dimensional analysis of the airway with cone-beam computed tomography. Am J Orthod Dentofac Orthop. 2011;140(5):607–15.
22. Weissheimer A, Macedo de Menezes L, Sameshima GT, Enciso R, Pham J, Grauer D. Imaging software accuracy for 3-dimensional analysis of the upper airway. Am J Orthod Dentofac Orthop. 2012;142(6):801–13.
23. Souza KR, Oltramari-Navarro PV, Navarro Rde L, Conti AC, Almeida MR. Reliability of a method to conduct upper airway analysis in cone-beam computed tomography. Braz Oral Res. 2013;27(1):48–54.
24. Mattos CT, Cruz CV, da Matta TC, Pereira Lde A, Solon-de-Mello Pde A, Ruellas AC, Sant'anna EF. Reliability of upper airway linear, area, and volumetric measurements in cone-beam computed tomography. Am J Orthod Dentofac Orthop. 2014;145(2):188–97.
25. Taylor M, Hans MG, Strohl KP, Nelson S, Broadbent BH. Soft tissue growth of the oropharynx. Angle Orthod. 1996;66(5):393–400.
26. Linder-Aronson S, Leighton BC. A longitudinal study of the development of the posterior nasopharyngeal wall between 3 and 16 years of age. Eur J Orthod. 1983;5(1):47–58.
27. Vogler RC, Ii FJ, Pilgram TK. Age-specific size of the normal adenoid pad on magnetic resonance imaging. Clin Otolaryngol Allied Sci. 2000;25(5):392–5.
28. Mislik B, Hänggi MP, Signorelli L, Peltomäki TA, Patcas R. Pharyngeal airway dimensions: a cephalometric, growth-study-based analysis of physiological variations in children aged 6–17. Eur J Orthod. 2014;36(3):331–9.
29. Laird A. Airway analysis in class I, II, and III dentoskeltal types: a CBCT study. Master thesis. Oklahoma City, OK. 2015.
30. Cohen AM, Vig PS. A serial growth study of the tongue and intermaxillary space. Angle Orthod. 1976;46(4):332–7.
31. Değerliyurt K, Ueki K, Hashiba Y, et al. The effect of mandibular setback or two-jaws surgery on pharyngeal airway among different genders. Int J Oral Maxillofac Surg. 2009;38(6):647–52.
32. Kim JS, Kim JK, Hong SC, Cho JH. Pharyngeal airway changes after sagittal split ramus osteotomy of the mandible: a comparison between genders. J Oral Maxillofac Surg. 2010;68(8):1802–6.
33. Moss-Salentijn L, Melvin L. Moss and the functional matrix. J Dent Res. 1997;76:1814–7.
34. Angle E. Treatment of malocclusion of the teeth. Philadelphia, PA: SS White Manufacturing Company; 1907.
35. Grauer D, Cevidanes LS, Styner MA, Ackerman JL, Proffit WR. Pharyngeal airway volume and shape from cone-beam computed tomography: relationship to facial morphology. Am J Orthod Dentofac Orthop. 2009;136(6):805–14.
36. Iwasaki T, Hayasaki H, Takemoto Y, Kanomi R, Yamasaki Y. Oropharyngeal airway in children with class III malocclusion evaluated by cone-beam computed tomography. Am J Orthod Dentofac Orthop. 2009;136(3):318–9.
37. Iwasaki T, Sato H, Suga H, Takemoto Y, Inada E, Saitoh I, Kakuno E, Kanomi R, Yamasaki Y. Relationships among nasal resistance, adenoids, tonsils, and tongue posture and maxillofacial form in class II and class III children. Am J Orthod Dentofac Orthop. 2017;151(5):929–40.
38. Primozic J, Farcnik F, Perinetti G, Richmond S, Ovsenik M. The association of tongue posture with the dentoalveolar maxillary and mandibular morphology in class III malocclusion: a controlled study. Eur J Orthod. 2013;35(3):388–93.
39. Ceylan I, Oktay H. A study on the pharyngeal size in different skeletal patterns. Am J Orthod Dentofac Orthop. 1995;108(1):69–75.
40. Kim YJ, Hong JS, Hwang YI, Park YH. Three-dimensional analysis of pharyngeal airway in preadolescent children with different anteroposterior skeletal patterns. Am J Orthod Dentofac Orthop. 2010;137:306–7.
41. El H, Palomo JM. An airway study of different maxillary and mandibular sagittal positions. Eur J Orthod. 2013;35(2):262–70.

42. Abu Alhaija ES, Al-Khateeb SN. Uvulo-glosso-pharyngeal dimensions in different anteroposterior skeletal patterns. Angle Orthod. 2005;75(6):1012–8.
43. Claudino LV, Mattos CT, Ruellas AC, Sant' Anna EF. Pharyngeal airway characterization in adolescents related to facial skeletal pattern: a preliminary study. Am J Orthod Dentofac Orthop. 2013;143(6):799–809.
44. de Freitas MR, Alcazar NM, Janson G, de Freitas KM, Henriques JF. Upper and lower pharyngeal airways in subjects with class I and class II malocclusions and different growth patterns. Am J Orthod Dentofac Orthop. 2006;130(6):742–5.
45. Shigeta Y, Ogawa T, Venturin J, Nguyen M, Clark GT, Enciso R. Gender- and age-based differences in computerized tomographic measurements of the oropharynx. Oral Surg Oral Med Oral Pathol Oral Radiol Endod. 2008;106:563–70.
46. Ucar FI, Uysal T. Orofacial airway dimensions in subjects with class I malocclusion and different growth patterns. Angle Orthod. 2011;81(3):460–8.
47. Jakobsone G, Stenvik A, Espeland L. The effect of maxillary advancement and impaction on the upper airway after bimaxillary surgery to correct class III malocclusion. Am J Orthod Dentofac Orthop. 2011;139(4):e369–76.
48. Shigeta Y, Ogawa T, Ando E, Clark GT, Enciso R. Influence of tongue/mandible volume ratio on oropharyngeal airway in Japanese male patients with obstructive sleep apnea. Oral Surg Oral Med Oral Pathol Oral Radiol Endod. 2011;111:239–43.
49. Akin M, Ucar FI, Chousein C, Sari Z. Effects of chin cup or facemask therapies on the orofacial airway and hyoid position in class III subjects. J Orofac Orthop. 2015;76:520–30.
50. Tecco S, Caputi S, Festa F. Evaluation of cervical posture following palatal expansion: a 12-month follow-up controlled study. Eur J Orthod. 2007;29(1):45–51.
51. Hershey HG, Stewart BL, Warren DW. Changes in nasal airway resistance associated with rapid maxillary expansion. Am J Orthod. 1976;69(3):274–84.
52. Hartgerink DV, Vig PS, Abbott DW. The effect of rapid maxillary expansion on nasal airway resistance. Am J Orthod Dentofac Orthop. 1987;92(5):381–9.
53. De Felippe NL, Bhushan N, Da Silveira AC, Viana G, Smith B. Long-term effects of orthodontic therapy on the maxillary dental arch and nasal cavity. Am J Orthod Dentofac Orthop. 2009;136(4):490–1.
54. Christie KF, Boucher N, Chung CH. Effects of bonded rapid palatal expansion on the transverse dimensions of the maxilla: a cone-beam computed tomography study. Am J Orthod Dentofac Orthop. 2010;137(4):79–85.
55. Baratieri C, Alves M Jr, de Souza MM, de Souza Araújo MT, Maia LC. Does rapid maxillary expansion have long-term effects on airway dimensions and breathing? Am J Orthod Dentofac Orthop. 2011;140(2):146–56.
56. Lee JW, Park KH, Kim SH, Park YG, Kim SJ. Correlation between skeletal changes by maxillary protraction and upper airway dimensions. Angle Orthod. 2011;81(3):426–32.
57. Hänggi MP, Teuscher UM, Roos M, Peltomäki TA. Long-term changes in pharyngeal airway dimensions following activator-headgear and fixed appliance treatment. Eur J Orthod. 2008;30(6):598–605.
58. Proffit WR. Forty-year review of extraction frequencies at a university orthodontic clinic. Angle Orthod. 1994;64(6):407–14.
59. Tweed CH. Indications for the extraction of teeth in orthodontic procedures. Am J Orthod Oral Surg. 1944;30:405–28.
60. Germec-Cakan D, Taner T, Akan S. Uvulo-glossopharyngeal dimensions in non-extraction, extraction with minimum anchorage, and extraction with maximum anchorage. Eur J Orthod. 2011;33(5):515–20.
61. Wang Q, Jia P, Anderson NK, Wang L, Lin J. Changes of pharyngeal airway size and hyoid bone position following orthodontic treatment of class I bimaxillary protrusion. Angle Orthod. 2012;82(1):115–21.
62. Valiathan M, El H, Hans MG, Palomo MJ. Effects of extraction versus non-extraction treatment on oropharyngeal airway volume. Angle Orthod. 2010;80(6):1068–74.
63. Al Maaitah E, El Said N, Abu Alhaija ES. First premolar extraction effects on upper airway dimension in bimaxillary proclination patients. Angle Orthod. 2012;82(5):853–9.
64. Achilleos S, Krogstad O, Lyberg T. Surgical mandible setback and changes in uvuloglossopharyngeal morphology and head posture: a short- and long-term cephalometric study in males. Eur J Orthod. 2000;22(4):383–94.
65. Kawakami M, Yamamoto K, Fujimoto M, Ohgi K, Inoue M, Kirita T. Changes in tongue and hyoid positions, and posterior airway space following mandibular set back surgery. J Craniomaxillofac Surg. 2005;33(2):107–10.
66. Park SB, Kim YI, Son WS, Hwang DS, Cho BH. Cone-beam computed tomography evaluation of short- and long-term airway change and stability after orthognathic surgery in patients with class III skeletal deformities: bimaxillary surgery and mandibular setback surgery. Int J Oral Maxillofac Surg. 2012;41(1):87–93.
67. Eggensperger N, Smolka W, Iizuka T. Long-term changes of hyoid bone position and pharyngeal airway size following mandibular setback by sagittal split ramus osteotomy. J Craniomaxillofac Surg. 2005;33(2):111–7.
68. Aydemir H, Memikoglu U, Karasu H. Pharyngeal airway space, hyoid bone position and head posture after orthognathic surgery in class III patients. Angle Orthod. 2012;82(6):993–1000.
69. Riley RW, Nelson B, Powell NB, Guilleminault C. Obstructive sleep apnea syndrome: a surgical protocol for dynamic upper airway reconstruction. J Oral Maxillofac Surg. 1993;51:742–7.

70. Hart PS, McIntyre BP, Kadioglu O, Currier GF, Sullivan SM, Li J, Shay C. Postsurgical volumetric airway changes in 2-jaw orthognathic surgery patients. Am J Orthod Dentofacial Orthop. 2015;147:536–46.
71. Lugaresi E, Plazzi G. Heavy snorer disease: from snoring to the sleep apnea syndrome-an overview. Respiration. 1997;64(1):11–4.
72. Friberg D. Heavy snorer's disease: a progressive local neuropathy. Acta Otolaryngol. 1999;119(8):925–33.
73. Young T, Peppard PE, Gottlieb DJ. Epidemiology of obstructive sleep apnea: a population health perspective. Am J Respir Crit Care Med. 2002;165(9):1217–39.
74. Martin SE, Mathur R, Marshall I, Douglas NJ. The effect of age, sex, obesity and posture on upper airway size. Eur Respir J. 1997;10(9):2087–90.
75. Ozbek MM, Memikoglu TU, Gogen H, Lowe AA, Baspinar E. Oropharyngeal airway dimensions and functional-orthopedic treatment in skeletal class II cases. Angle Orthod. 1998;68(4):327–36.

Diagnostic Value of 3D Imaging in Clinical Orthodontics

7

Juan Martin Palomo, Hakan El, Neda Stefanovic, and Mohamed Bazina

Abstract

The digital volumetric tomography era has begun, and we now have access to significant additional diagnostic information. When moving from 2D to 3D, distances and angles turn into areas and volumes, and more information may take orthodontics to the next level, increasing the scope of what can be done clinically. This chapter shows some simple ways of incorporating 3D analysis into a busy orthodontic office, without the need for special software.

J. M. Palomo (✉)
Department of Orthodontics, Craniofacial Imaging Center, School of Dental Medicine, Case Western Reserve University, Cleveland, OH, USA
e-mail: jmp5@case.edu

H. El
Department of Orthodontics, School of Dental Medicine, Hacettepe University, Ankara, Turkey

N. Stefanovic
Department of Orthodontics, Faculty of Dental Medicine, University of Belgrade, Belgrade, Serbia

M. Bazina
Division of Orthodontics, College of Dentistry, University of Kentucky, Lexington, KY, USA

O. Kadioglu, G. F. Currier (eds.), *Craniofacial 3D Imaging*,
https://doi.org/10.1007/978-3-030-00722-5_7

7.1 Introduction

Cone beam computed tomography (CBCT) has completely revolutionized orthodontic imaging allowing the clinician to see the patient as what they really are, three-dimensional (3D) structures. However, the clinician should not treat 3D images like 2D images. A 3D image allows the individual to evaluate valuable information, not possible with traditional 2D imaging, and this may significantly impact the type of treatment plan and potentially the results. This chapter describes how a clinician can use all three dimensions for a more comprehensive diagnosis.

7.2 Patient Head Positioning

The first effort, in terms of radiological examination for orthodontic diagnosis and treatment planning, is to precisely position the head as was initially started with the invention of the Bolton-Broadbent cephalometer in 1925 [1]. Since then similar apparatus have been used in cephalometric radiology to orient the head for a standardized position between acquisitions. The change in head position, especially in 2D radiological examinations, is known to complicate landmark identification and thus can cause significant measurement error [2]. Currently, in order to keep pace with the rapidly changing technology, orthodontists are transitioning from 2D radiographs to 3D CBCT images. This is an understandable change since the field of interest of orthodontics itself is oriented toward 3D. Although CBCT technology has many shortcomings, the application of strict head positioning for 2D lateral head films has been mostly eliminated, but still there is a requirement to compare older 2D films with the new 3D images to justify this transition [3, 4]. Therefore, this alone means that we cannot completely omit the head positioning step before acquiring a CBCT image.

CBCT image acquisition starts with the positioning of the patient's head while seated, standing, or in a supine position, depending upon the CBCT machine being used. Most have used head postures in the literature while taking extraoral photographs or during a radiologic examination such as the natural head position [5] or Frankfort horizontal (FH) plane parallel to the floor [6, 7]. Natural head position is considered the most balanced, natural position of the head when someone views an object at eye level [8]. Several methods have been described to obtain the natural head position for 2D [9–12] and 3D [13–18] image acquisition. However, regardless of which head position is used, the head should be oriented in three planes of space in order to evaluate all soft tissue, skeletal, and dental characteristics in a standardized way. While this suggestion should be applied strictly for 2D image acquisition systems, the same does not hold true for 3D imaging as several post-processing software programs can give the operator the ability to orient the head in the desired position.

The Ackerman-Proffit classification can be used as a reference in order to position the head in three planes of space [19]. This method is based on the maneuverability of an airplane in the air and can be described as 6 degrees of freedom. As seen in Fig. 7.1, all six of the movements occur on or around the x, y, and z Cartesian coordinate system. Straight positional movements on the x-, y-, and z-axes are mediolateral, superoanterior, and anteroposterior movements, respectively. Furthermore, rotational movements of the head such as pitch, yaw, and roll can occur around the x-, y-, and z-axes, respectively.

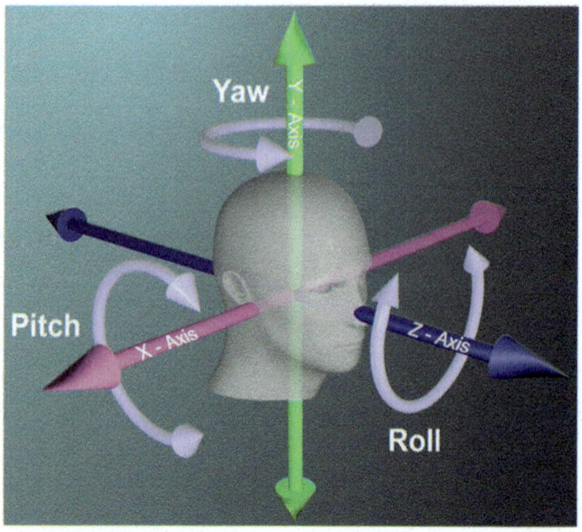

Fig. 7.1 Positioning of the head with 6-degrees of freedom. Straight movements on the x-, y-, and z-axes may cause positioning errors on coronal, sagittal, and/or axial planes. Pitch, yaw, and roll types of rotations occur around the x-, y-, and z-axes, respectively, and may lead to anatomic changes, especially on cervical vertebrae and upper airway that cannot be corrected with software orientation

Most distinct head positioning errors that may be encountered in the sagittal plane are linear movements alongside the z-axis and rotational movements around the x-axis (pitch). Alongside the z-axis, protruded/retruded positions of the head may cause anterior/posterior sloping of the cervical vertebral column, respectively. Furthermore, the patient can bite in a different way than usual [20]. This change in the cervical vertebral column and occlusion cannot be corrected with the current segmentation programs during the post-processing stage of images. Craniocervical angulation and posture have been used mainly for the temporomandibular joint, craniofacial growth, and several malocclusion studies [20–22]. Therefore, it is important to obtain the correct position of the cervical vertebral column. The anteroposterior position of the head can be adjusted using the headrest, laser aligners (shading on older CBCT devices), and chin cup or lip rest of the CBCT device. However, for orthodontic purposes, it is not advisable for the patient to lean against the chin cup or the lip rest piece [23]. It may cause a change in the occlusion or, even worse, a deformation of the relevant soft tissue if the patient rests on them firmly. Therefore, instead of using the chin cup or

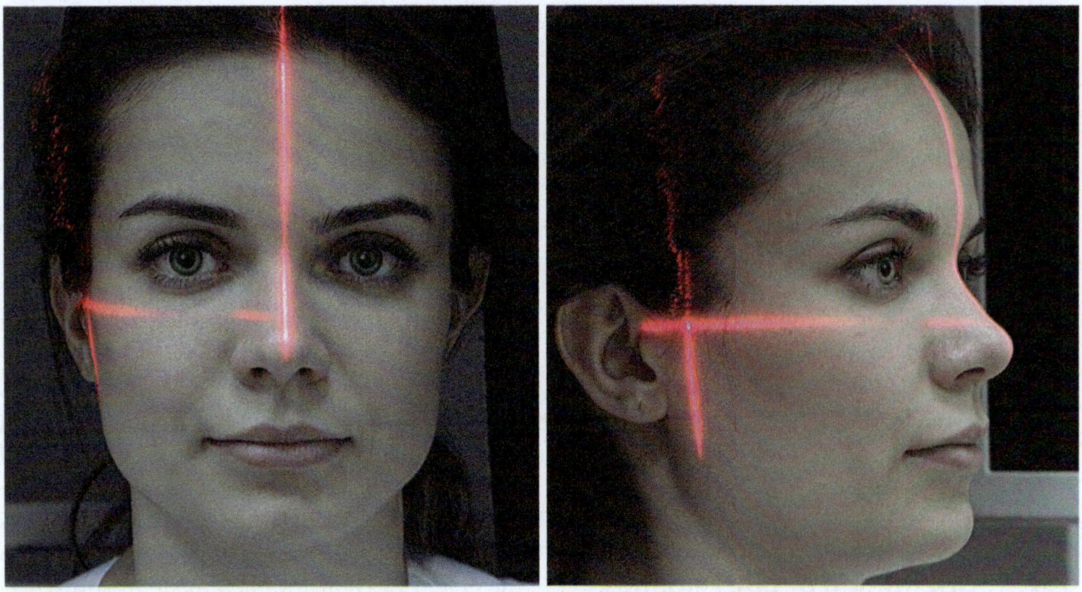

Fig. 7.2 Laser aligners of the CBCT device. Vertical line is mainly used to define the midsagittal plane and to correct roll and yaw types of rotations of the head. Horizontal line is mainly used to define Frankfort horizontal plane (or true horizontal plane in case of the natural head position) and to correct pitch type of rotations such as overextension or overflexion

the lip rest, head straps around the forehead or, if available, ear rods in a reversed direction (without inserting the rods into the acoustic meatus) can be used to stabilize the head [8]. Laser aligners, on the other hand, can be considered as an industrial standard for contemporary CBCT devices and help the operator significantly during positioning by dropping a cross-reference mark on the head (Fig. 7.2). Around the *x*-axis, pitch errors, such as overflexion or overextension of the head, can also cause measurement errors. These positional inaccuracies may similarly lead to a change in the cervical vertebral column and occlusion as well as the oropharyngeal region [24]. If the head is overextended or overflexed, the change in the position of the hyoid bone due to suprahyoid muscles is especially known to cause a change in the dimensions of the oropharyngeal airway [25, 26]. Yet again, pitch errors are corrected using the laser aligners and scout images prior to image acquisition. Since the c-arm (gantry) of the CBCT device is adjusted to the height of the patient (either sitting or standing) before image acquisition, it is less likely to encounter a linear positional error on the *y*-axis. The vertical position of the head can easily be adjusted with the horizontal marker of the laser aligner according to the desired head position. Yaw rotations around the *y*-axis may cause a double image if a lateral cephalogram is going to be obtained from CBCT scans, but this can be easily adjusted with the orientation feature of the segmentation programs (see "Orientation of CBCT Images for the Orthodontic Patient") [27].

On the other hand, the positional errors that may be encountered in the coronal plane are rotational movements around the *x*-, *y*-, and *z*-axes. It is much less likely for patients to position their heads linearly to one side on the *x*-axis. Even so, this can be detected with the vertical marker of the laser aligners and corrected accordingly. The vertical laser marker is positioned on the midsagittal plane of the head. Using the vertical marker also aids in correcting the yaw and roll rotations. Especially yaw rotation is known to cause more linear measurement errors as compared to pitch and roll rotations [6, 28]. However, it should also be noted that since a 3D image is under question, minimal positional errors will have no effect on the accuracy of linear or angular measurements as all points will move in the same direction [27] and can be corrected during the software orientation stage. The horizontal laser marker on the coronal plane can also be effectively used to correct the roll type of rotations.

It is clear in the literature that minimal changes in the orientation of the head do not affect the accuracy and reliability of CBCT measurements as long as a stable and a repeatable head position is obtained [23]. The most important consideration here is to avoid extreme head positions that may affect the true anatomical representation of the region of interest [23, 28]. Furthermore, it is advised to standardize or orient the head position, especially to compare previous 2D records with the new 3D images.

7.3 Orientation of CBCT Images for the Orthodontic Patient

Head orientation plays an important role when measuring distances and angles on any radiographic evaluation of the orthodontic patients [29–33]. Since the current patient positioning tools in CBCT machines are not sufficient to produce a reproducible head position that can be used for longitudinal assessments of the patients, image analysis software also provides tools to adjust and correct the patients' head position after image acquisition. Keep in mind that after image acquisition, any extracranial reference that might be used during acquisition is not transferred to the 3D volume.

In 3D imaging, dental and skeletal displacements and bone remodeling can be quantified by 3D linear and angular measurements or by 3D linear surface distances (color-coded maps). Although 3D linear distances are a simplification of the complex nature of morphologic changes, they provide relevant clinical evaluations of changes in space. Clinical questions require more precise information regarding the location and amount of changes in the three dimensions of space (*x*-, *y*-, and *z*-axes). Quantification of directional differences in each plane of space can be obtained by decomposing the distances between projections of the 3D landmarks in the *x*, *y*, and *z* coordinate system. For any study sample comprising scans with different head orientations, the understanding of the direction of the changes for all patients depends on the establishment of a common coordinate system. The inconsistency of head orientation across patients in a study sample can lead to inconsistent measurements.

Multiple orientation methods have been proposed in the literature. The CWRU orientation method uses five anatomical structures and one plane. In the coronal view the axial plane is set through the center of the left and right optical foramina to adjust the roll of the volume. In the axial view the coronal plane is set through the center of left foramen ovale to the center of right foramen ovale to adjust any jaw of the volume. In the sagittal view, the axial plane overlaps McRae's plane denoted by the anterior border of the foramen magnum (Basion) to the posterior border of the foramen magnum (Opisthion) to adjust the pitch of the volume (Fig. 7.3) [34]. Ruellas et al. recommended the use of transporionic plane, the Frankfort horizontal plane, and the midsagittal plane, which are coincident with the *x*-, *y*-, and *z*-axes, respectively (Fig. 7.4). They found the method to be reproducible and considered this orientation closest to natural head position [35]. In the same study, they evaluated how head orientation interfered with the amount of directional change in three-dimensional space and concluded that the 3D distances were not affected by head orientation, while the amount of directional change in each plane of space was strongly influenced by head orientation [35].

Extra care should be taken during patient positioning and head orientation after image acquisition if the image evaluation involves the use of measurements based on the 3D components in the *x*-, *y*-, and *z*-axes or if two-dimensional images are reconstructed from a 3D data set. The head orientation should be saved so that every image of every subject has the same reference planes.

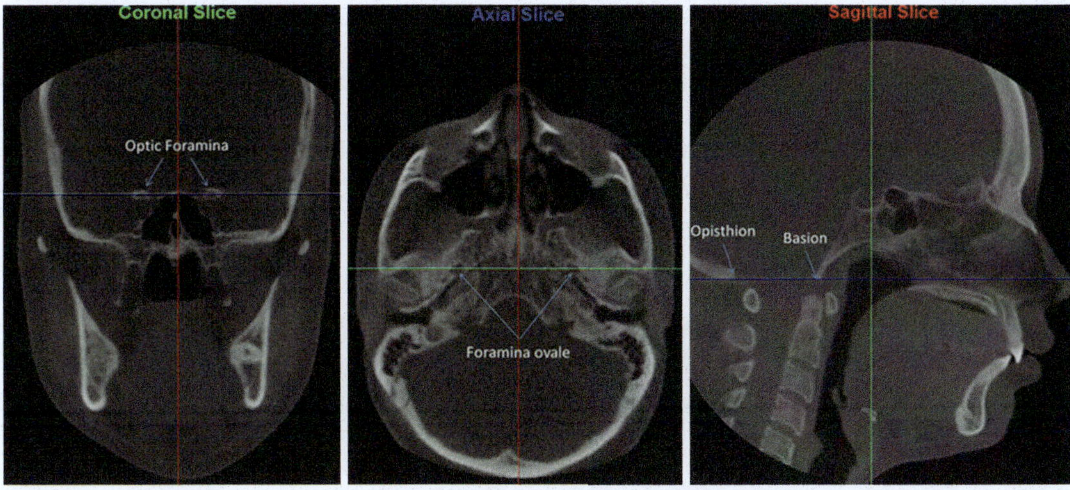

Fig. 7.3 CBCT head orientation using the three planes of space (following Wu et al. orientation method)

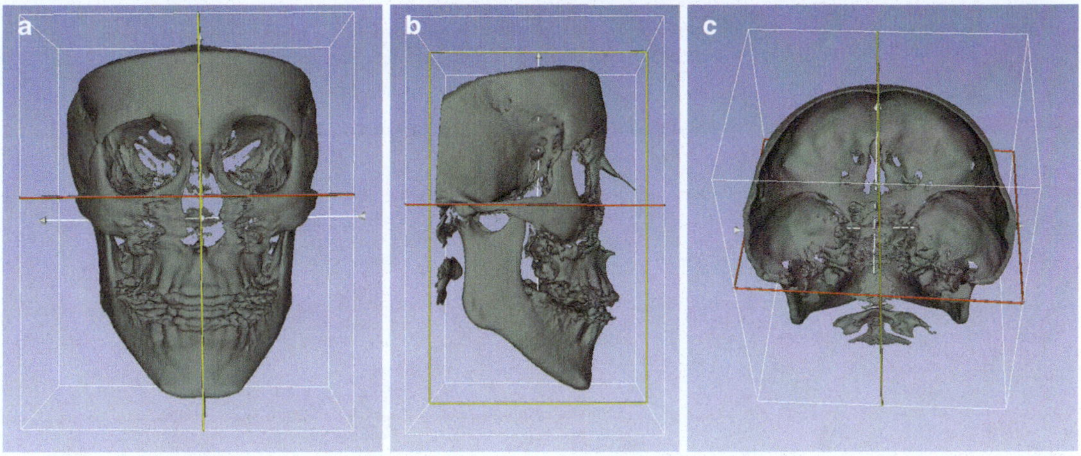

Fig. 7.4 Figures illustrating Ruellas et al.'s head orientation method: (**a**) frontal view; (**b**) lateral view; and (**c**) superior view. The midsagittal plane of the 3D model was oriented vertically and coincident with the yellow (sagittal) plane; the Frankfort horizontal plane was oriented horizontally to coincide with the red (axial) plane; and the transporionic line was oriented to match with the intersection of the axial and coronal planes

7.4 Acquisition of 2D Orthodontic Records from 3D CBCT Scans

7.4.1 Acquisition of Cephalograms

2D cephalograms in the orthodontic practice from the most used to the least are lateral, anterior-posterior (AP), posterior-anterior (PA), and submentovertex (SMV) radiographs for diagnostic purposes and evaluation of treatment progress. Cephalometry has been a very valuable tool that changed the way we think about orthodontic diagnosis and treatment planning since the day it was introduced [36]. It has given great insight from evaluating craniofacial growth to a better understanding of skeletal discrepancies, from calculating response to our treatments to assessing long-term stability after orthodontic treatment [37]. Although there are attempts to improve novel 3D analyses to better understand the skeletal and dental relationships, there still exists significant differences between conventional 2D and 3D cephalometry [38]. When 2D cephalometry is considered to be under development for a period longer than 80 years, it is understandable why orthodontists still prefer to generate 2D cephalograms from 3D images for analysis purposes.

It has been emphasized that 3D cephalometry produces similar results to direct skull measurements [3, 39]. Moreover, 2D cephalograms that are generated from CBCTs are comparable with measurements obtained directly from dry skulls and from conventional cephalograms of patients [37, 40, 41]. The main reason for this phenomena

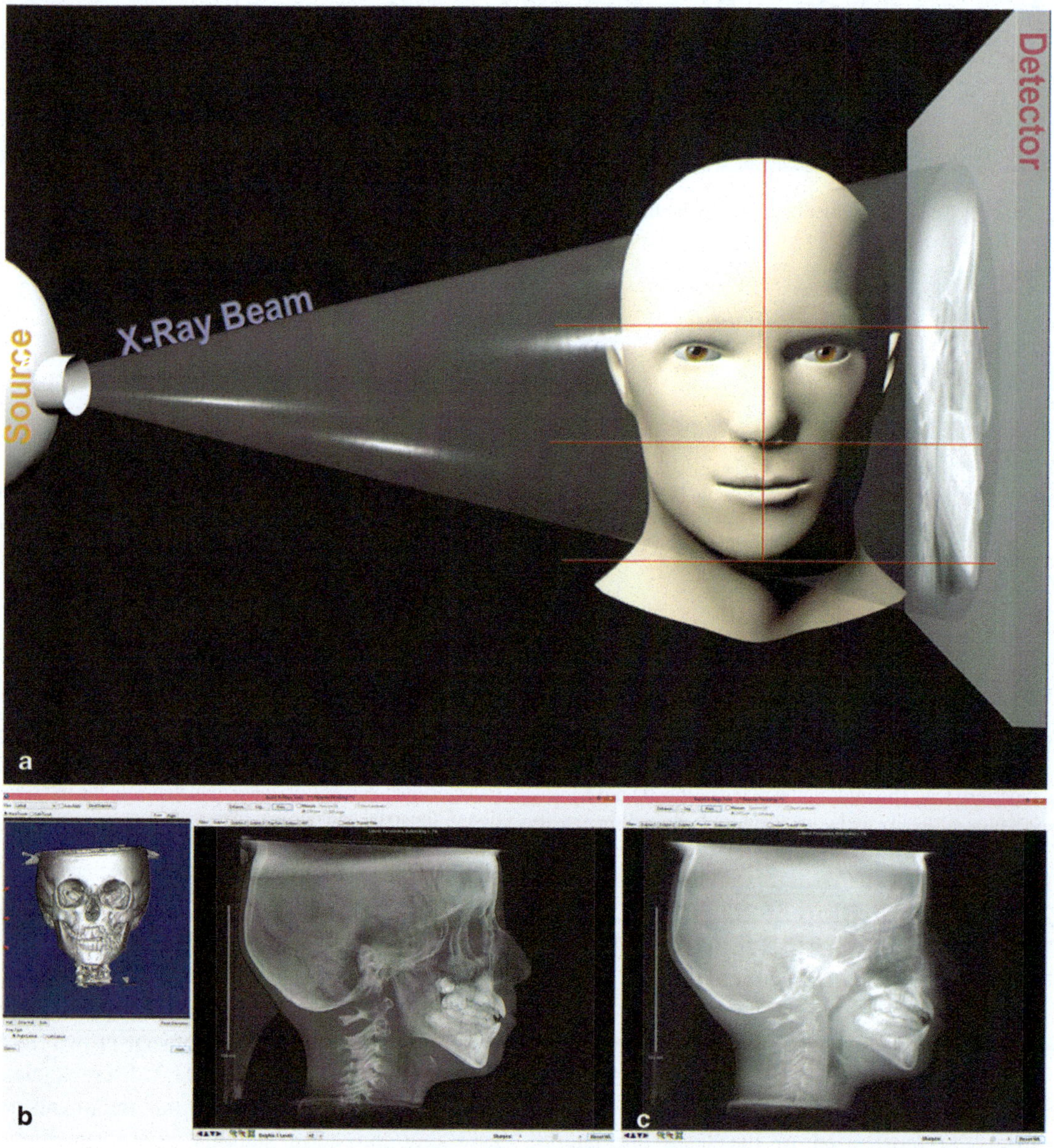

Fig. 7.5 (**a**) Illustration of perspective projection. Please note that the image on the detector is magnified causing a slight change over the proportions of the head. (**b**) Simulated perspective projection and the 2D lateral cephalogram obtained with the preadjusted setting (Dolphin 1) of Dolphin 3D imaging software (Dolphin Imaging & Management Solutions, Chatsworth, CA) with fictitious built-in 7% magnification. (**c**) The same lateral cephalogram with raysum settings providing a closer look to conventional 2D lateral head films

is the ability to eliminate the radiographic distortion and magnification factor in CBCT images that are seen in conventional 2D radiograms. Radiographic distortion may occur if the midsagittal plane of the head is not parallel to the image sensor and not perpendicular to the x-ray beam [42]. As discussed in this chapter earlier, this can easily be corrected for an acquired 3D volumetric image with the orientation tools of various software. Indeed, post-image processing orientation of the head correctly is the very first step of obtaining 2D cephalograms in order to minimize duplication and distortion of anatomic structures.

Magnification, on the other hand, presents another error factor in measurement for conventional 2D films. X-rays, emanating from a source, have a divergent pattern [43]. This pattern is the main reason for an object to be magnified on the sensor (Fig. 7.5a). Furthermore, since a 3D object as the cranium is under question, it has been reported that magnification of craniofacial structures varies from 0% to 24%, depending upon the related structure's proximity to either the sensor or the x-ray source [44]. The structures closer to x-ray source tend to magnify more compared to the structures closer to the sensor. The magnification factor of a conventional 2D cephalogram is calculated by using the simple formula below where a is the distance between the sensor and the midsagittal plane of the face and b is the distance between the x-ray source and the midsagittal plane of the face:

$$\text{Magnification rate} = \frac{a}{b} \times 100$$

The magnification factor of conventional lateral cephalometric radiographs varies from 0.6% to 7.5% depending upon the device used [37, 43]. It is particularly important to know the magnification of a conventional cephalometric film if a comparison/superimposition is going to be performed with a CBCT-generated cephalogram. Most of the segmentation software in the market today present a feature where the operator can choose to simulate a perspective or an orthogonal x-ray projection while adjusting the magnification rate. Therefore, in order to create a 2D cephalogram from CBCT data for comparison with an older 2D conventional film, perspective projection and the relevant magnification setting should be adjusted (Fig. 7.5b). Programs also offer several reconstruction parameters to create 2D images by means of multiplanar reformation (MPR) such as the ray sum method (Fig. 7.5c).

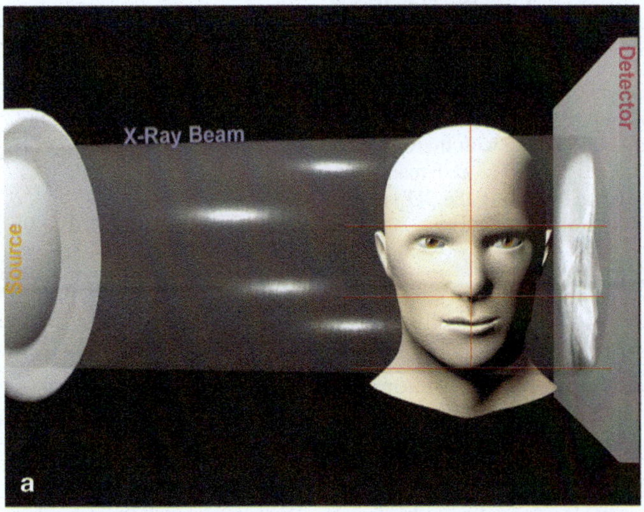

Fig. 7.6 (**a**) Illustration of orthogonal projection. Please note that the image on the detector reflects more concordant proportions of the head without magnification. (**b**) Simulated orthogonal projection and the 2D lateral cephalogram obtained with the preadjusted setting (Dolphin 1) of Dolphin 3D imaging software (Dolphin Imaging & Management Solutions, Chatsworth, CA) with 0% magnification. (**c**) The same lateral cephalogram with raysum settings providing a closer look to conventional 2D lateral head films

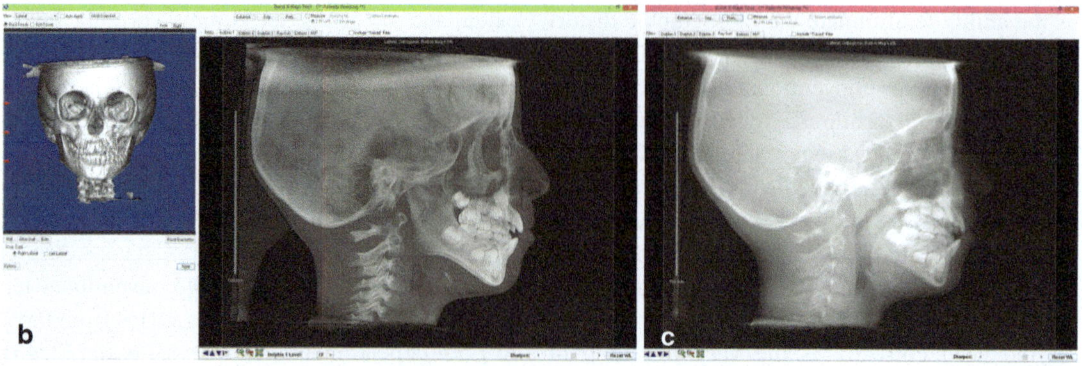

Fig. 7.6 (continued)

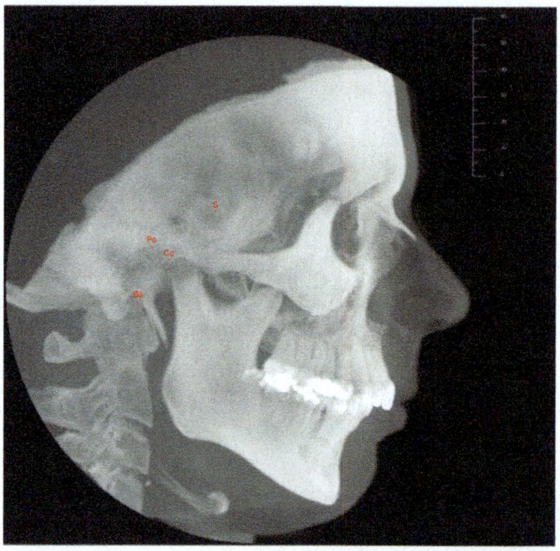

Fig. 7.7 Lateral cephalometric film obtained directly from 3D volumetric view using the maximum intensity projection (MIP) and grayscale setting with Invivo 5 software (Anatomage, San Jose, CA). The red marks are inserted on the relative landmarks using the 3D volume render view

The ray sum processing algorithm, rather than averaging the data, simply adds all values to create the closest result to a conventional radiograph [45]. As a result, a more reliable and a comparable 2D sample can be obtained [37]. However, if the aim is only to generate a 2D image from 3D data, it is recommended to use the orthogonal projection feature. Orthogonal projection simulates as if the entire x-ray beam travels parallel to each other and perpendicular to the midsagittal plane (Fig. 7.6a–c). This way it can be assumed that no magnification occurs and comparable results to direct skull measurements can be obtained [37]. Furthermore, volumetric images can also be used for purposes of cephalometric analysis giving the operator the ability to work with actual dimensions of the skull. For this purpose, maximum intensity projection (MIP) and the grayscale view can be preferred instead of ray sum. However, MIP images are achieved by displaying only the highest attenuation value from the data. Therefore, deeper structures may not be detected properly. One way to overcome this problem is to insert marks prior to analysis on the deeper landmarks such as Sella, Basion, etc. on the 3D volumetric view and use these inserted marks while performing a cephalometric analysis (Fig. 7.7).

The preparation of AP (Fig. 7.8) and SMV (Fig. 7.9) films are also a straightforward procedure. For the production of such films, the operator only needs to select the proper feature, borders, projection type (perspective/orthogonal), and the preferred imaging modality.

7.4.2 Acquisition of Panoramic Radiographs

Panoramic radiographs provide a full coverage of the dentition, giving valuable diagnostic information about the axial inclinations of the teeth, the surrounding structures, attachments, anomalies/

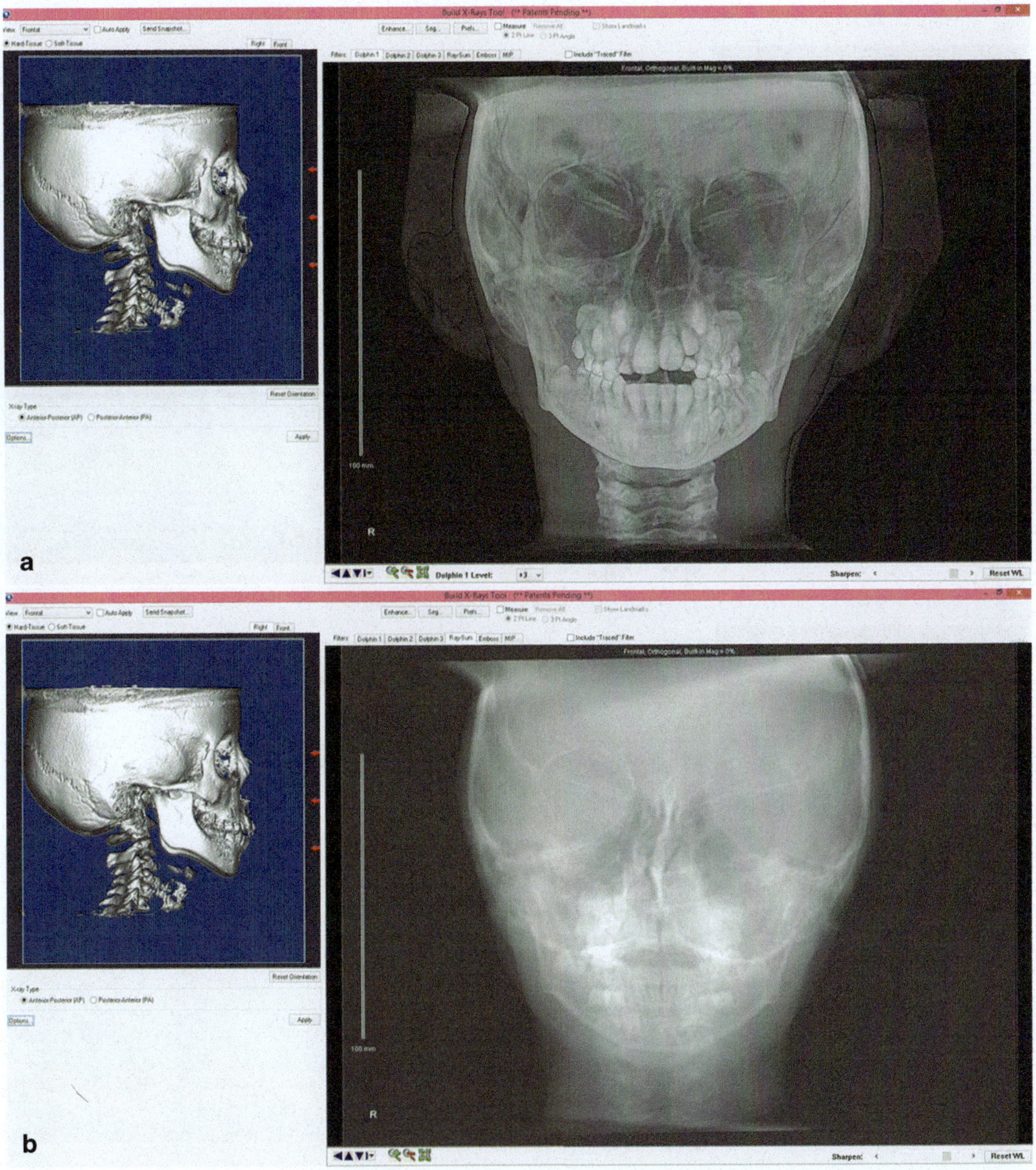

Fig. 7.8 (**a**) Simulated orthogonal projection and the 2D anterior-posterior (AP) cephalogram obtained with the preadjusted setting (Dolphin 1) of Dolphin 3D imaging software with 0% magnification. Note that the orientation of the head changed to sagittal compared to coronal orientation while obtaining lateral cephalograms. (**b**) The same AP cephalogram with raysum settings providing a closer look to conventional 2D AP head films

pathologies, and general information about temporomandibular joints. Therefore, it becomes one of the routinely used orthodontic records. Although the CBCT provides an immersive experience for the operator, evaluating all previously mentioned structures in one 2D general view is still appealing to orthodontists.

CBCT-generated panoramic radiographs are obtained by identifying an arch curve using the axial slices (Fig. 7.10a). The orientation of the

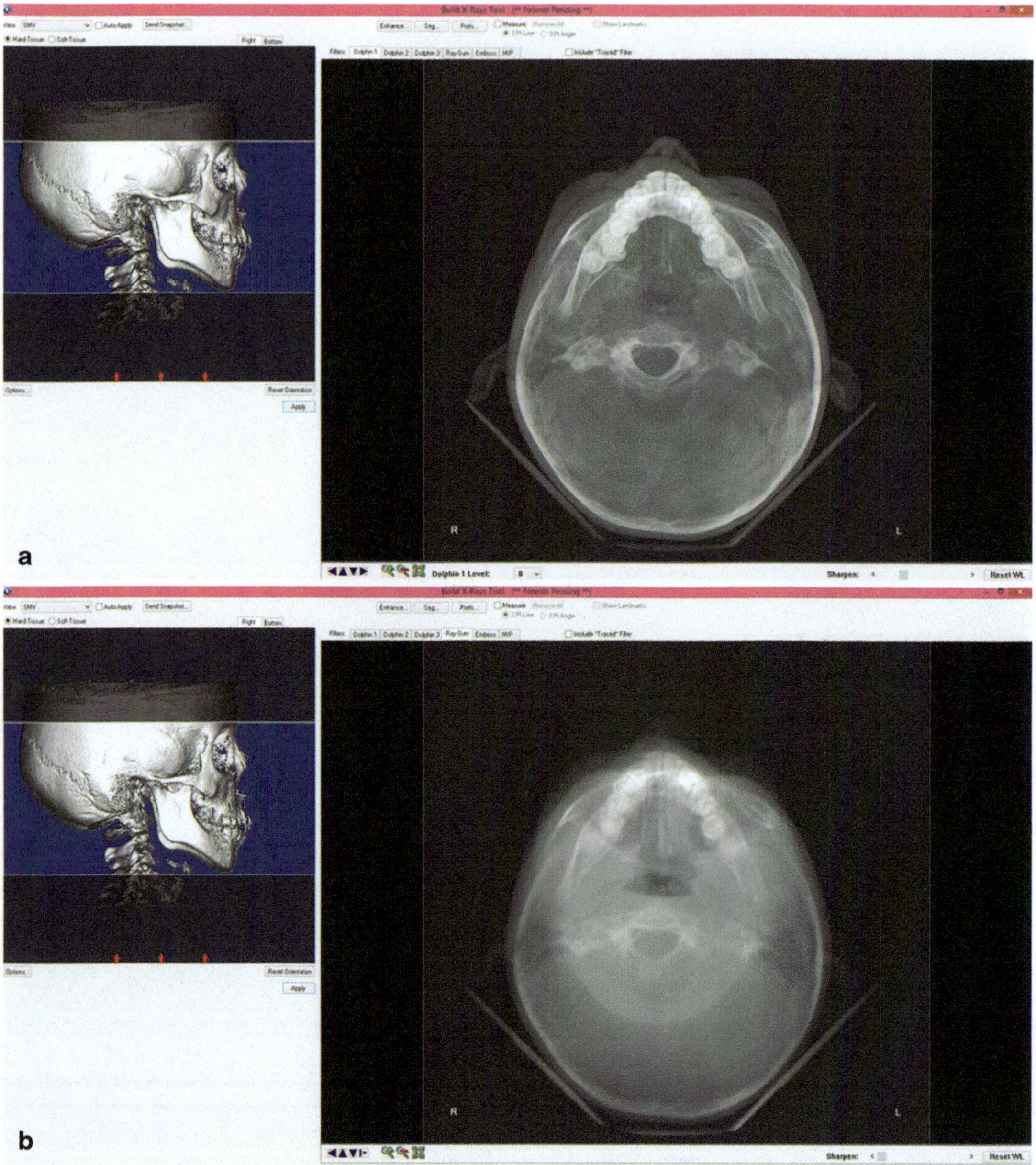

Fig. 7.9 (**a**) Simulated orthogonal projection and (**b**) the 2D submentovertex (SMV) cephalogram obtained with the preadjusted setting (Dolphin 1) of Dolphin 3D imaging software with 0% magnification. Note that the orientation of the head is same compared to previous figure, but the projection trajectory changed from anterior-posterior direction to the inferior-superior direction (small red arrows represent projection trajectory)

head, especially eliminating the yaw and roll type of rotations, is important since this will affect the axial slice views, restrain forming a better projection trajectory, and cause a change in the axial inclinations of the teeth. The second step is to define the superior and inferior limits of the panoramic view. Middle cranial fossa and/or the base of the orbit as the superior limit and base of the mandible and/or hyoid bone as the inferior limit usually meet the requirements (Fig. 7.10a). Following, it is recommended to position the axial slice indicator line on the occlusal plane in

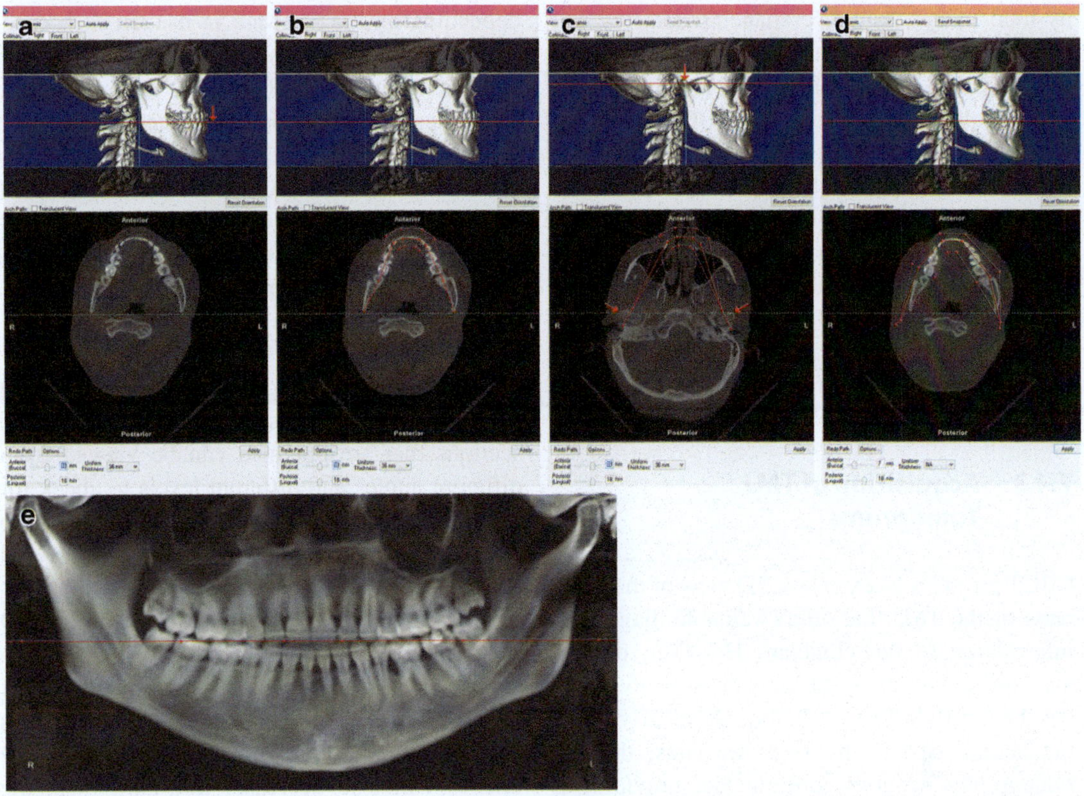

Fig. 7.10 Showing the necessary steps to acquire a CBCT-generated panoramic radiograph. (**a**) The highlighted region on top image represents the superior and inferior borders of the panoramic film. The head is oriented to coincide the axial slice (red line marked with arrow) with the occlusal plane so that all the teeth on dental arch is visible. (**b**) Defining path using the dental arch as a reference (bottom image). Note that the condyles are not present in the current axial view. (**c**) The line representing the axial slice (red line marked with arrow) has been carried superiorly where the condyles are located. The path formed previously has been extended posteriorly to also include the condyles (bottom image). (**d**) Adjusting the focal trough in order to include region of interest completely. (**e**) Resultant CBCT-generated panoramic image

order to create a path, which will serve as the projection trajectory. If there is a clockwise or counterclockwise rotation of the occlusal plane on the oriented image, then reorienting the image (pitch rotation) to make the occlusal plane parallel to the ground can be advised (Fig. 7.10b). Forming the path using the view of the occlusal plane on the axial slice makes it impossible to locate the TMJs. However, this is not a concern because after creating the dental arch form, the axial slice indicator line can be carried superiorly to also visualize and add the condylar and coronoid processes to the previously drawn path by carrying the posterior points accordingly (Fig. 7.10c). This way, the lateral boundaries of the panoramic radiograph are also determined. The final and most important step to obtain a panoramic radiograph is to adjust the simulated focal trough. The focal trough is the anterior and posterior boundaries around the originally drawn path in which the structures are well defined. Most of the programs allow the operator to increase or decrease the focal trough size. The decision to increase or decrease its size can be given by inspecting the axial slices carefully starting from the upper limit to the lower limit. This way, one can see what will be included in the final view and what will not. All teeth with their root apices, mandibular bone, maxillary alveolar bone, and condyles must fall fully in the focal trough for a diagnostic quality final image. Any discrepancies detected can be corrected by changing the path points and/or changing the size of the focal trough (Fig. 7.10d).

It should be kept in mind that CBCT-generated panoramic radiographs (Fig. 7.10e) cannot be trusted completely as they present distortions such as shortening, elongation, and magnification, especially if the curve is not identified properly. Since the reconstructed panoramic image is susceptible to operator error, it is important to confirm the panoramic radiographic findings with 2D MPR views of the volume [46, 47].

7.4.3 Acquisition of TMJ Tomograms

CBCT provides an extensive 3D view of the bony parts of the TMJ that offers valuable diagnostic information to the clinician. However, due to anatomic contiguities, it is only possible to evaluate the condyle-fossa relationship clearly from the lateral aspects in 3D view. Thus, the 2D tomographic sections come to aid, providing a better understanding of the condyle-fossa relationship and the anatomy [48].

After orienting the head, using the 3D sagittal view, the operator selects the upper and lower limits where the condyle-fossa complex is fully contained (Fig. 7.11a). Condyles rest in the fossa slightly rotated mediolaterally. When an imaginary line is drawn from the lateral pole to the medial pole, the lines generally intersect on the anterior border of foramen magnum close to the Basion point [49]. This is important, because while adjusting the slice orientation to obtain the best possible tomographic sections of the TMJ, these anatomic landmarks come in handy. Consequently, the line representing the axial slice is carried onto the sagittal view to the point where the condyle is the widest in the mediolateral direction on the axial view so that a perpendicular projection trajectory can be formed. Thus, condyles can be evaluated without getting affected due to the inclined placement in the articular fossa. The tomographic sections can be taken from the coronal (Fig. 7.11a), sagittal (Fig. 7.11b), and circular (Fig. 7.11c) directions, depending on the software in use. In each case, the lines representing the sections can be adjusted to pass from the mediolateral, anteroposterior long axes of the condyles or according to the needs of the operator. If a circular projection is selected, then 0° and 90° intervals will represent the condyle-fossa relationship from two different perpendicular views, either from sagittal and coronal, respectively, or from coronal and sagittal, respectively (Fig. 7.11c). If it is desired to evaluate the condyle-fossa relationship at different levels, then a linear projection (sagittal or coronal) along the condyles must be selected. Finally, the sections that are obtained can be viewed from different angles and different levels to check for symmetry, shape, size, and condylar position to gather information about the anatomy and condyle-fossa relationship.

As can be understood, we still need the 2D images during the transition to the third dimension, which is the most important advantage of the CBCT. With the development of 3D analyses and segmentation programs, the need for 2D images will also decrease. But until then, the orthodontic practitioner should be familiar with 2D image acquisition methods.

7.5 Visualization of Impacted Teeth

CBCT is not a routine orthodontics record yet due to its respectively higher radiation doses compared to cephalograms and panoramic and periapical films [50]. Impacted or ectopically erupting teeth, especially the maxillary canines, have been traditionally evaluated using the tube shift method (parallax technique) [51]. This method, however, gives the clinicians a relative idea about the position and difficulty of the impaction. Therefore, maybe one of the situations that require a 3D evaluation, without question, is ectopically erupting or impacted teeth [52]. It has been shown that CBCT gives the ability to think three dimensionally for complexly positioned canines and thereby render the treatment easier, reducing treatment time [53].

Following third molars, maxillary permanent canines are the second most frequently impacted teeth ranging from 0.8% to 2.8%, and they are more often palatally located [54]. In such scenarios, using a smaller field of view (FOV) and lower resolution settings are recommended [55, 56]. A field of view (FOV) of about 10 cm or less generally produces excellent results as compared

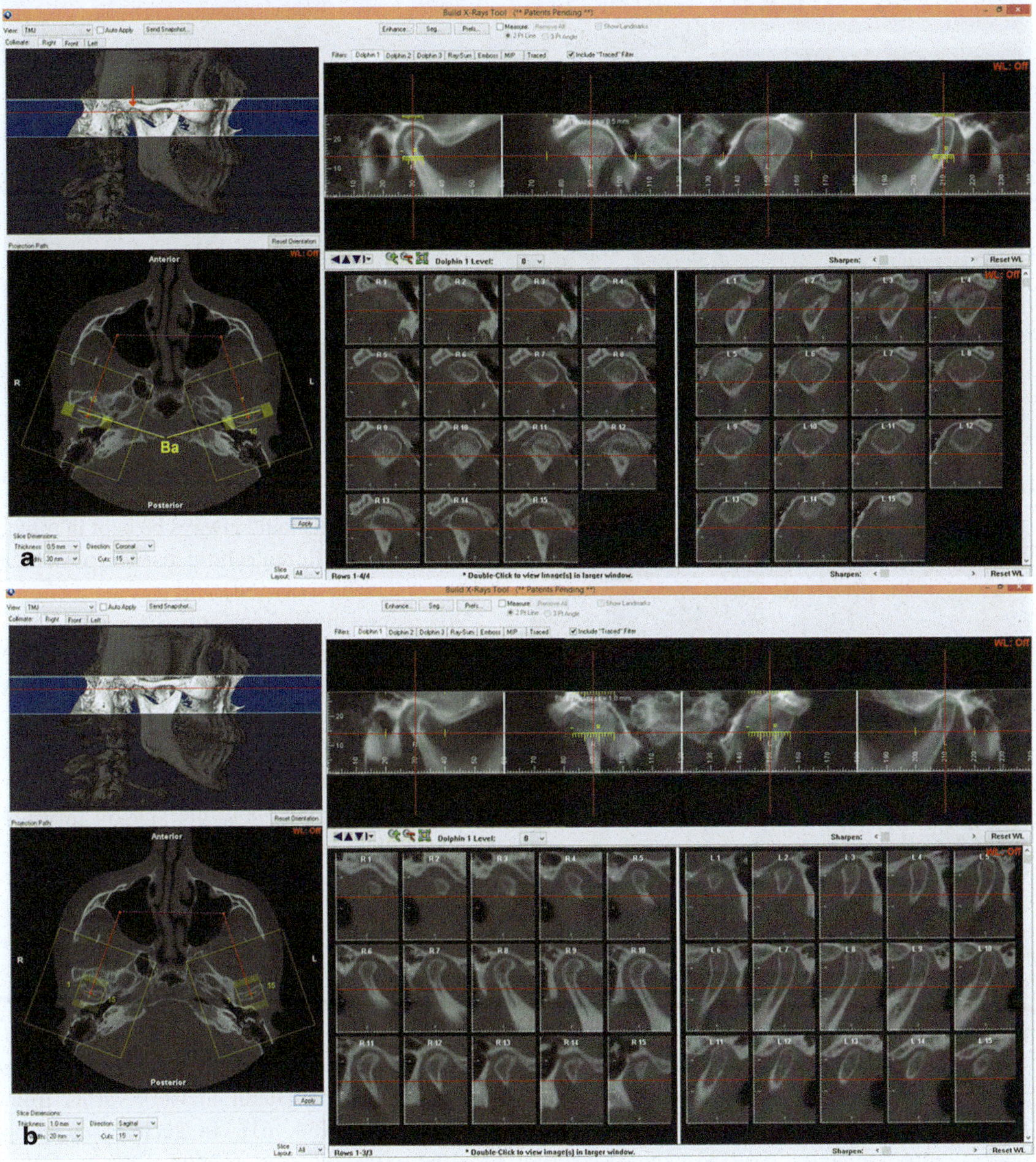

Fig. 7.11 (**a**) The highlighted region on top left image represents the superior and inferior borders of the region of interest (condyles). The axial slice line (red line marked with arrow) is on the condyles where they are widest mediolaterally by inspecting the axial cross section (bottom left image). The lines (yellow) passing from the lateral and medial poles of the condyles meet approximately on the anterior border of foramen magnum (Ba). Slices are formed according to desired direction as (**a**) coronal, (**b**) sagittal, or (**c**) circular. Also the thickness, width, and number of cross cuts can be adjusted by the operator

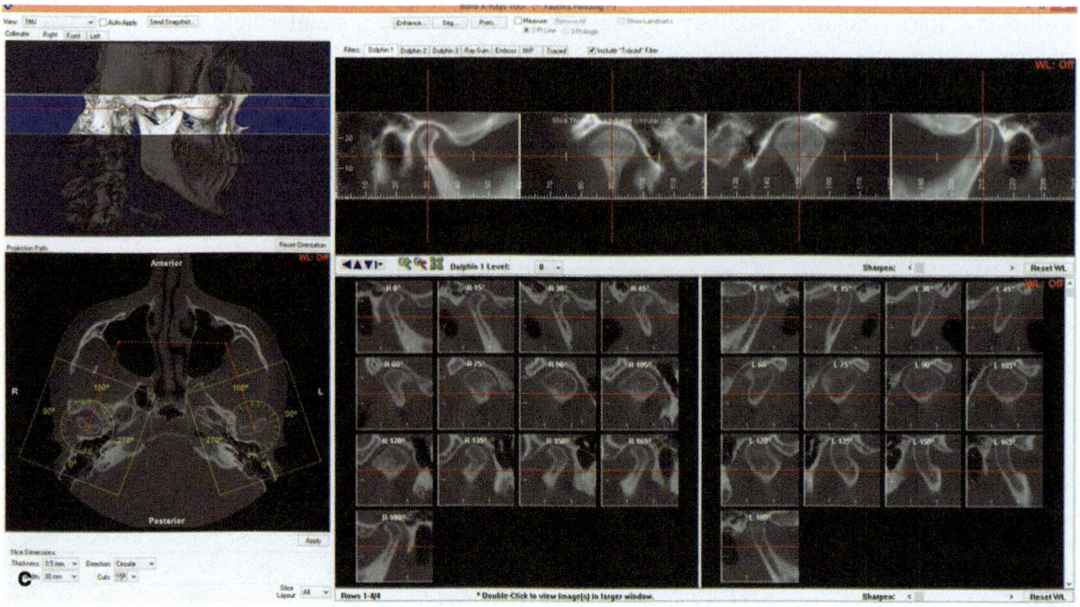

Fig. 7.11 (continued)

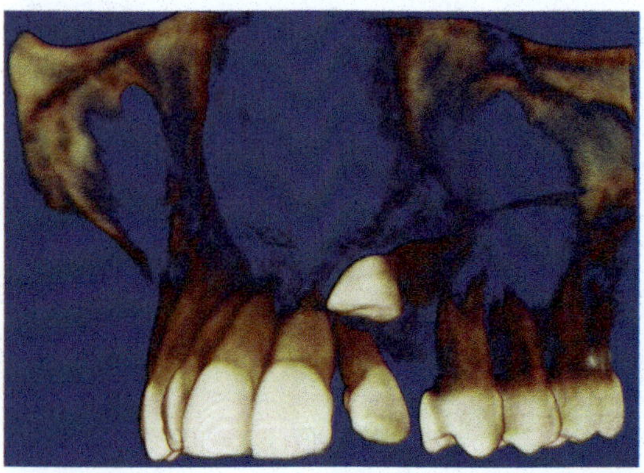

Fig. 7.12 Volume rendering view of an impacted canine

to a panoramic radiograph in evaluation of the position and the geometry of the impaction, relationships with adjacent teeth and to decide on the force directions to be applied to erupt these teeth into the dental arch. The mentioned FOV produces approximately 18–333 microsieverts (mSv) of effective radiation dose, depending on the selected resolution and the type of device in use which comes close to a combination of panoramic radiographs (6–50 mSv) and cephalograms (2–10 mSv) [57]. But still, the concern for ionizing radiation should always be kept in mind, especially for children whose tissues present a higher radiosensitivity [58].

When screening is performed for an impaction, the volume rendering view is the one to give a general idea about the situation (Fig. 7.12). It must be kept in mind that the 3D surface rendering mode is only for visualization purposes, not for diagnosis and analysis [59]. Using this view informs the clinician about the general position and location of the impacted teeth, possible damages to adjacent teeth, the path to move the teeth into the dental arch most efficiently, and how a minimally invasive surgical approach can be applied [60]. One of the least desirable situations is root resorption that can be seen on teeth adjacent to the impaction. According to several studies, the rate of maxillary canines to cause root resorption on lateral incisors varies from 48% to 66.7% [54, 61]. In such cases, it is required to use the cross-section views to evaluate the amount of root resorption. For this purpose, it is necessary to reorient the image considering the long axis of the tooth. First, the axial slice where the suspected crown of the tooth is the widest mesiodistally is determined

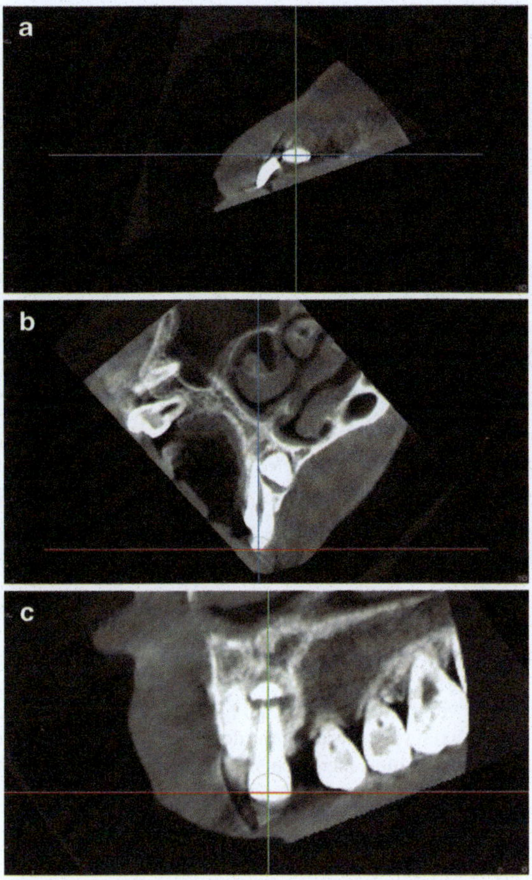

Fig. 7.13 (**a**) Orientation of the axial slice to coincide the widest mesiodistal width of the maxillary lateral incisor to the coronal slice (blue line). (**b**) Orientation of the sagittal slice to coincide the long axis of the maxillary lateral incisor to the coronal slice (blue line). (**c**) Orientation of the coronal slice to coincide the long axis of the maxillary lateral incisor to the sagittal slice (green line). It is apparent on the sagittal slice (**b**) that on the labial side of the root of the maxillary lateral incisor, there is a wedge-shaped severe resorption due to the impacted canine

(Fig. 7.13a). For this purpose, the axial slice can be rotated (yaw) to match the widest mesiodistal width of the corresponding crown with the line representing the coronal slice. Then using the sagittal view, the corresponding long axis of the tooth can be coincided (pitch) with the line representing the coronal slice (Fig. 7.13b). Finally, on the coronal view, the tooth's long axis is aligned (roll) with the line representing the sagittal slice (Fig. 7.13c). Now, the tooth under investigation can be evaluated in detail along its long axis from every aspect by moving the respective lines (axial, coronal, and sagittal) to visualize the amount of resorption. The resorption amount can be evaluated by using the grading system suggested by Ericson and Kurol as no resorption, slight resorption, moderate resorption, or severe resorption [62].

It is also necessary to evaluate the position of the canine to establish a proper treatment protocol. There are several classifications today, most of which have been adopted from previous 2D studies to judge the severity of the impaction. For this purpose, CBCT-generated 2D panoramic films and 3D volume rendering views can be used. The most used evaluations include the following: the distance of canine crown and root from the occlusal plane and/or the palatal plane [63, 64], the mesiodistal space available for canines [65], the canine overlap with adjacent teeth (sector) [66, 67], deviation from the occlusal arch [67], and the angulation of canines with respect to the midline, lateral incisor, and occlusal plane [53] (Fig. 7.14a–d). All of these tasks can be performed with ease using the linear and angular measurement tools of available segmentation programs.

It is apparent that 3D visualization of impacted teeth has improved the diagnostic capabilities of the orthodontist and thus led to faster treatment times by means of better treatment planning. However, although CBCT is considered a golden standard in terms of impacted teeth, it should be preferred primarily for complex impactions to lower the risks of ionizing radiation.

7.6 Assessment of Root Resorption

External root resorption is a relatively common unpredictable and idiopathic adverse effect of orthodontic treatment. Fortunately, in most cases, we see apical root shortening or surface resorption that does not decrease the functional capacity or longevity of the affected tooth [68]. In less than 5% of orthodontically treated anterior teeth, there is loss of more than one-third of the original root length [69].

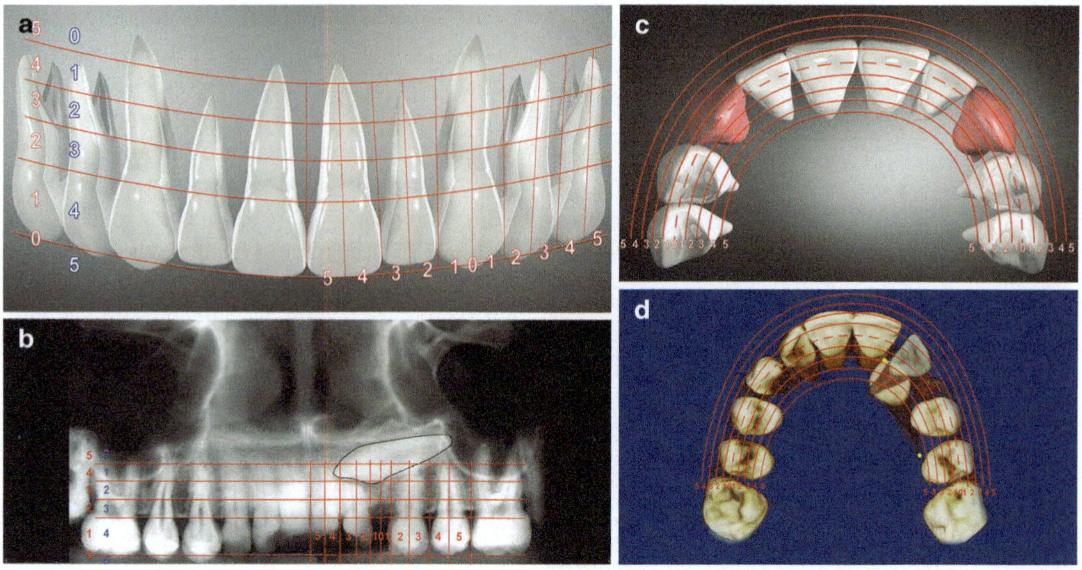

Fig. 7.14 (**a**) The grid-like scale used to grade the difficulty of canine impaction [67]. The horizontal and vertical lines are used to define the vertical and horizontal position of the impacted canine, respectively. The evaluation is done separately for both the cusp tip and the root tip. For the horizontal regions, red numbers and blue numbers are for the cusp tip and root tip, respectively. For the vertical regions, the numbering system is used for both the cusp tip and root tip. Depending on the anatomic location of the cusp and root tips, a number is given on a 0–5 scale. The sum of scores decides the anticipated difficulty of treatment. (**b**) Canine cusp tip is located in the fourth region both horizontally and vertically. The root tip is located on area 0 vertically and on area 5 horizontally. (**c**) The lines on the axial view represent the deviation from the occlusal arch with 2 mm. increments. (**d**) The canine tip for the same patient is located on area 2 and root tip on area 5 (yellow spots)

Panoramic and periapical radiographs, as well as lateral cephalograms, have been used for many decades to detect root resorption. However, the inability to see anything but root shortening or mesiodistal surface resorption diminishes their diagnostic value [70]. Moreover, the projection of 3D structures onto a 2D medium, superimposition of anatomical structures, improper patient positioning, geometric distortion, differential magnification, and lack of reproducibility further limit the information obtained from these records, especially in terms of the evaluation of quantitative root resorption [70–78]. Many anatomical and pathological details cannot be seen on traditional radiographs [73]. It has also been reported that conventional radiographs underestimate, or overestimate, the amount of resorption [69, 77, 79–83]. Finally, 2D intraoral radiography is not reliable for detecting early stages of external root resorption [81, 84–86].

Low radiation doses, multidirectional presentation, 1:1 ratio, high spatial resolution, and relative affordability are some of the reasons why cone beam computed tomography (CBCT) has become an essential diagnostic tool in dentistry [51, 87–89]. CBCT provides distortion-free slice images of roots and more accurate information about root resorption than 2D radiographs [90–92]. We can manipulate CBCT in the axial, sagittal, and coronal planes, reconstruct images in such a way that eliminates the overlying noise, zoom in and out of specific areas, and adjust brightness and contrast, which enables us to obtain more accurate and reliable measurements of anatomical structures [81, 93], in this case roots (Fig. 7.15). The difference between CBCT measurements and actual root lengths has been established to be around 0.1 mm, which is clinically insignificant [89, 94]. Furthermore, with the CBCT one can see highly detailed images and easily detect even the early stages of root resorption [73]. It has been reported that CBCT images give false-positive results in less than 10% of cases with no lesions, as opposed to 2D modali-

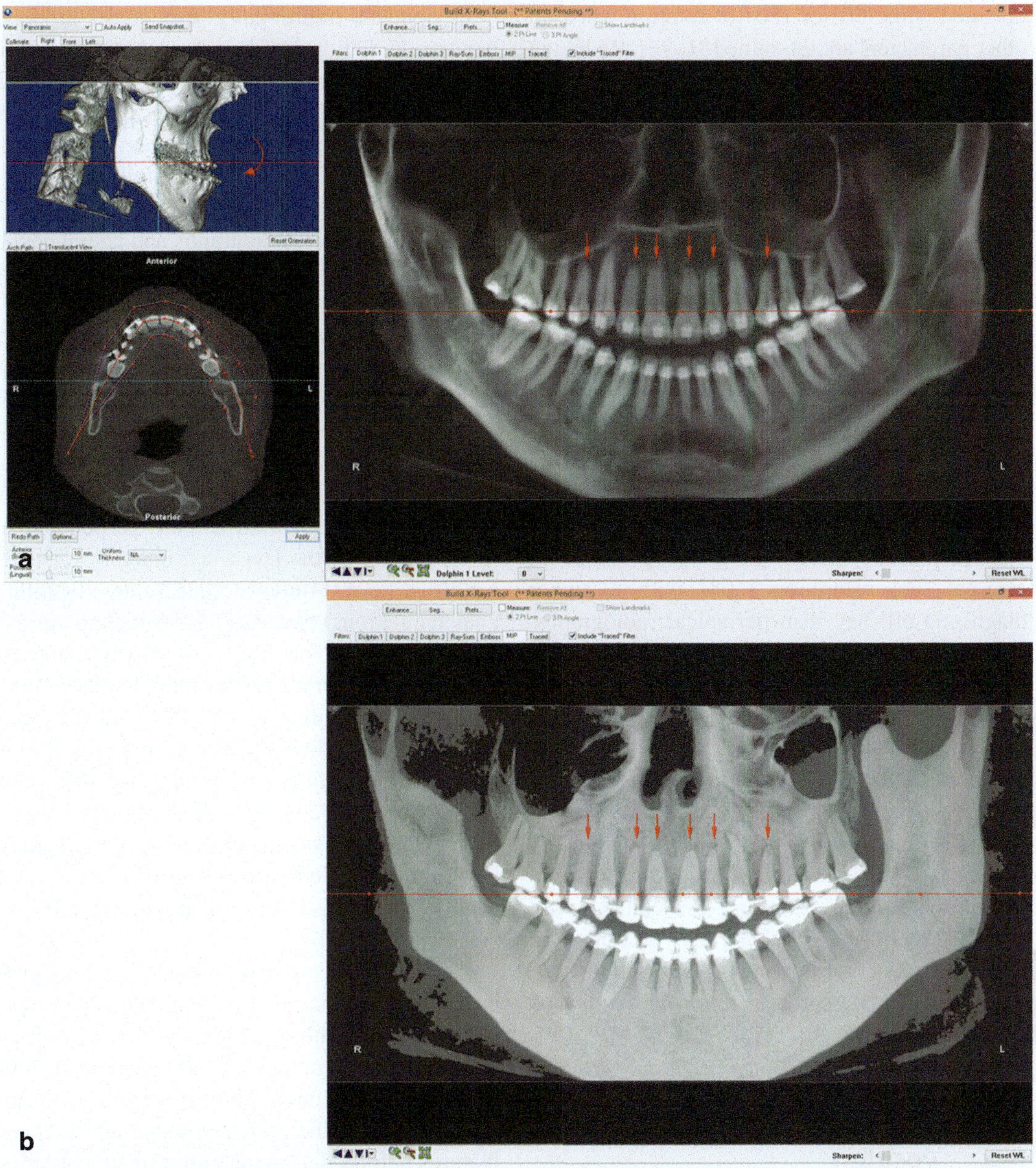

Fig. 7.15 Assessing root resorption. (**a**) In order to avoid misrepresentation due to the inclination, image orientation is done according to the long axis of teeth (top left). Root resorption visible on the Dolphin Imaging standard panoramic reconstruction from a DICOM dataset (top right). (**b**) More detailed information can also be obtained using MIP view

ties, where false positives are present in slightly over 20% of cases [73].

The quality of CBCT images is affected by the scanning unit, field of view (FOV), scanning time, tube amperage, tube voltage, voxel size, and spatial resolution [95]. Different voxel sizes have no effect on the accuracy and reliability of measurements of teeth and root lengths [89]. Different commercially available settings of a CBCT scanner can give similar accuracy and spatial resolution in the whole field of view [94]. There is no difference in the precision of volumes obtained with 360° or 180° of rotation when all other parameters are the same [81, 96]. Reducing the number of image projections does not result in reduced dimensional accuracy of 3D measurements [97]. No statistical differences in the diagnostic ability of the small and medium FOV CBCT machines have been found [98].

We can say that linear CBCT measurements are highly accurate [99–102]. CBCT images are considered exceptionally precise in identifying and qualifying root resorption [83, 103]. Their sensitivity and specificity are excellent, which makes them superior to digital intraoral and panoramic radiographs in detecting external root resorption [73, 81, 92, 104–107]. A systematic review and meta-analysis from 2017 concluded that CBCT could be a reliable diagnostic tool in clinical practice for detecting the presence of external root resorption [108] with a higher diagnostic efficacy than periapical radiographs. Furthermore, to put things in perspective, we need to mention the fact that orthodontic treatment planning probably would not vary much if external apical root resorption were diagnosed on 2D or CBCT images but would differ significantly if buccal or lingual root resorption were identified, which is only possible using CBCT [109]. Likewise, enhanced information regarding root resorption associated with impacted teeth [98, 103, 104] may be critical in changing extraction treatment plans and eventually extracting a resorbed lateral incisor instead of a healthy premolar [109].

7.7 Assessment of Tooth Position

Almost half a century ago, Andrews [110] defined the six keys of normal occlusion and explained the importance of proper mesiodistal and labiolingual inclination of the teeth. Moreover, proper axial inclination and root parallelism are important for good occlusion, incisal function, and treatment stability [111–115]. Therefore, assessment of tooth position is an essential part of orthodontic diagnosis, treatment planning, and treatment progress and outcome evaluation. Orthodontists traditionally use study models, panoramic radiographs, and cephalograms for this purpose. Root parallelism and mesiodistal tooth angulations are commonly assessed with panoramic radiographs, while lateral cephalograms are used for evaluating labiolingual angulation of anterior teeth [115–121].

Both panoramic radiographs and lateral cephalograms are very convenient but not precise and reliable enough for assessing tooth position [120, 122]. Panoramic radiography is technique and operator sensitive [123] and gives a distorted 2D representation of a 3D object [124]. Most errors occur in patient positioning [125]. Discrepancies between optimal and actual beam directions are especially present in the premolar area [126]. An x-ray beam that is not horizontally perpendicular to the surface of the jaw when imaging adjacent teeth that have different torque values (labiolingual angulations) creates a false perception of the root tip (mesiodistal angulation), especially in the canine and premolar regions. Increased lingual root torque usually appears as more mesial root tip and labial root torque as more distal root tip. This phenomenon is inconsistent and extremely variable [119, 127]. Limitations of lateral cephalograms come from relative inaccuracy in tracing cephalometric points on incisors and canines, as well as from radiographic superimpositions [117, 122].

Cone beam computed tomography scans give us the DICOM datasets from which we can single out axial, sagittal, and coronal slices, as well as reconstruct 3D images of individual teeth and craniofacial structures. This allows orthodontists to fully explore the patient's dental and skeletal features before, during, and after treatment [116, 120, 128]. When assessing tooth position using CBCT datasets, conventional axial, sagittal, and coronal slices, as well as custom sections, are used (Fig. 7.16) [116, 120, 128–131]. However, it has been suggested that 3D volume renderings might provide a more powerful and simplified tool for the visualization of root angulation and proximity but would not be an effective way for measuring mesiodistal tooth indications because of the difficulty to accurately select and localize measuring points in the same plane [129].

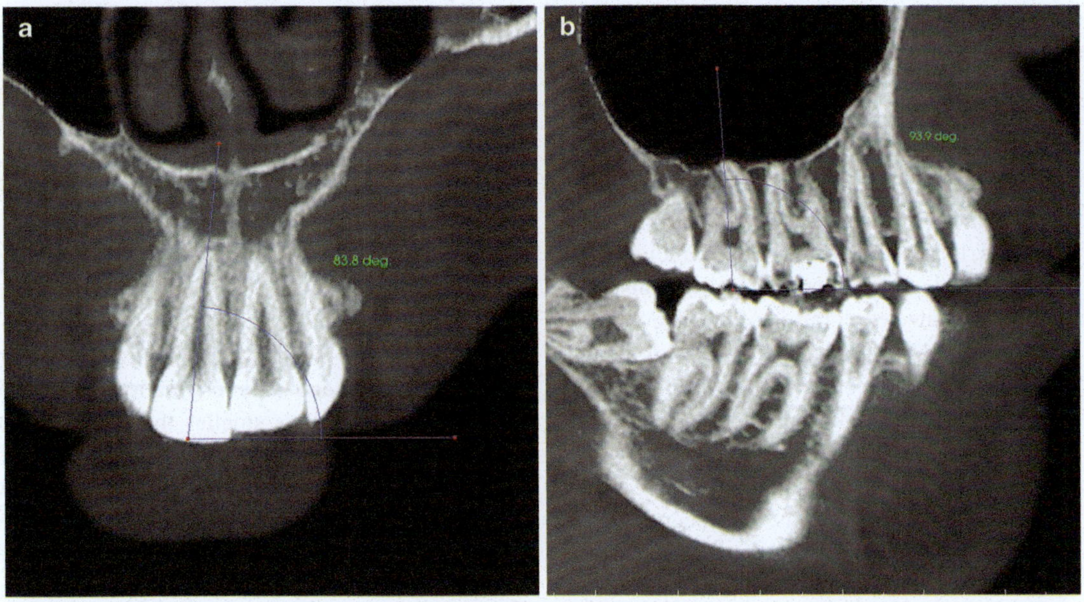

Fig. 7.16 Mesiodistal angulations are measured using the occlusal plane as a reference line. The occlusal plane is oriented parallel to the lower border of the display window for the sagittal and coronal views. The long axis of the single-rooted tooth is defined by the midpoint of the incisal edge or the facial cusp tip and the root apex (**a**). The long axis of the multi-rooted tooth is determined by the occlusal aspect of the buccal groove and the depth of the bi- or trifurcation (**b**)

7.8 Assessment of Alveolar Bone Heights and Volume

Orthodontic tooth movement is always related to the remodeling of the surrounding periodontal ligament (PDL) and alveolar bone. Bone remodeling is achieved through bone resorption and bone apposition formation. Optimal force systems initiate adequate biological responses of the PDL and the alveolar bone [132, 133]. However, heavy orthodontic forces in a limited area can cause contact between a tooth and the alveolar cortical plate, which may lead to the resorption of the cortical bone, subsequent alveolar bone defects (fenestrations and dehiscences), root exposure, and decreased bony support for the teeth [134–138]. Fenestrations are isolated areas where the root is denuded of the bone with only periosteum and gingiva over it, and dehiscences are bony defects in which denuded areas involve the alveolar bone margin [139]. Transverse movements can create dehiscences and fenestrations in the buccal and lingual cortical plates [135, 136, 140–142]. Excessive palatal movement of maxillary incisors brings their roots closer to the palatal cortex, which bends and remodels to some extent until contact is made but with further movement leads to penetration of the cortical plate accompanied by bone loss, root resorption, and finally relapse. Orthodontic proclination of maxillary incisors can also lead to dehiscences [143]. Too much lingual movement of mandibular incisors may cause irreversible distraction of the lingual cortex and decrease bone support [134–136, 140–142, 144–146]. There is no consensus on whether excessive proclination of mandibular incisors causes periodontal recession [147–154]. However, inflammation during orthodontic treatment and initial lack of bony support will possibly damage teeth and the periodontium [145, 148, 149, 155–160]. In summation, the amount and type of orthodontic tooth movement and patient features, such as boundary conditions before treatment and bone ability to remodel, affect bone quality and quantity after orthodontic treatment [109]. This is why analyzing alveolar bone characteristics, measuring the height, thickness, and volume of the alveolar bone, may be valuable in certain patients, like those presenting with thin alveolar bone phenotypes or pre-existing periodontal disease, as well as cases requiring tooth movement past another tooth or obstruction. Patients in need of extensive orthodontic movement close to or beyond alveolar limits, like the borderline non-extraction cases or orthodontic-surgical patients, would also benefit from detailed exploration of the alveolar bone [109, 116, 161]. Undiagnosed defects of the buccal alveolar bone before orthodontic treatment may increase the risk for dental relapse [162, 163] or gingival recession [156–158, 160].

Traditional 2D radiography, mostly periapical and panoramic radiographs, has been used for this purpose for quite some time, but superimposition of structures and lack of information from the third dimension make it very difficult to obtain precise values and visualize bony defects, such as dehiscences and fenestrations [146, 164, 165]. High spatial resolution, relative affordability, low radiation doses, and the fact that it provides distortion-free images that can be viewed in all three planes of space as well as in a 3D reconstruction make cone beam computed tomography the diagnostic tool of choice once more [51, 81, 87–89, 93]. The accuracy and reliability of alveolar bone measurements on CBCT images and its suitability for examining alveolar bone morphology have been confirmed by many authors [89, 94, 101, 103, 166–173]. Nevertheless, possible overestimations of up to 2 mm (bone thickness greater than voxel size) and underestimations of up to 1 mm (bone thickness smaller than voxel size) have also been reported, as well as limits of agreement in the mandibular incisor area of up to 2 mm [169, 171, 174]. These inconsistencies are probably a consequence of research design and sample variations, but still they suggest that the accuracy of alveolar bone CBCT measurements might depend on different factors. Comparing different numbers of projections (153, 306, and 612 projections) or different scanning times

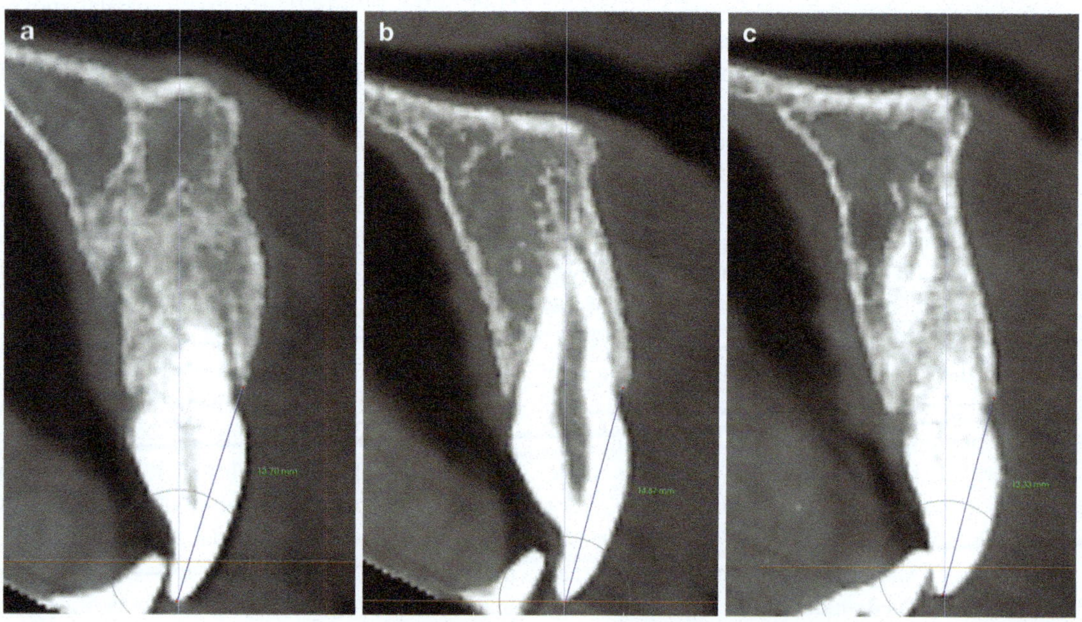

Fig. 7.17 Measuring facial alveolar bone height on CBCT sections from the center of the mesial (**a**), middle (**b**), and distal (**c**) third and facial bone thickness on CBCT sections from the center of the cervical (**d**), middle (**e**), and apical (**f**) third of the maxillary central incisor. These measurements can also be obtained on posterior teeth, as well as in the palatal/lingual region

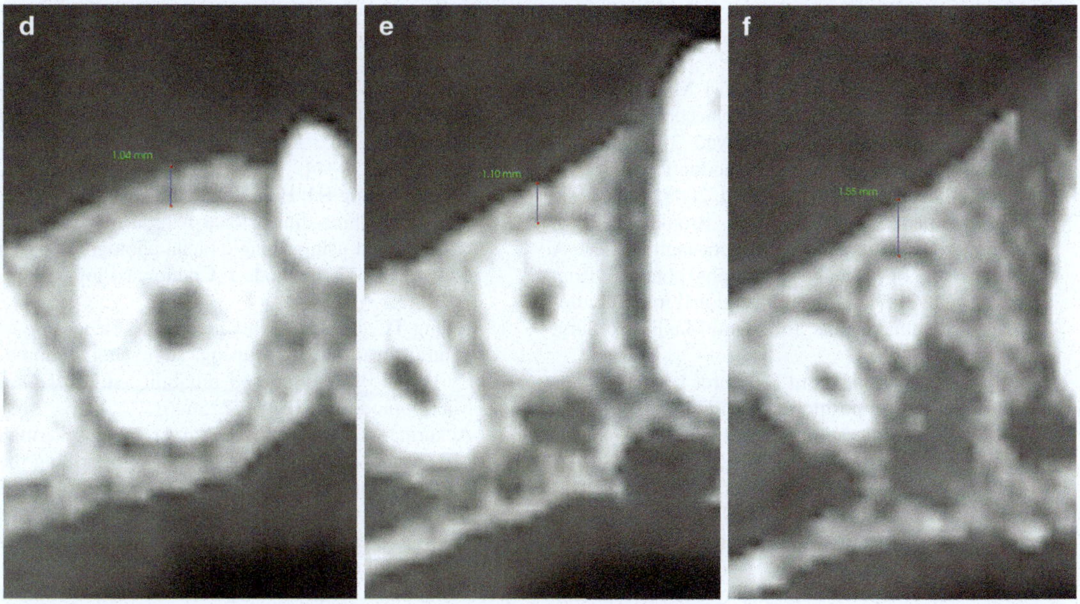

Fig. 7.17 (continued)

showed no statistical differences in measurement accuracy [97, 175]. Reducing the rotation arc of the CBCT scanner from 360° to 180° has resulted in comparable measurements of periapical lesions [81, 96]. Voxel size alone has almost no effect on linear measurement accuracy [89, 176–178], but some authors claim it might when soft tissues are present [171, 174].

When analyzing alveolar bone morphology, orthodontists and other dental specialists measure the alveolar bone heights and thicknesses in order to identify bone defects (Fig. 7.17). One can obtain accurate and reliable buccal alveolar bone dimensions from CBCT images; however, measuring bone height is more accurate then measuring bone thickness. This may be due to easier landmark identification. Landmarks for buccal bone height are placed on the incisal edge or cusp of a tooth and the border between the alveolar crest and gingival soft tissue, whereas for buccal bone thickness one needs to distinguish between the cementum and bone that have similar radiodensities [165, 167, 172, 175]. Looking into dehiscences and fenestration, when a defect is found on a CBCT image, it is a true dehiscence about 50% of the time and a true fenestration about 25% of the time. When no defect is found on a CBCT image, most likely there was no defect to be found in the first place. The low prevalence of dehiscences and fenestrations makes identification of true negatives more important than identification of actual defects. This is why the CBCT is still a good tool in diagnosing these buccal bony defects [165].

7.9 Chapter Conclusions

A relatively new technology, such as CBCT and 3D imaging, sees annual improvements of hardware and software. The clinical basics of what needs to be analyzed may remain the same, but the methodology above described may change.

It seems that 3D imaging is here to stay with its incorporation in clinical orthodontics not only important but perhaps essential for the currently practicing orthodontist.

References

1. Palomo JM, Hunt DW Jr, Hans MG, Broadbent BH Jr. A longitudinal 3-dimensional size and shape comparison of untreated class I and class II subjects. Am J Orthod Dentofac Orthop. 2005;127:584–91.

2. Vig PS, Hall DJ. The inadequacy of cephalometric radiographs for airway assessment. Am J Orthod. 1980;77:230–3.
3. Periago DR, Scarfe WC, Moshiri M, Scheetz JP, Silveira AM, Farman AG. Linear accuracy and reliability of cone beam CT derived 3-dimensional images constructed using an orthodontic volumetric rendering program. Angle Orthod. 2008;78:387–95.
4. van Vlijmen OJ, Berge SJ, Bronkhorst EM, Swennen GR, Katsaros C, Kuijpers-Jagtman AM. A comparison of frontal radiographs obtained from cone beam CT scans and conventional frontal radiographs of human skulls. Int J Oral Maxillofac Surg. 2009;38:773–8.
5. Lundstrom A, Lundstrom F, Lebret LM, Moorrees CF. Natural head position and natural head orientation: basic considerations in cephalometric analysis and research. Eur J Orthod. 1995;17:111–20.
6. Shokri A, Miresmaeili A, Farhadian N, Falah-Kooshki S, Amini P, Mollaie N. Effect of changing the head position on accuracy of transverse measurements of the maxillofacial region made on cone beam computed tomography and conventional posterior-anterior cephalograms. Dentomaxillofac Radiol. 2017;46:20160180.
7. Cevidanes L, Oliveira AE, Motta A, Phillips C, Burke B, Tyndall D. Head orientation in CBCT-generated cephalograms. Angle Orthod. 2009;79:971–7.
8. Cassi D, De Biase C, Tonni I, Gandolfini M, Di Blasio A, Piancino MG. Natural position of the head: review of two-dimensional and three-dimensional methods of recording. Br J Oral Maxillofac Surg. 2016;54:233–40.
9. Ferrario VF, Sforza C, Germano D, Dalloca LL, Miani A Jr. Head posture and cephalometric analyses: an integrated photographic/radiographic technique. Am J Orthod Dentofac Orthop. 1994;106:257–64.
10. Lundstrom F, Lundstrom A. Natural head position as a basis for cephalometric analysis. Am J Orthod Dentofac Orthop. 1992;101:244–7.
11. Dvortsin DP, Ye Q, Pruim GJ, Dijkstra PU, Ren Y. Reliability of the integrated radiograph-photograph method to obtain natural head position in cephalometric diagnosis. Angle Orthod. 2011;81:889–94.
12. Usumez S, Orhan M. Inclinometer method for recording and transferring natural head position in cephalometrics. Am J Orthod Dentofac Orthop. 2001;120:664–70.
13. Damstra J, Fourie Z, Ren Y. Simple technique to achieve a natural position of the head for cone beam computed tomography. Br J Oral Maxillofac Surg. 2010;48:236–8.
14. Xia JJ, McGrory JK, Gateno J, Teichgraeber JF, Dawson BC, Kennedy KA, et al. A new method to orient 3-dimensional computed tomography models to the natural head position: a clinical feasibility study. J Oral Maxillofac Surg. 2011;69:584–91.
15. de Paula LK, Ackerman JL, Carvalho Fde A, Eidson L, Cevidanes LH. Digital live-tracking 3-dimensional minisensors for recording head orientation during image acquisition. Am J Orthod Dentofac Orthop. 2012;141:116–23.
16. Weber DW, Fallis DW, Packer MD. Three-dimensional reproducibility of natural head position. Am J Orthod Dentofac Orthop. 2013;143:738–44.
17. Kim DS, Yang HJ, Huh KH, Lee SS, Heo MS, Choi SC, et al. Three-dimensional natural head position reproduction using a single facial photograph based on the POSIT method. J Craniomaxillofac Surg. 2014;42:1315–21.
18. Hsung TC, Lo J, Li TS, Cheung LK. Recording of natural head position using stereophotogrammetry: a new technique and reliability study. J Oral Maxillofac Surg. 2014;72:2256–61.
19. Ackerman JL, Proffit WR, Sarver DM, Ackerman MB, Kean MR. Pitch, roll, and yaw: describing the spatial orientation of dentofacial traits. Am J Orthod Dentofac Orthop. 2007;131:305–10.
20. Solow B, Sonnesen L. Head posture and malocclusions. Eur J Orthod. 1998;20:685–93.
21. Sonnesen L, Bakke M, Solow B. Temporomandibular disorders in relation to craniofacial dimensions, head posture and bite force in children selected for orthodontic treatment. Eur J Orthod. 2001;23:179–92.
22. Solow B, Siersbaek-Nielsen S. Cervical and craniocervical posture as predictors of craniofacial growth. Am J Orthod Dentofac Orthop. 1992;101:449–58.
23. El-Beialy AR, Fayed MS, El-Bialy AM, Mostafa YA. Accuracy and reliability of cone-beam computed tomography measurements: influence of head orientation. Am J Orthod Dentofac Orthop. 2011;140:157–65.
24. Lenza MG, Lenza MM, Dalstra M, Melsen B, Cattaneo PM. An analysis of different approaches to the assessment of upper airway morphology: a CBCT study. Orthod Craniofac Res. 2010;13:96–105.
25. Kawakami M, Yamamoto K, Fujimoto M, Ohgi K, Inoue M, Kirita T. Changes in tongue and hyoid positions, and posterior airway space following mandibular setback surgery. J Craniomaxillofac Surg. 2005;33:107–10.
26. Jiang C, Yi Y, Jiang C, Fang S, Wang J. Pharyngeal airway space and hyoid bone positioning after different orthognathic surgeries in skeletal class II patients. J Oral Maxillofac Surg. 2017;75:1482–90.
27. van Vlijmen OJ, Berge SJ, Swennen GR, Bronkhorst EM, Katsaros C, Kuijpers-Jagtman AM. Comparison of cephalometric radiographs obtained from cone-beam computed tomography scans and conventional radiographs. J Oral Maxillofac Surg. 2009;67:92–7.
28. Hassan B, van der Stelt P, Sanderink G. Accuracy of three-dimensional measurements obtained from cone beam computed tomography surface-rendered images for cephalometric analysis: influence of patient scanning position. Eur J Orthod. 2009;31:129–34.

29. Moorrees CAA, Kean MR. Natural head position, a basic consideration in the interpretation of cephalometric radiographs. Am J Orthod. 1958;45:785–91.
30. Foster TD, Howat AP, Naish PJ. Variation in cephalometric reference lines. Br J Orthod. 1981;8:183–7.
31. Moyers RE, Bookstein FL. The inappropriateness of conventional cephalometrics. Am J Orthod. 1979;75:599–617.
32. Solow B, Tallgren A. Natural head position in standing subjects. Acta Odontol Scand. 1971;29:591–607.
33. Kumar VLJ. Effect of cone beam CT study orientation on synthesized 2D radiographs from Dolphin 3D software. Kansas City, MO: American Association of Oral and Maxillofacial Radiology; 2006.
34. Wu RPJ, Landers M, et al. Anatomically based cranial landmarks for three-dimensional superimposition. Orthodontics. Cleveland, OH: Case Western Reserve University; 2012.
35. Ruellas AC, Tonello C, Gomes LR, Yatabe MS, Macron L, Lopinto J, et al. Common 3-dimensional coordinate system for assessment of directional changes. Am J Orthod Dentofac Orthop. 2016;149:645–56.
36. Hans MG, Palomo JM, Valiathan M. History of imaging in orthodontics from broadbent to cone-beam computed tomography. Am J Orthod Dentofac Orthop. 2015;148:914–21.
37. Kumar V, Ludlow J, Soares Cevidanes LH, Mol A. In vivo comparison of conventional and cone beam CT synthesized cephalograms. Angle Orthod. 2008;78:873–9.
38. Li N, Hu B, Mi F, Song J. Preliminary evaluation of cone beam computed tomography in three-dimensional cephalometry for clinical application. Exp Ther Med. 2017;13:2451–5.
39. Baumgaertel S, Palomo JM, Palomo L, Hans MG. Reliability and accuracy of cone-beam computed tomography dental measurements. Am J Orthod Dentofac Orthop. 2009;136:19–25. discussion 25-18
40. Ludlow JB, Gubler M, Cevidanes L, Mol A. Precision of cephalometric landmark identification: cone-beam computed tomography vs conventional cephalometric views. Am J Orthod Dentofacial Orthop. 2009;136:312.e311. 310; discussion 312–13
41. Kumar V, Ludlow JB, Mol A, Cevidanes L. Comparison of conventional and cone beam CT synthesized cephalograms. Dentomaxillofac Radiol. 2007;36:263–9.
42. Spolyar JL. Head positioning error in cephalometric radiography—an implant study. Angle Orthod. 1987;57:77–88.
43. Rino Neto J, de Paiva JB, Queiroz GV, Attizzani MF, Miasiro JH. Evaluation of radiographic magnification in lateral cephalograms obtained with different X-ray devices: experimental study in human dry skull. Dental Press J Orthod. 2013;18:17.e11-17.
44. Jacobson A, Caufield PW. Introduction to radiographic cephalometry. Philadelphia, PA: Lea & Febiger; 1985.
45. Dalrymple NC, Prasad SR, Freckleton MW, Chintapalli KN. Informatics in radiology (infoRAD): introduction to the language of three-dimensional imaging with multidetector CT. Radiographics. 2005;25:1409–28.
46. Palomo JM, Valiathan M, Hans MG. 3D orthodontic diagnosis and treatment planning. In: Kapila S, editor. Cone beam computed tomography in orthodontics: indications, insights, and innovations. Ames, IA: John Wiley & Sons Inc.; 2014. p. 221–46.
47. Valai-Kasim SA, Krishnaswamy NR, Tom B, Thavarajah R. Rotational panoramic radiographs-unusual triple images. J Clin Exp Dent. 2015;7:e183–6.
48. Okeson JP. Management of temporomandibular disorders and occlusion. Elsevier Mosby: St. Louis, MO; 2013.
49. Okeson JP. Functional anatomy and biomechanics of the masticatory system. In: Okeson JP, editor. Management of temporomandibular disorders and occlusion. St. Louis, MO: Elsevier Mosby; 2013. p. 2–20.
50. Comission E. Cone beam CT for dental and maxillofacial radiology: evidence-based guidelines. Luxembourg: SEDENTEXCT; 2012. p. 172.
51. Kau CH, Richmond S, Palomo JM, Hans MG. Three-dimensional cone beam computerized tomography in orthodontics. J Orthod. 2005;32:282–93.
52. Smith BR, Park JH, Cederberg RA. An evaluation of cone-beam computed tomography use in postgraduate orthodontic programs in the United States and Canada. J Dent Educ. 2011;75:98–106.
53. Alqerban A, Jacobs R, van Keirsbilck PJ, Aly M, Swinnen S, Fieuws S, et al. The effect of using CBCT in the diagnosis of canine impaction and its impact on the orthodontic treatment outcome. J Orthod Sci. 2014;3:34–40.
54. Walker L, Enciso R, Mah J. Three-dimensional localization of maxillary canines with cone-beam computed tomography. Am J Orthod Dentofac Orthop. 2005;128:418–23.
55. Garib DG, Calil LR, Leal CR, Janson G. Is there a consensus for CBCT use in orthodontics? Dental Press J Orthod. 2014;19:136–49.
56. Proffit WR, Sarver DM, Ackerman JL. Orthodontics diagnosis: the problem-oriented approach. In: Proffit WR, Fields HW, Sarver DM, editors. Contemporary orthodontics. St. Louis, MO: Mosby Elsevier; 2013. p. xii, 754.
57. Radiology AAoOaM. Clinical recommendations regarding use of cone beam computed tomography in Orthodontics. Position statement by the American Academy of Oral and Maxillofacial Radiology. Oral Surg Oral Med Oral Pathol Oral Radiol. 2013;116:238–57.
58. Farman AG. Image gently: enhancing radiation protection during pediatric imaging. Oral Surg Oral Med Oral Pathol Oral Radiol. 2014;117:657–8.
59. Pauwels R, Araki K, Siewerdsen JH, Thongvigitmanee SS. Technical aspects of dental CBCT: state of the art. Dentomaxillofac Radiol. 2015;44:20140224.

60. Machado GL. CBCT imaging – a boon to orthodontics. Saudi Dent J. 2015;27:12–21.
61. Ericson S, Kurol PJ. Resorption of incisors after ectopic eruption of maxillary canines: a CT study. Angle Orthod. 2000;70:415–23.
62. Ericson S, Kurol J. Incisor root resorptions due to ectopic maxillary canines imaged by computerized tomography: a comparative study in extracted teeth. Angle Orthod. 2000;70:276–83.
63. Oz AZ, Oz AA, El H, Palomo JM. Maxillary sinus volume in patients with impacted canines. Angle Orthod. 2017;87:25–32.
64. Power SM, Short MB. An investigation into the response of palatally displaced canines to the removal of deciduous canines and an assessment of factors contributing to favourable eruption. Br J Orthod. 1993;20:215–23.
65. Cernochova P, Krupa P, Izakovicova-Holla L. Root resorption associated with ectopically erupting maxillary permanent canines: a computed tomography study. Eur J Orthod. 2011;33:483–91.
66. Ericson S, Kurol J. Early treatment of palatally erupting maxillary canines by extraction of the primary canines. Eur J Orthod. 1988;10:283–95.
67. Kau CH, Pan P, Gallerano RL, English JD. A novel 3D classification system for canine impactions—the KPG index. Int J Med Robot. 2009;5:291–6.
68. Brezniak N, Wasserstein A. Root resorption after orthodontic treatment: part 1. Literature review. Am J Orthod Dentofac Orthop. 1993;103:62–6.
69. Levander E, Malmgren O. Evaluation of the risk of root resorption during orthodontic treatment: a study of upper incisors. Eur J Orthod. 1988;10:30–8.
70. Chan EK, Darendeliler MA. Exploring the third dimension in root resorption. Orthod Craniofac Res. 2004;7:64–70.
71. Brezniak N, Goren S, Zoizner R, Dinbar A, Arad A, Wasserstein A, et al. A comparison of three methods to accurately measure root length. Angle Orthod. 2004;74:786–91.
72. Cohenca N, Simon JH, Mathur A, Malfaz JM. Clinical indications for digital imaging in dento-alveolar trauma. Part 2: root resorption. Dent Traumatol. 2007;23:105–13.
73. Creanga AG, Geha H, Sankar V, Teixeira FB, McMahan CA, Noujeim M. Accuracy of digital periapical radiography and cone-beam computed tomography in detecting external root resorption. Imaging Sci Dent. 2015;45:153–8.
74. Katona TR. Flaws in root resorption assessment algorithms: role of tooth shape. Am J Orthod Dentofacial Orthop. 2006;130:698.e619–27.
75. Leach HA, Ireland AJ, Whaites EJ. Radiographic diagnosis of root resorption in relation to orthodontics. Br Dent J. 2001;190:16–22.
76. Patel S, Dawood A, Mannocci F, Wilson R, Pitt FT. Detection of periapical bone defects in human jaws using cone beam computed tomography and intraoral radiography. Int Endod J. 2009;42:507–15.
77. Sameshima GT, Asgarifar KO. Assessment of root resorption and root shape: periapical vs panoramic films. Angle Orthod. 2001;71:185–9.
78. Webber RL, Messura JK. An in vivo comparison of diagnostic information obtained from tuned-aperture computed tomography and conventional dental radiographic imaging modalities. Oral Surg Oral Med Oral Pathol Oral Radiol Endod. 1999;88:239–47.
79. Apajalahti S, Peltola JS. Apical root resorption after orthodontic treatment – a retrospective study. Eur J Orthod. 2007;29:408–12.
80. Dudic A, Giannopoulou C, Leuzinger M, Kiliaridis S. Detection of apical root resorption after orthodontic treatment by using panoramic radiography and cone-beam computed tomography of super-high resolution. Am J Orthod Dentofac Orthop. 2009;135:434–7.
81. Durack C, Patel S, Davies J, Wilson R, Mannocci F. Diagnostic accuracy of small volume cone beam computed tomography and intraoral periapical radiography for the detection of simulated external inflammatory root resorption. Int Endod J. 2011;44:136–47.
82. Janson GR, De Luca Canto G, Martins DR, Henriques JF, De Freitas MR. A radiographic comparison of apical root resorption after orthodontic treatment with 3 different fixed appliance techniques. Am J Orthod Dentofac Orthop. 2000;118:262–73.
83. Patel S, Dawood A, Wilson R, Horner K, Mannocci F. The detection and management of root resorption lesions using intraoral radiography and cone beam computed tomography – an in vivo investigation. Int Endod J. 2009;42:831–8.
84. Andreasen FM, Sewerin I, Mandel U, Andreasen JO. Radiographic assessment of simulated root resorption cavities. Endod Dent Traumatol. 1987;3:21–7.
85. Chapnick L. External root resorption: an experimental radiographic evaluation. Oral Surg Oral Med Oral Pathol. 1989;67:578–82.
86. Goldberg F, De Silvio A, Dreyer C. Radiographic assessment of simulated external root resorption cavities in maxillary incisors. Endod Dent Traumatol. 1998;14:133–6.
87. Mah J, Hatcher D. Three-dimensional craniofacial imaging. Am J Orthod Dentofac Orthop. 2004;126:308–9.
88. Noujeim M, Prihoda T, Langlais R, Nummikoski P. Evaluation of high-resolution cone beam computed tomography in the detection of simulated interradicular bone lesions. Dentomaxillofac Radiol. 2009;38:156–62.
89. Sherrard JF, Rossouw PE, Benson BW, Carrillo R, Buschang PH. Accuracy and reliability of tooth and root lengths measured on cone-beam computed tomographs. Am J Orthod Dentofac Orthop. 2010;137:S100–8.
90. Castro IO, Alencar AH, Valladares-Neto J, Estrela C. Apical root resorption due to orthodontic treat-

ment detected by cone beam computed tomography. Angle Orthod. 2013;83:196–203.
91. Lund H, Grondahl K, Grondahl HG. Cone beam computed tomography for assessment of root length and marginal bone level during orthodontic treatment. Angle Orthod. 2010;80:466–73.
92. Ren H, Chen J, Deng F, Zheng L, Liu X, Dong Y. Comparison of cone-beam computed tomography and periapical radiography for detecting simulated apical root resorption. Angle Orthod. 2013;83:189–95.
93. Dalmau E, Zamora N, Tarazona B, Gandia JL, Paredes V. A comparative study of the pharyngeal airway space, measured with cone beam computed tomography, between patients with different craniofacial morphologies. J Craniomaxillofac Surg. 2015;43:1438–46.
94. Ballrick JW, Palomo JM, Ruch E, Amberman BD, Hans MG. Image distortion and spatial resolution of a commercially available cone-beam computed tomography machine. Am J Orthod Dentofac Orthop. 2008;134:573–82.
95. Kamburoglu K, Murat S, Kolsuz E, Kurt H, Yuksel S, Paksoy C. Comparative assessment of subjective image quality of cross-sectional cone-beam computed tomography scans. J Oral Sci. 2011;53:501–8.
96. Lennon S, Patel S, Foschi F, Wilson R, Davies J, Mannocci F. Diagnostic accuracy of limited-volume cone-beam computed tomography in the detection of periapical bone loss: 360 degrees scans versus 180 degrees scans. Int Endod J. 2011;44:1118–27.
97. Brown AA, Scarfe WC, Scheetz JP, Silveira AM, Farman AG. Linear accuracy of cone beam CT derived 3D images. Angle Orthod. 2009;79:150–7.
98. Alqerban A, Jacobs R, Souza PC, Willems G. In-vitro comparison of 2 cone-beam computed tomography systems and panoramic imaging for detecting simulated canine impaction-induced external root resorption in maxillary lateral incisors. Am J Orthod Dentofac Orthop. 2009;136:764.e761–11. discussion 764–5
99. Lascala CA, Panella J, Marques MM. Analysis of the accuracy of linear measurements obtained by cone beam computed tomography (CBCT-NewTom). Dentomaxillofac Radiol. 2004;33:291–4.
100. Marmulla R, Wortche R, Muhling J, Hassfeld S. Geometric accuracy of the NewTom 9000 Cone Beam CT. Dentomaxillofac Radiol. 2005;34:28–31.
101. Misch KA, Yi ES, Sarment DP. Accuracy of cone beam computed tomography for periodontal defect measurements. J Periodontol. 2006;77:1261–6.
102. Mischkowski RA, Pulsfort R, Ritter L, Neugebauer J, Brochhagen HG, Keeve E, et al. Geometric accuracy of a newly developed cone-beam device for maxillofacial imaging. Oral Surg Oral Med Oral Pathol Oral Radiol Endod. 2007;104:551–9.
103. Alqerban A, Jacobs R, Fieuws S, Nackaerts O, Willems G. Comparison of 6 cone-beam computed tomography systems for image quality and detection of simulated canine impaction-induced external root resorption in maxillary lateral incisors. Am J Orthod Dentofac Orthop. 2011;140:e129–39.
104. Alqerban A, Jacobs R, Fieuws S, Willems G. Comparison of two cone beam computed tomographic systems versus panoramic imaging for localization of impacted maxillary canines and detection of root resorption. Eur J Orthod. 2011;33:93–102.
105. Bernardes RA, de Paulo RS, Pereira LO, Duarte MA, Ordinola-Zapata R, de Azevedo JR. Comparative study of cone beam computed tomography and intraoral periapical radiographs in diagnosis of lingual-simulated external root resorptions. Dent Traumatol. 2012;28:268–72.
106. D'Addazio PS, Campos CN, Ozcan M, Teixeira HG, Passoni RM, Carvalho AC. A comparative study between cone-beam computed tomography and periapical radiographs in the diagnosis of simulated endodontic complications. Int Endod J. 2011;44:218–24.
107. Kamburoglu K, Kursun S, Yuksel S, Oztas B. Observer ability to detect ex vivo simulated internal or external cervical root resorption. J Endod. 2011;37:168–75.
108. Yi J, Sun Y, Li Y, Li C, Li X, Zhao Z. Cone-beam computed tomography versus periapical radiograph for diagnosing external root resorption: a systematic review and meta-analysis. Angle Orthod. 2017;87:328–37.
109. Kapila SD, Nervina JM. CBCT in orthodontics: assessment of treatment outcomes and indications for its use. Dentomaxillofac Radiol. 2015;44:20140282.
110. Andrews LF. The six keys to normal occlusion. Am J Orthod. 1972;62:296–309.
111. Edwards JG. The prevention of relapse in extraction cases. Am J Orthod. 1971;60:128–44.
112. Graber TM. Postmortems in posttreatment adjustment. Am J Orthod. 1966;52:331–52.
113. Hatasaka HH. A radiographic study of roots in extraction sites. Angle Orthod. 1976;46:64–8.
114. Holdaway RA. Bracket angulation as applied to the edgewise appliance. Angle Orthod. 1952;22:227–36.
115. Mayoral G. Treatment results with light wires studied by panoramic radiography. Am J Orthod. 1982;81:489–97.
116. Cattaneo PM, Salih RA, Melsen B. Labio-lingual root control of lower anterior teeth and canines obtained by active and passive self-ligating brackets. Angle Orthod. 2013;83:691–7.
117. Gracco A, Luca L, Bongiorno MC, Siciliani G. Computed tomography evaluation of mandibular incisor bony support in untreated patients. Am J Orthod Dentofac Orthop. 2010;138:179–87.
118. Hasund A, Ulstein G. The position of the incisors in relation to the lines NA and NB in different facial types. Am J Orthod. 1970;57:1–14.
119. Lucchesi MV, Wood RE, Nortje CJ. Suitability of the panoramic radiograph for assessment of mesiodistal angulation of teeth in the buccal segments of the mandible. Am J Orthod Dentofac Orthop. 1988;94:303–10.

120. Peck JL, Sameshima GT, Miller A, Worth P, Hatcher DC. Mesiodistal root angulation using panoramic and cone beam CT. Angle Orthod. 2007;77:206–13.
121. Ursi WJ, Almeida RR, Tavano O, Henriques JF. Assessment of mesiodistal axial inclination through panoramic radiography. J Clin Orthod. 1990;24:166–73.
122. Baumrind S, Frantz RC. The reliability of head film measurements. 2. Conventional angular and linear measures. Am J Orthod. 1971;60:505–17.
123. McKee IW, Glover KE, Williamson PC, Lam EW, Heo G, Major PW. The effect of vertical and horizontal head positioning in panoramic radiography on mesiodistal tooth angulations. Angle Orthod. 2001;71:442–51.
124. Xie Q, Soikkonen K, Wolf J, Mattila K, Gong M, Ainamo A. Effect of head positioning in panoramic radiography on vertical measurements: an in vitro study. Dentomaxillofac Radiol. 1996;25:61–6.
125. Schiff T, D'Ambrosio J, Glass BJ, Langlais RP, McDavid WD. Common positioning and technical errors in panoramic radiography. J Am Dent Assoc. 1986;113:422–6.
126. Scarfe WC, Nummikoski P, McDavid WD, Welander U, Tronje G. Radiographic interproximal angulations: implications for rotational panoramic radiography. Oral Surg Oral Med Oral Pathol. 1993;76:664–72.
127. Samawi SS, Burke PH. Angular distortion in the orthopantomogram. Br J Orthod. 1984;11:100–7.
128. Cattaneo PM, Treccani M, Carlsson K, Thorgeirsson T, Myrda A, Cevidanes LH, et al. Transversal maxillary dento-alveolar changes in patients treated with active and passive self-ligating brackets: a randomized clinical trial using CBCT-scans and digital models. Orthod Craniofac Res. 2011;14:222–33.
129. Bouwens DG, Cevidanes L, Ludlow JB, Phillips C. Comparison of mesiodistal root angulation with posttreatment panoramic radiographs and cone-beam computed tomography. Am J Orthod Dentofac Orthop. 2011;139:126–32.
130. Pontes LF, Cecim RL, Machado SM, Normando D. Tooth angulation and dental arch perimeter-the effect of orthodontic bracket prescription. Eur J Orthod. 2015;37:435–9.
131. Rhoden FK, Maltagliati LA, de Castro Ferreira Conti AC, Almeida-Pedrin RR, Filho LC, de Almeida Cardoso M. Cone beam computed tomography-based evaluation of the anterior teeth position changes obtained by passive self-ligating brackets. J Contemp Dent Pract. 2016;17:623–9.
132. Henneman S, Von den Hoff JW, Maltha JC. Mechanobiology of tooth movement. Eur J Orthod. 2008;30:299–306.
133. Wise GE, King GJ. Mechanisms of tooth eruption and orthodontic tooth movement. J Dent Res. 2008;87:414–34.
134. Sarikaya S, Haydar B, Ciger S, Ariyurek M. Changes in alveolar bone thickness due to retraction of anterior teeth. Am J Orthod Dentofac Orthop. 2002;122:15–26.
135. Wainwright WM. Faciolingual tooth movement: its influence on the root and cortical plate. Am J Orthod. 1973;64:278–302.
136. Wehrbein H, Bauer W, Diedrich P. Mandibular incisors, alveolar bone, and symphysis after orthodontic treatment. A retrospective study. Am J Orthod Dentofac Orthop. 1996;110:239–46.
137. Zachrisson BU, Alnaes L. Periodontal condition in orthodontically treated and untreated individuals. I. Loss of attachment, gingival pocket depth and clinical crown height. Angle Orthod. 1973;43:402–11.
138. Zachrisson BU, Alnaes L. Periodontal condition in orthodontically treated and untreated individuals. II Alveolar bone loss: radiographic findings. Angle Orthod. 1974;44:48–55.
139. Newman MG, Takei HH, Carranza FA. Carranza's clinical periodontology. Philadelphia, PA: W.B. Saunders Company; 2002.
140. Handelman CS. The anterior alveolus: its importance in limiting orthodontic treatment and its influence on the occurrence of iatrogenic sequelae. Angle Orthod. 1996;66:95–109. discussion 109–110
141. Wehrbein H, Fuhrmann RA, Diedrich PR. Periodontal conditions after facial root tipping and palatal root torque of incisors. Am J Orthod Dentofac Orthop. 1994;106:455–62.
142. Wehrbein H, Fuhrmann RA, Diedrich PR. Human histologic tissue response after long-term orthodontic tooth movement. Am J Orthod Dentofac Orthop. 1995;107:360–71.
143. Fuhrmann R. Three-dimensional interpretation of periodontal lesions and remodeling during orthodontic treatment. Part III. J Orofac Orthop. 1996;57:224–37.
144. DeAngelis V. Observations on the response of alveolar bone to orthodontic force. Am J Orthod. 1970;58:284–94.
145. Ericsson I, Thilander B, Lindhe J, Okamoto H. The effect of orthodontic tilting movements on the periodontal tissues of infected and non-infected dentitions in dogs. J Clin Periodontol. 1977;4:278–93.
146. Meikle MC. The dentomaxillary complex and overjet correction in class II, division 1 malocclusion: objectives of skeletal and alveolar remodeling. Am J Orthod. 1980;77:184–97.
147. Allais D, Melsen B. Does labial movement of lower incisors influence the level of the gingival margin? A case-control study of adult orthodontic patients. Eur J Orthod. 2003;25:343–52.
148. Artun J, Grobety D. Periodontal status of mandibular incisors after pronounced orthodontic advancement during adolescence: a follow-up evaluation. Am J Orthod Dentofac Orthop. 2001;119:2–10.
149. Artun J, Urbye KS. The effect of orthodontic treatment on periodontal bone support in patients with advanced loss of marginal periodontium. Am J Orthod Dentofac Orthop. 1988;93:143–8.
150. Djeu G, Hayes C, Zawaideh S. Correlation between mandibular central incisor proclination and gingival recession during fixed appliance therapy. Angle Orthod. 2002;72:238–45.

151. Dorfman HS. Mucogingival changes resulting from mandibular incisor tooth movement. Am J Orthod. 1978;74:286–97.
152. Pearson LE. Gingival height of lower central incisors, orthodontically treated and untreated. Angle Orthod. 1968;38:337–9.
153. Ruf S, Hansen K, Pancherz H. Does orthodontic proclination of lower incisors in children and adolescents cause gingival recession? Am J Orthod Dentofac Orthop. 1998;114:100–6.
154. Wingard CE, Bowers GM. The effects of facial bone from facial tipping of incisors in monkeys. J Periodontol. 1976;47:450–4.
155. Grant DA, Stern IB, Listgarten MA, Orban BJ, Gottlieb B. Periodontics in the tradition of Gottlieb and Orban. St. Louis, MO: Elsevier Mosby; 1988.
156. Melsen B, Allais D. Factors of importance for the development of dehiscences during labial movement of mandibular incisors: a retrospective study of adult orthodontic patients. Am J Orthod Dentofac Orthop. 2005;127:552–61. quiz 625
157. Wennstrom JL. Mucogingival considerations in orthodontic treatment. Semin Orthod. 1996;2:46–54.
158. Wennstrom JL, Lindhe J, Sinclair F, Thilander B. Some periodontal tissue reactions to orthodontic tooth movement in monkeys. J Clin Periodontol. 1987;14:121–9.
159. Wennstrom JL, Stokland BL, Nyman S, Thilander B. Periodontal tissue response to orthodontic movement of teeth with infrabony pockets. Am J Orthod Dentofac Orthop. 1993;103:313–9.
160. Yared KF, Zenobio EG, Pacheco W. Periodontal status of mandibular central incisors after orthodontic proclination in adults. Am J Orthod Dentofac Orthop. 2006;130:6.e1–8.
161. Baysal A, Ucar FI, Buyuk SK, Ozer T, Uysal T. Alveolar bone thickness and lower incisor position in skeletal class I and class II malocclusions assessed with cone-beam computed tomography. Korean J Orthod. 2013;43:134–40.
162. Ericsson I, Thilander B. Orthodontic relapse in dentitions with reduced periodontal support: an experimental study in dogs. Eur J Orthod. 1980;2:51–7.
163. Rothe LE, Bollen AM, Little RM, Herring SW, Chaison JB, Chen CS, et al. Trabecular and cortical bone as risk factors for orthodontic relapse. Am J Orthod Dentofac Orthop. 2006;130:476–84.
164. Lang NP, Hill RW. Radiographs in periodonties. J Clin Periodontol. 1977;4:16–28.
165. Leung CC, Palomo L, Griffith R, Hans MG. Accuracy and reliability of cone-beam computed tomography for measuring alveolar bone height and detecting bony dehiscences and fenestrations. Am J Orthod Dentofac Orthop. 2010;137:S109–19.
166. Fu JH, Yeh CY, Chan HL, Tatarakis N, Leong DJ, Wang HL. Tissue biotype and its relation to the underlying bone morphology. J Periodontol. 2010;81:569–74.
167. Ganguly R, Ruprecht A, Vincent S, Hellstein J, Timmons S, Qian F. Accuracy of linear measurement in the Galileos cone beam computed tomography under simulated clinical conditions. Dentomaxillofac Radiol. 2011;40:299–305.
168. Ludlow JB, Laster WS, See M, Bailey LJ, Hershey HG. Accuracy of measurements of mandibular anatomy in cone beam computed tomography images. Oral Surg Oral Med Oral Pathol Oral Radiol Endod. 2007;103:534–42.
169. Patcas R, Müller L, Ullrich O, Peltomäki T. Accuracy of cone-beam computed tomography at different resolutions assessed on the bony covering of the mandibular anterior teeth. Am J Orthod Dentofac Orthop. 2012;141:41–50.
170. Shiratori LN, Marotti J, Yamanouchi J, Chilvarquer I, Contin I, Tortamano-Neto P. Measurement of buccal bone volume of dental implants by means of cone-beam computed tomography. Clin Oral Implants Res. 2012;23:797–804.
171. Sun Z, Smith T, Kortam S, Kim DG, Tee BC, Fields H. Effect of bone thickness on alveolar bone-height measurements from cone-beam computed tomography images. Am J Orthod Dentofac Orthop. 2011;139:e117–27.
172. Timock AM, Cook V, McDonald T, Leo MC, Crowe J, Benninger BL, et al. Accuracy and reliability of buccal bone height and thickness measurements from cone-beam computed tomography imaging. Am J Orthod Dentofac Orthop. 2011;140:734–44.
173. Tomasi C, Bressan E, Corazza B, Mazzoleni S, Stellini E, Lith A. Reliability and reproducibility of linear mandible measurements with the use of a cone-beam computed tomography and two object inclinations. Dentomaxillofac Radiol. 2011;40:244–50.
174. Wood R, Sun Z, Chaudhry J, Tee BC, Kim DG, Leblebicioglu B, et al. Factors affecting the accuracy of buccal alveolar bone height measurements from cone-beam computed tomography images. Am J Orthod Dentofac Orthop. 2013;143:353–63.
175. Cook VC, Timock AM, Crowe JJ, Wang M, Covell DA Jr. Accuracy of alveolar bone measurements from cone beam computed tomography acquired using varying settings. Orthod Craniofac Res. 2015;18(Suppl 1):127–36.
176. Kamburoglu K, Kolsuz E, Kurt H, Kilic C, Ozen T, Paksoy CS. Accuracy of CBCT measurements of a human skull. J Digit Imaging. 2011;24:787–93.
177. Medelnik J, Hertrich K, Steinhauser-Andresen S, Hirschfelder U, Hofmann E. Accuracy of anatomical landmark identification using different CBCT- and MSCT-based 3D images: an in vitro study. J Orofac Orthop. 2011;72:261–78.
178. Stratemann SA, Huang JC, Maki K, Miller AJ, Hatcher DC. Comparison of cone beam computed tomography imaging with physical measures. Dentomaxillofac Radiol. 2008;37:80–93.

8 Treatment Planning, Outcome Assessment, and Upper Airway Imaging Using CBCT in Clinical Orthodontics

Juan Martin Palomo, Hakan El, Neda Stefanovic, Rany Bous, and Tarek Elshebiny

Abstract

The cone-beam computed tomography scan provides additional information that can be used not only for diagnosis but also for treatment planning and outcome assessment. This chapter will introduce some novel methods of 3D analysis not possible with 2D radiography, which can be easily incorporated right away in clinic orthodontics. Some of these measurements, such as the transverse analysis, have been shown to significantly improve the quality of treatment. In other words, the more we know, the more control we have, and the better result we can achieve.

J. M. Palomo (✉)
Department of Orthodontics, Craniofacial Imaging Center, School of Dental Medicine, Case Western Reserve University, Cleveland, OH, USA
e-mail: jmp5@case.edu

H. El
Department of Orthodontics, School of Dental Medicine, Hacettepe University, Ankara, Turkey

N. Stefanovic
Department of Orthodontics, Faculty of Dental Medicine, University of Belgrade, Belgrade, Serbia

R. Bous · T. Elshebiny
Department of Orthodontics, School of Dental Medicine, Case Western Reserve University, Cleveland, OH, USA

O. Kadioglu, G. F. Currier (eds.), *Craniofacial 3D Imaging*,
https://doi.org/10.1007/978-3-030-00722-5_8

8.1 Introduction

Cone-beam computed tomography (CBCT) is about more information. Unlike traditional two-dimensional radiography, all the information provided by a 3D CBCT cannot be seen at once and at the same time. A CBCT image provides layers of information, which not only display the current state of the patient but also give insight that may be essential in the orthodontic treatment planning. The additional information is also useful when assessing changes and evaluating results, so outcome assessments also can be more comprehensive. This chapter will address certain techniques that go beyond the regular diagnosis and provide further information for the treatment planning and outcome assessment of the orthodontic patient.

8.2 Imaging of Upper Airway

Imaging of the upper airway using CBCT is a very promising area which may be able to provide sound scientific data and added diagnostic value for treatment planning [1]. The pathological conditions present in the upper airway have far-reaching effects on the dentofacial region [2] and have been discussed over a period of more than a hundred years [3, 4]. However, this crucial and vital function is often overlooked in the name of better esthetic results and the establishment of a better occlusion.

Currently, with the state-of-the-art devices, hard and soft tissues, as well as the upper airway, can be displayed, providing a different perspective on the treatment plans of our patients. The upper airway has been evaluated using many techniques among which lateral cephalograms have been one of the most preferred modalities by orthodontists. However, the evaluation of the upper airway from only the sagittal view on a superimposed image does not properly portray the airway and has led to certain problems [5]. CBCT imaging, on the other hand, has helped us to identify the nature of airway-related problems, especially for obstructive sleep apnea syndrome (OSAS) by also visualizing the coronal view [6, 7].

In contrast to solid structures, the upper airway is a functional space surrounded by soft and hard tissues. The radiodensity of air under standard pressure and temperature using conventional CT has measured as −1000 Hounsfield units (HU) [8]. The reason for this low estimate is due to the very low attenuation of X-rays while passing from a hollow structure. The same holds true for CBCT devices, but there are some limitations due to beam hardening artifacts, more scatter radiation compared to conventional CT, the limited dynamic range of the X-ray area detectors, and the inability to show the actual HU values [9]. These unfavorable conditions cause the upper airway to receive an HU roughly in between −500 and −1000. This is particularly important since a well-defined segmentation of the upper airway depends on defining the correct threshold intervals. Thresholding is for discriminating the image according to gray values and is a highly preferred method for segmentation purposes [10]. All voxels above and below a defined limit are grouped together to discriminate the region of interest (ROI) apart from other tissues. This way, the upper airway can be segmented like a solid structure using correct threshold settings.

In terms of airway segmentation, current software can be categorized as manual, automatic, or semiautomatic segmentation programs. Manual segmentation programs give the greatest operator control by allowing the user to define the boundaries of the airway slice by slice. However, it is a very time-consuming procedure and not intended for clinical use. Automatic segmentation depends on correct thresholding capabilities and thus more prone to errors, but it can be performed within a couple of minutes. Semiautomatic segmentation provides both capabilities to segment the airway. Unfortunately, at this point, there is not a consensus for volumetric measurements between different programs due to the use of different algorithms. However, the correlation of volumetric measurements and intraoperator reliability is high [11]. The more important consideration here is to define the minimum cross-sectional area (MCA) and not the volume.

As the name suggests, MCA is the most constricted axial slice area within a defined airway volume (Fig. 8.1). Several studies also show that the MCA is a variable that can be used to explain the airway volume [12]. When the upper airway is assumed as a long cylindrical pipe, it is more significant to define the radius for determining the airflow resistance than knowing the volume of this canal according to Hagen-Poiseuille equation: [13, 14].

$$R_{AW} = \frac{8\mu l}{\pi r^4}$$

where R_{AW} is the airway resistance, μ dynamic viscosity, l length, and r radius. As can be seen from this equation, radius is inversely proportional in fourth power to airway resistance. This means that even if there is a small decrease in radius, this will be reflected as a dramatic increase in airway resistance that makes defining the MCA, especially for OSAS patients, a mandatory task.

On the other hand, this equation cannot be fully trusted as the upper airway, more specifically the oropharynx part, constantly displays a change in shape (or radius if we come to think of it as a cylinder) during inspiration and expiration phases [15]. Since CBCT is a static imaging technique, then a question comes to mind: in which phase of respiration should we acquire the image in order to standardize the records? In individuals with normal respiration, the upper airway contracts during early inspiration due to the formation of negative intraluminal pressure and begins to expand with the activation of dilator muscles toward the end of inspiration. In the early stages of expiration, the airway reaches its maximum dimensions due to the positive intraluminal pressure and narrows toward the end [16]. Since it is more important to define the areas of constriction, it may be suggested to acquire the images during the end of the expiration period.

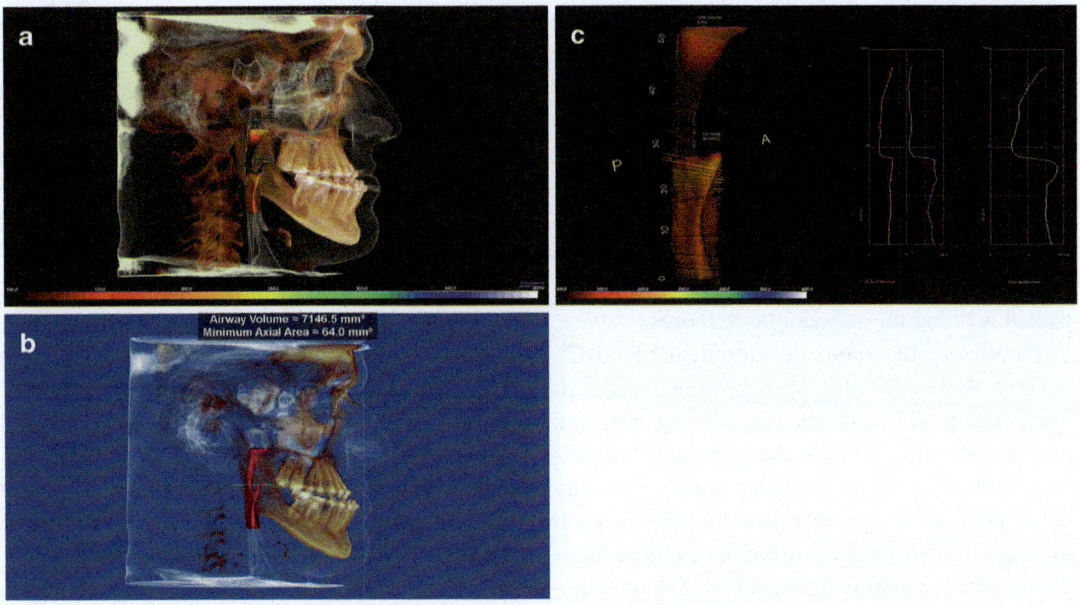

Fig. 8.1 Visualization of the oropharyngeal airway. (**a**) Determination of volume and minimally constricted axial area between defined limits using InVivo5 software. The color-coding also helps to visualize areas of constriction. (**b**) Determination of volume and minimally constricted axial area between defined limits using Dolphin Imaging software. Note that the same patient with the same head orientation is used for both programs. Although volumetric differences are more distinct (8.2 cc for InVivo5 and 7.2 cc for Dolphin), the MCA readings are closer (65.3 mm^2 for InVivo5 and 64 mm^2 for Dolphin) and at the same anatomical level for both programs. (**c**) A new feature of InVivo5 also provides users to visualize mediolateral and anterior-posterior widths and the cross-sectional areas all along the defined limits

One of the other most important considerations of upper airway imaging is the positioning of the patient. It has been reported that there is a 20–40% more contraction in the upper airway in supine position as compared to upright or sitting position [17]. Furthermore during sleep, the collapsibility of pharyngeal region tends to increase as compared to wakefulness [18]. Therefore, in the case of an OSAS screening, supine position is recommended during image acquisition in order to better evaluate the constricted regions. For other purposes, the sitting or upright positions will certainly give an idea about the constricted regions for the orthodontist to establish a treatment plan.

3D upper airway imaging has been a great addition to the orthodontists' armamentarium. It should always be kept in mind that we cannot diagnose or perform risk assessment for OSAS with radiographic imaging techniques. However, defining the areas of constriction in the upper airway and knowing what modifications can occur during the implementation of various orthodontic/orthopedic/orthognathic treatments can provide valuable information in the treatment planning of patients with an anatomical predisposition to OSAS.

8.3 3D Surgery

One of the other main fields to support the use of CBCT for patients presenting with skeletal deformities is the planning of orthognathic surgery. Preparation of a patient for orthognathic surgery has always presented a challenge for the clinicians. The use of 2D radiographs and analyses of a 3D entity have always left a question mark, especially in asymmetrical situations. Furthermore, in order to confirm the planning on 2D radiographs and to prepare the splints to guide the surgeon during the surgery, mandatory use of articulators [19] has presented another challenge and with loss of time and work power.

The reliability and accuracy of CBCT devices have been tested with several studies and found to reflect true dimensions of anatomical structures [20–22]. Although the resolution of maxillary and mandibular bones is in the range of acceptable/good quality for a surgical simulation, teeth resolutions are not sufficient for producing a surgical splint to be able to use in the mouth. Furthermore, it is almost an impossible task with the current segmentation programs to separate upper and lower teeth apart since CBCT images are taken while the patient is biting for orthodontic purposes.

Parallel to the rapid development of CBCT devices, there have also been favorable developments in the 3D field, such as 3D scanners and printers becoming more widespread and accessible. Following these developments, software developers have incorporated the 3D surgical modules to their programs. Patient's dental models are now scanned with the aid of 3D laser scanners, and the digital models are placed accurately overlaying the teeth on CBCT images [23]. Thus, the fabrication of a reliable surgical splint is possible by using the highly detailed upper and lower teeth that can be separated.

CBCT can capture the facial soft tissue and display it correctly to an extent. However, the resultant image lacking many features of the face makes it expressionless and hard to create a realistic simulation. Stereophotogrammetry is a technique to measure accurately certain anthropometric dimensions [24]. This method refers to combining multiple views of photos to form a 3D image [25]. Structured light technique is also a system used for capturing 3D information based on the triangulation principle [25]. There are systems available as of today incorporating the benefits of stereophotogrammetry and structured light technique in one device in order to capture photo-realistic 3D facial soft tissue images [26]. Also, some manufacturers are adding this functionality to their CBCT devices so that in addition to a CBCT scan, a 3D facial photo is also captured, simultaneously. The 3D photos acquired from such devices can easily be overlaid manually or automatically on the CBCT scan soft tissue profile to create a virtual patient [23]. This image fusion process includes merging facial soft tissues, facial skeleton, and dentition and is an accurate and a realistic tool for treatment planning [27]. Segmentation programs also offer a feature called "2D photo wrapping." With this method, the operator can wrap a 2D frontal photograph of the patient to his/her CBCT-generated soft tissue profile. However, although this method creates a photo-realistic look, some deformation may occur inevitably. Hence, 2D wrapping must be handled with a specific technique and diligence.

After creating the "virtual patient," 3D surgical procedures can be applied according to the specific software's user guides. Generally, it is a step-by-step procedure consisting of defining the borders of the maxilla, mandible (corpus and ramus), and teeth, deciding on which surgical cuts to perform (maxillary, mandibular, or both) and performing the movements of the jaws with 6 degrees of freedom (yaw, pitch, roll, superoinferior, anteroposterior, and lateral) according to the cephalometric surgical plan. In the final stage, 3D digital splint(s) can be created depending on single or double jaw surgery, and a hard copy can be printed using 3D printers (Fig. 8.2).

It is apparent that 3D digital world has much to offer for the clinicians. Incorporation of 3D surgery modules to various software is allowing the operator to work more accurately and decreasing significantly the amount of time spent for surgical preparation. With the advances in CBCT devices and related industry, it can be assumed that we are approaching the end of an important era; the era of articulators.

8.4 Temporary Anchorage Device Planning

Temporary anchorage devices (TADs), also known as orthodontic mini-implants, miniscrews, or miniscrew implants, have become very popular over the past 20 years for providing skeletal anchorage and improving orthodontic mechanics [28–31]. As the name suggests, TADs are temporarily fixed to the jaw bone in order to enhance orthodontic anchorage and are removed after they have served their purpose [28]. Stability of TADs during orthodontic treatment is essential for their clinical effectiveness in minimizing anchorage loss [31–34]. Before placing a TAD, detailed clini-

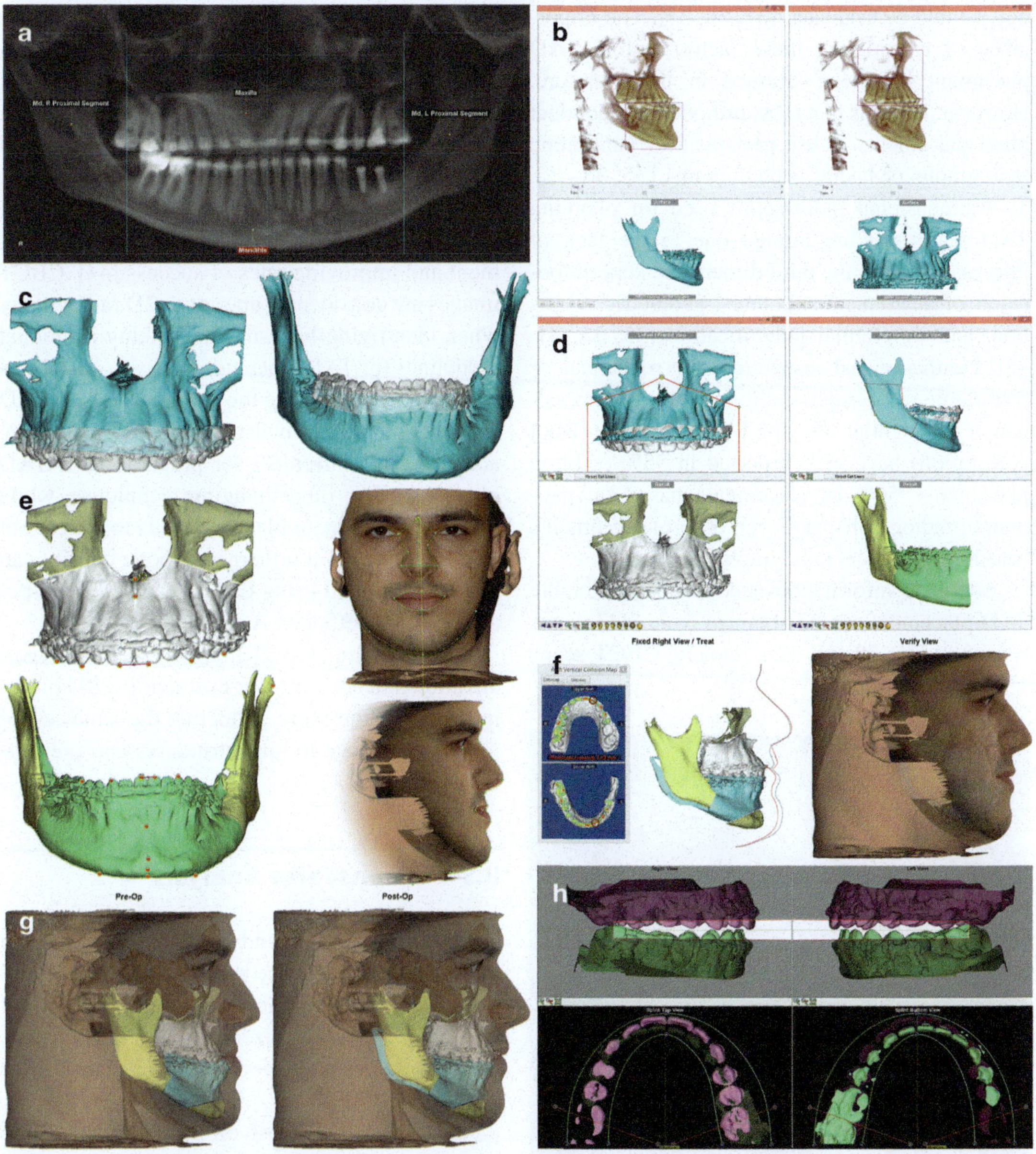

Fig. 8.2 Workflow of Dolphin Imaging software for 3D surgery. (**a**) 2D segmentation of the hard tissues using the panoramic view. (**b**) 3D segmentation of the mandible (top and bottom left images) and maxilla (top and bottom right images). (**c**) Laser-scanned upper and lower models (.STL files) are inserted on respective jaws, adjusted and unwanted parts cropped out. (**d**) Definition of surgical cuts. (**e**) 3D landmark identification for hard and soft tissues. (**f**) 3D surgery performed according to the treatment plan. Soft and hard tissue changes are verified. It is also possible to visualize new teeth contacts using the vertical collision map. (**g**) Comparison of pre- and post-op soft and hard tissue changes. (**h**) Preparation of splints

cal and radiographic assessment of the placement site should be performed [35]. Surrounding bone quality, such as bone density, depth and cortical bone thickness, characteristics of the mucosa and attached gingiva, soft tissue thickness and mobility, proximity of the roots, interradicular distances, proximity of nerves and blood vessels, the location of the inferior alveolar nerve, and sinus morphology, should be evaluated [32, 36, 37]. The importance of considering these factors before TAD placement has been stressed in the literature. However, it seems that most orthodontists conduct blind placements or use periapical or panoramic radiographs or lateral cephalograms [35, 38–42]. Two-dimensional radiographs used for planning TAD placement are limited due to the lack of information from the third dimension, superimposition of anatomical structures, geometric distortion, and differential magnification [35, 38, 41, 42]. Neither visual inspection nor periapical or panoramic radiographs exhibit reliability for evaluation of potential sites for TAD placement. Root perforations have been reported in 55% of blind placements, 50% of placements based on panoramic radiographs, and even 60% of placements based on periapical radiographs [35].

Adequate information needed for predictable TAD placement can be obtained from 3D images (Fig. 8.3) [35, 38, 41, 42]. Namely, CBCT scans can be used for evaluating the quality and quantity of the cortical and underlying trabecular bone, which is very important for TAD stability. This is especially important when placing TADs near complex anatomical structures where quality and quantity of bone might be compromised [40, 43]. Determining these alveolar bone features could be valuable for finding optimal sites for TAD placement and improving rates of success [44]. CBCT images are considered superior to 2D radiography when identifying best sites and giving positional guidelines for TAD placement [32, 38, 45–52], as well as for determining the proximity of the TAD to the root [53]. Small-volume CBCT is more accurate and convenient for preparation of TAD placement than other imaging techniques, while using it in treatment planning leads to fewer root perforations [35]. Furthermore, bone density can also be measured using CBCT scans if the X-ray attenuation coefficient is corrected [53].

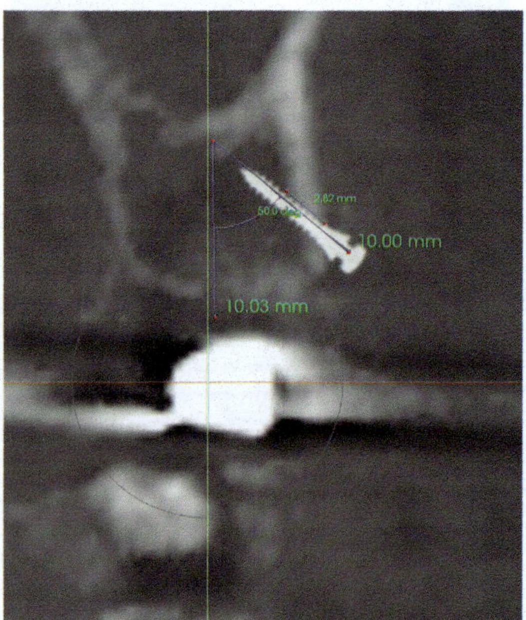

Fig. 8.3 Linear (cortical bone thickness) and angular (TAD angulation) parameters important for TAD stability were measured on the coronal view of the maxillary molar area. Cortical bone thickness was measured on the facial side as the faciolingual distance between the cortical and cancellous bone. TAD angulation was determined by the angle between the long axis of the TAD and the long axis of the tooth

In summation, CBCT images give better visualization and accuracy for evaluating TAD placement sites, therefore reducing the chances of potential damage to vital structures and preventing failure [38].

8.5 Transverse Analysis

Thorough diagnosis and treatment planning in all three planes of space is the key to a successful orthodontic treatment. The introduction of CBCT has revolutionized our diagnostic tools, by allowing us to inspect all three planes more accurately [54].

Until recently, most clinicians would base their diagnosis on lateral cephalograms, which are a 2D representation of a 3D structure, and that caused the transverse dimension to be overlooked and lacking in accurate diagnostic measures. However, orthodontists agree that transverse deficiencies are an essential component of diagnosis, and it has been incorporated in the American Board of Orthodontics (ABO) Cast-Radiograph Evaluation (C-R evaluation), where the buccolingual inclinations of posterior teeth are measured by comparing height differences between buccal and lingual cusps on dental casts [55]. Measures such as dental casts and posteroanterior cephalograms have been used, but these measures lacked reliability and consistency [56]. Other clinicians base their decision on the presence or absence of crossbites, which is flawed because the crossbite could be of dental nature or an absence of it could be due to an excessive molar angulation compensating for an underlying skeletal deficiency. Moreover, experienced clinicians can diagnose the transverse dimension with a combination of measures that evaluate the presence of crossbites, the degree of crowding, arch width, perceived buccolingual inclinations of teeth, and the shape and height of the palatal vault [57, 58]. However, this requires experience, and the multiple measurements increase the potential for errors.

Enlow and Hans discussed dentoalveolar compensations and stated that "Intrinsic adjustments during growth are an important biological concept, as they allow regional parts to stay in a state of functional and structural equilibrium." [59] Moreover, Solow discussed how transverse skeletal jaw discrepancies are partly compensated for through adjustments of the buccolingual molar angulations [60]. In a CBCT study, it was shown that where a transverse deficiency exists without crossbites, it was usually due to molar inclinations beyond one standard of deviation from the mean, exhibiting a compensation for skeletal transverse deficiencies [61].

Using CBCTs, one is able to inspect the transverse dimensions more accurately and develop objective, reliable diagnostic norms and guidelines to help in diagnosis and treatment planning. The significant superiority of CBCTs over posteroanterior cephalograms was proven when both techniques were compared to dry skull measurements [62]. Moreover, Streit has shown low agreement among experienced clinicians when diagnosing "transverse deficiencies" using photographs, models, and frontal radiographs alone [63]. This supports the need for developing more reliable, standard measures to diagnose the transverse dimensions, which become even more important in identifying whether the deficiency is skeletal or dental, especially when dental compensations mask underlying skeletal deficiencies.

A few methods have been developed to utilize CBCTs to measure buccolingual inclinations of molars with the goal of developing norms and guidelines to aid in treatment decisions that include the transverse dimensions (such as whether to use dental versus skeletal expansion to correct deficiencies). Although the different methods used different landmarks, they all had high reliability [63–65].

8.5.1 The Case Western Reserve University (CWRU) Method

To address the need for measuring the buccolingual inclination of teeth, CWRU used a sample of 78 to develop a technique to measure buccolingual inclinations using CBCT and establish norms for "The CWRU Transverse Analysis" by averaging the buccolingual inclinations of first molars in individuals showing ideal posterior occlusion and intermaxillary relationships. The average maxillary first molar angulation was 100 ± 4°, while that of mandibular first molars was 77 ± 5°. These numbers were chosen to be the molar angulation norms for ideal occlusion. The method is simple to use and quick and can be done in any commercially available DICOM viewer that permits angular measurements. It provides an objective and quantifiable method to analyze the transverse dimension [63, 64, 66–69].

First, the head should be oriented with internal landmarks (the CWRU orientation method), as described in the previous Chap. 7 (Sec. 7.3). Then, the buccolingual inclinations of each maxillary first molar are measured through the angle outlined between the palatal long axis of the tooth (the line joining the mesiopalatal cusp tip with the palatal root apex) and a tangent to the inferior border of the nasal cavity (Fig. 8.4). The buccolingual inclination of each mandibular first molar is measured through the angle formed between the long axis of the tooth (the line connecting the central groove with the apex of the mesial root) and a tangent to the inferior border of the mandible (Fig. 8.5).

This method has shown an intra-rater reliability of 98.7% and an inter-rater reliability of 89.2% [69].

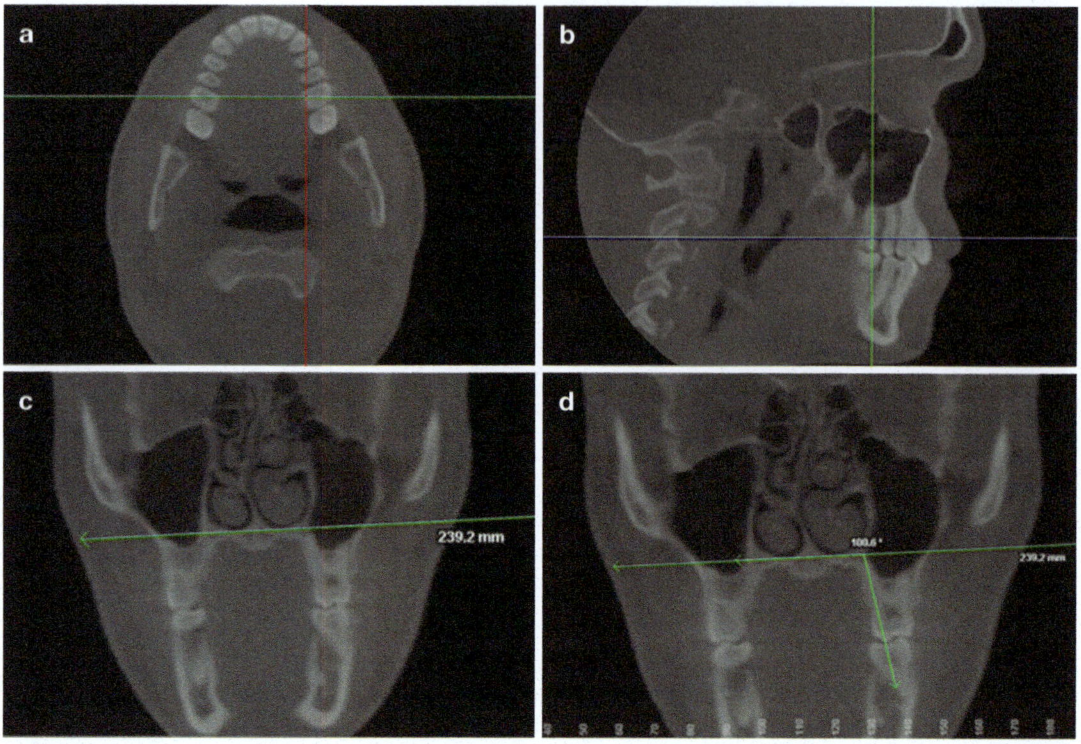

Fig. 8.4 Measuring maxillary molar angulation. (**a**) In axial view, position the line representing the sagittal plane through the maxillary first molars. (**b**) In sagittal view, position the line representing the coronal slice along the mesiopalatal cusp tip and the palatal root apex. (**c**) In coronal view, draw a reference line tangent to the nasal floor in coronal view. (**d**) Measure the inclination which is the angle between the reference line and the line passing through the long axis of the molar (the line passing through the apex of the palatal root and the mesiopalatal cusp tip)

8.5.2 Interpretation of the Measurements

The angulations are categorized as normal, deficient, or excessive. Excessive angulations indicate that the molars are tipped buccally beyond one standard of deviation from the norm, while deficient angulations indicate a lingual inclination below one standard of deviation from the norm.

For example, a maxillary molar with an excessive buccal inclination (above one standard of deviation) and/or a mandibular molar with a deficient inclination (below one standard of deviation) can indicate a possible dental compensation for a skeletally narrow maxilla. This suggests that using a rapid palatal expander (RPE) and expanding the maxilla can be beneficial for harmonious occlusion. On the other hand, a deficient maxillary molar inclination could possibly be a compensation for a wider maxilla, and thus RPE should be avoided.

A decision tree (Fig. 8.6) was developed to aid in decision-making. Normal molar angulations in absence of crossbites indicate a harmonious transverse relationship, and thus the transverse dimension should be maintained, while those with normal angulations in presence of crossbites exhibit a transverse skeletal deficiency without dental compensations, indicating a need for RPE [69].

For subjects that exhibit dental compensations for a skeletal transverse deficiency (excessive maxillary molar angulation, deficient mandibular molar angulation, or a combination), RPE is the treatment of choice to achieve a balanced occlusion, regardless of the presence or absence of crossbites. Meanwhile, subjects with deficient maxillary molar angulations should be treated

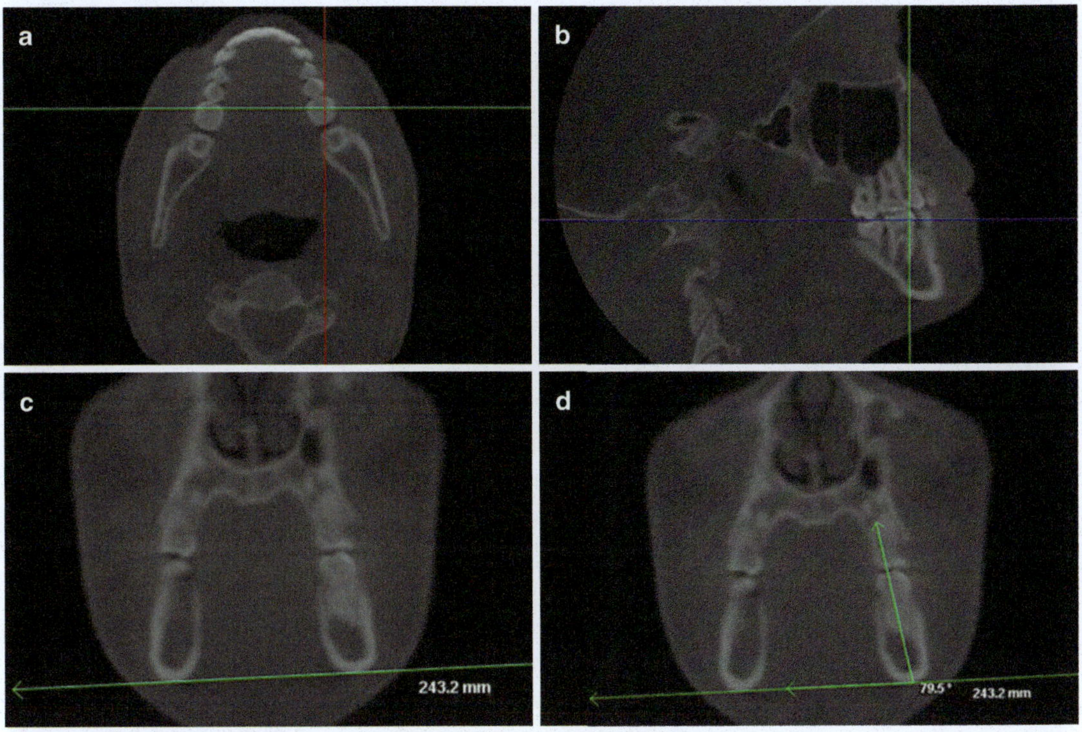

Fig. 8.5 Measuring mandibular molar angulation. (**a**) In axial view, position the line representing the sagittal plane through the mandibular first molars. (**b**) In sagittal view, position the line representing the coronal slice along the long axis of the molar (mesial cusp tip to mesial root apex). (**c**) In coronal view, draw a reference line tangent to the inferior border of the mandible in the coronal view. (**d**) Measure the inclination which is the angle between the reference line and the long axis of the molar (the line passing the root apex and central fossa of the molar)

with dental expansion (utilizing archwire expansion (AWE) or cross-elastics), regardless of the presence or absence of crossbites.

8.5.3 Objectivity and Clinical Implications

To measure the clinical implications and the effect of utilizing the CWRU transverse analysis on the quality of treatment, Mostafa et al. conducted a study that utilized the American Board of Orthodontics (ABO) Cast-Radiograph evaluation (C-R evaluation), which is the grading system for the ABO clinical examination, to study the clinical significance of this method objectively [69]. The CWRU transverse analysis was performed retrospectively, and the sample was divided into two groups, based upon whether the subjects followed what the CWRU transverse analysis would have suggested or not. The results showed a statistically significant improvement in the C-R evaluation score in the group that followed the CWRU transverse analysis method. This significance was even stronger in the buccolingual inclination component of the C-R evaluation, indicating that utilizing the analysis significantly enhances the overall quality of orthodontic treatment, as measured by the ABO criteria (Table 8.1) [69].

Case A shows an example of correctly utilizing the CWRU transverse analysis. Figure 8.7 shows the pre-treatment pictures and molar angulations, where maxillary right and left molars are excessively tipped buccally, at 105.9° and 105.4°, respectively, which is over one standard of deviation above the norm of $100 \pm 4°$ and the mandibular molars lingually tipped at 68.1° and 65.9° (below the norm of $77 \pm 5°$), indicating a dental compensation for a maxillary transverse skeletal deficiency.

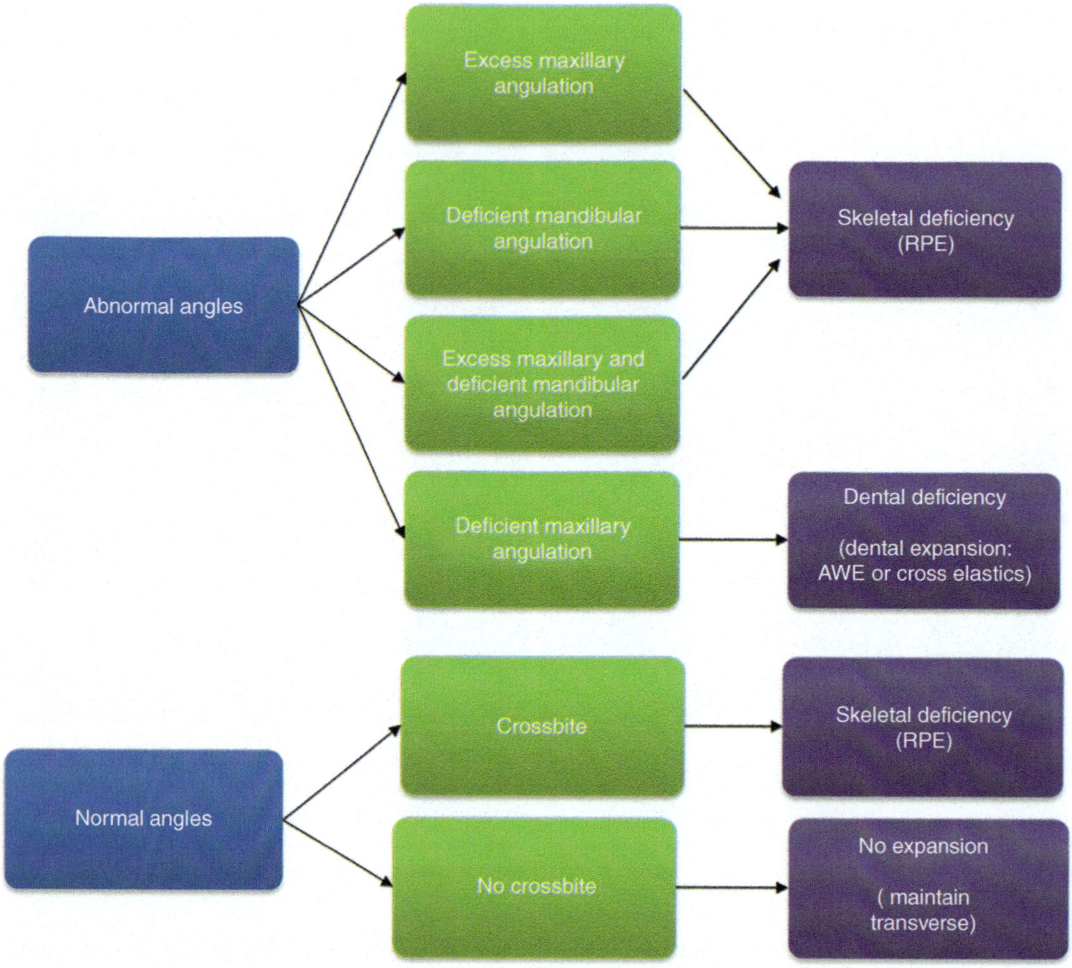

Fig. 8.6 Decision tree describing different molar angulations and suggested treatment plan

Table 8.1 Independent sample Mann-Whitney *U*-test between the CWRU's transverse analysis, the ABO's (*C-R Eval*), BL component of (*C-R Eval*), and active treatment duration

	Followed, $n = 46$		Did not follow, $n = 39$		
	Mean ± SD	Range	Mean ± SD	Range	*P*-value
ABO (*C-R Eval*)	18.7 ± 5.9	7–34	21.5 ± 6.2	12–39	0.041*
BL component of (*C-R Eval*) for posterior teeth	2.9 ± 1.7	0–7	4.6 ± 2.3	1–10	0.001*
Treatment duration (months)	25 ± 5.4	13–40	27.3 ± 6.4	16–48	0.106^{NS}

$n = 85$, followed = 46, did not follow = 39
NS not statistically significant
*Statistically significant at *P*-value of <=0.05, using independent sample Mann-Whitney-*U* test.

Although the patient did not have a posterior crossbite, the treatment plan included a RPE to account for the skeletal transverse deficiency. Figure 8.8 shows the posttreatment results, with improved posterior segment relationships. Maxillary molar angulations are improved, as they are now within one standard of deviation from the norm, at 103.2° and 102.7°. The treatment duration was 20 months, and the final ABO C-R evaluation score was 15, indicating a good

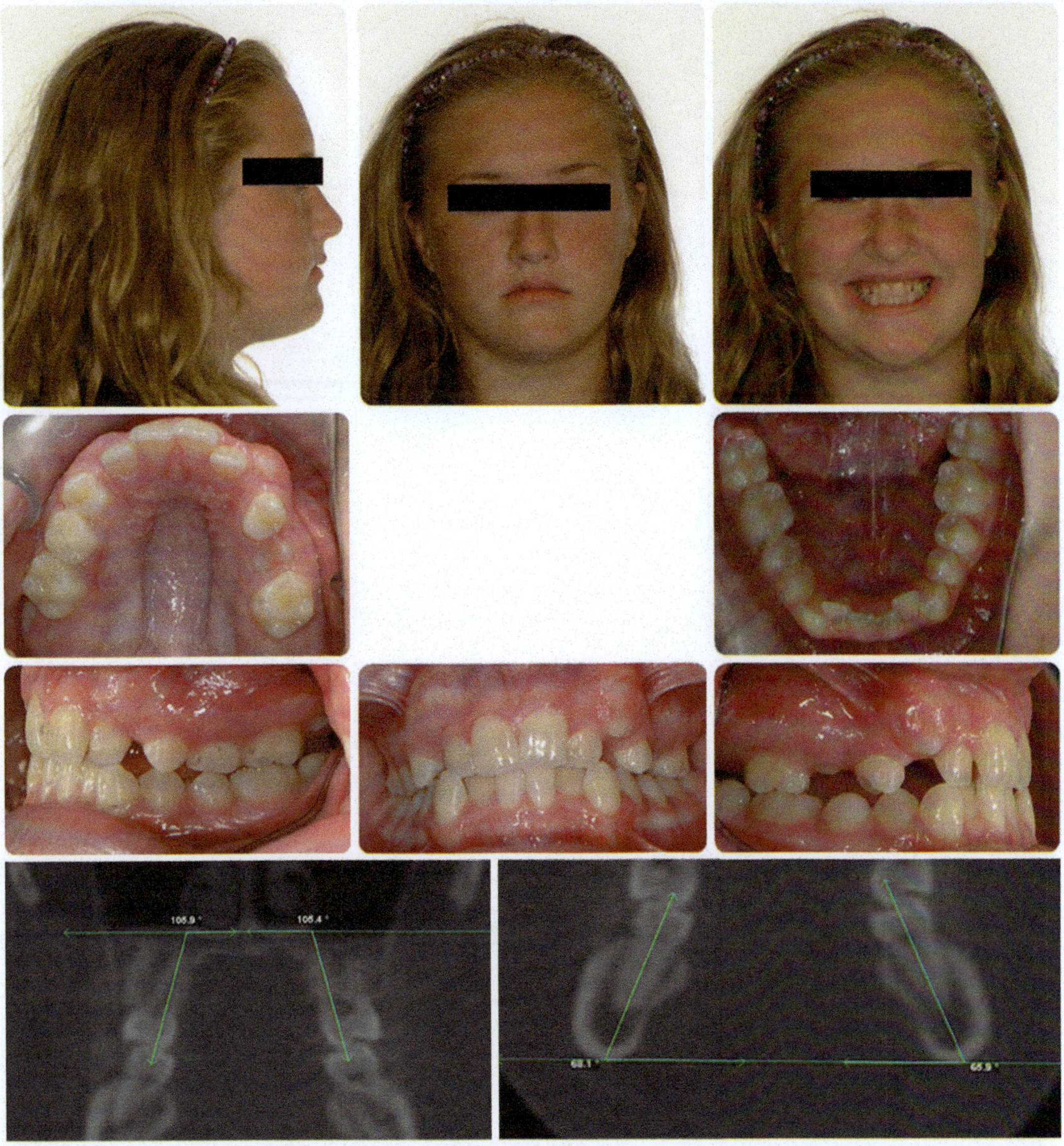

Fig. 8.7 Case A—pre-treatment pictures and molar angulations

final treatment result. None of the molars scored any points on the buccolingual inclination aspect of the C-R evaluation.

On the other hand, Case B shows a subject where the CWRU transverse analysis was not followed. The patient had a right unilateral crossbite, but the maxillary right first molar had a deficient angulation of 91.5°. Moreover, the mandibular right first molar had an excessive angle (91.2°) (Fig. 8.9). This indicates that the crossbite was dental or even a compensation for a wider maxilla. Here, RPE would not be indicated. However, he was treated with RPE, but the crossbite was not resolved and the maxillary molar angulations became even more deficient (Fig. 8.10). The total final C-R evaluation score was 37, which indicates a poor final result, which can be largely attributed to a misdiagnosis of the transverse dimension. If the CWRU transverse analysis was followed, the mechanics would have

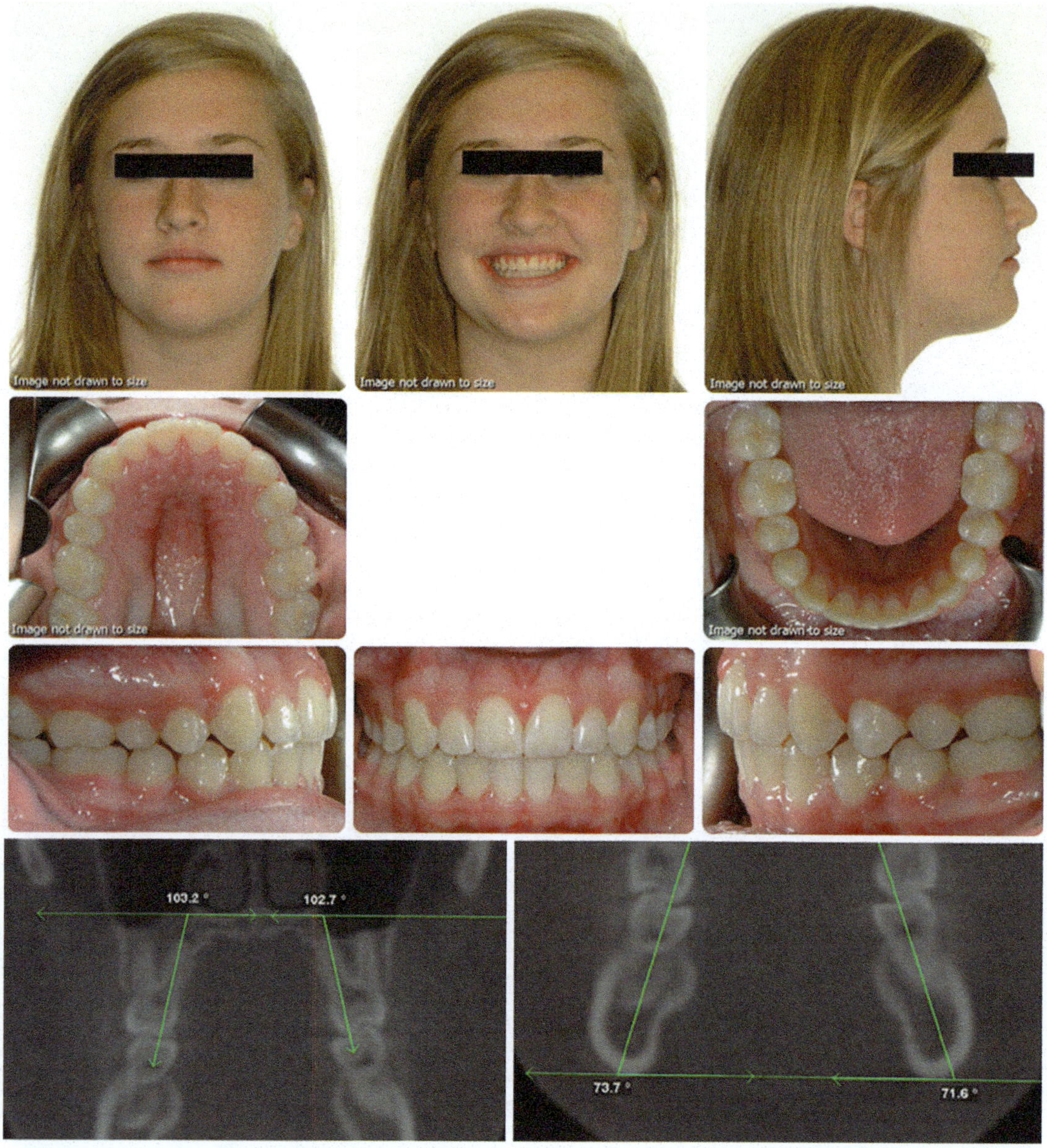

Fig. 8.8 Case A—posttreatment pictures and molar angulations

included dental expansion rather than RPE, and the final C-R evaluation score may have been better.

The transverse dimension is an important aspect of orthodontic case diagnosis. Vanarsdall emphasized the critical importance of the skeletal differential between the width of the maxilla and the width of the mandible, where he stated that: "Undiagnosed transverse discrepancy leads to the adverse periodontal response, unstable dental camouflage, and less than optimal dentofacial esthetics." [70] The CWRU transverse analysis is another tool that can be added to the orthodontist's armamentarium, supplementing other diagnostic information for evaluating the nature of the transverse discrepancy. It is a relatively simple method that, when used appropriately, can significantly enhance the quality of orthodontic treatment.

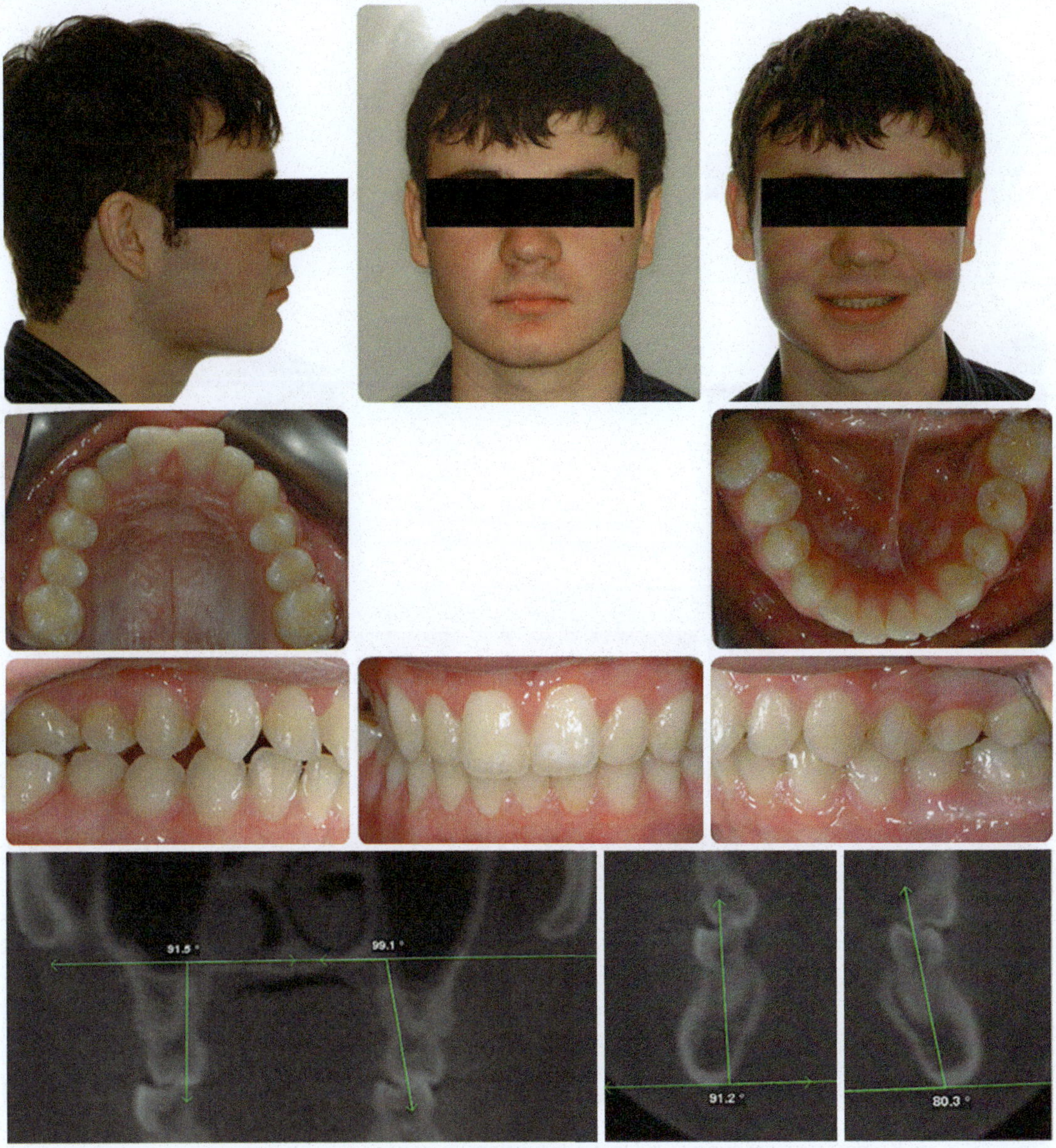

Fig. 8.9 Case B—pre-treatment pictures and molar angulations

8.6 Three-Dimensional Superimposition Methods Using CBCT

In 1931 Birdsall Holly Broadbent published a technique for superimposition of successive cephalometric films to study the physical changes that occurred during facial growth with time [71]. This imaging approach also became a standard part of orthodontic records for confirming diagnosis and evaluating treatment outcomes.

Several methods have been proposed for superimposing serial cephalograms in 2D [72, 73]. Cephalometric superimpositions allow clinicians to evaluate growth and treatment by evaluation of changes in the maxillary and mandibular displacement through a general superimposition on the cranial base, evaluation of changes in the maxillary dentoalveolar complex through local

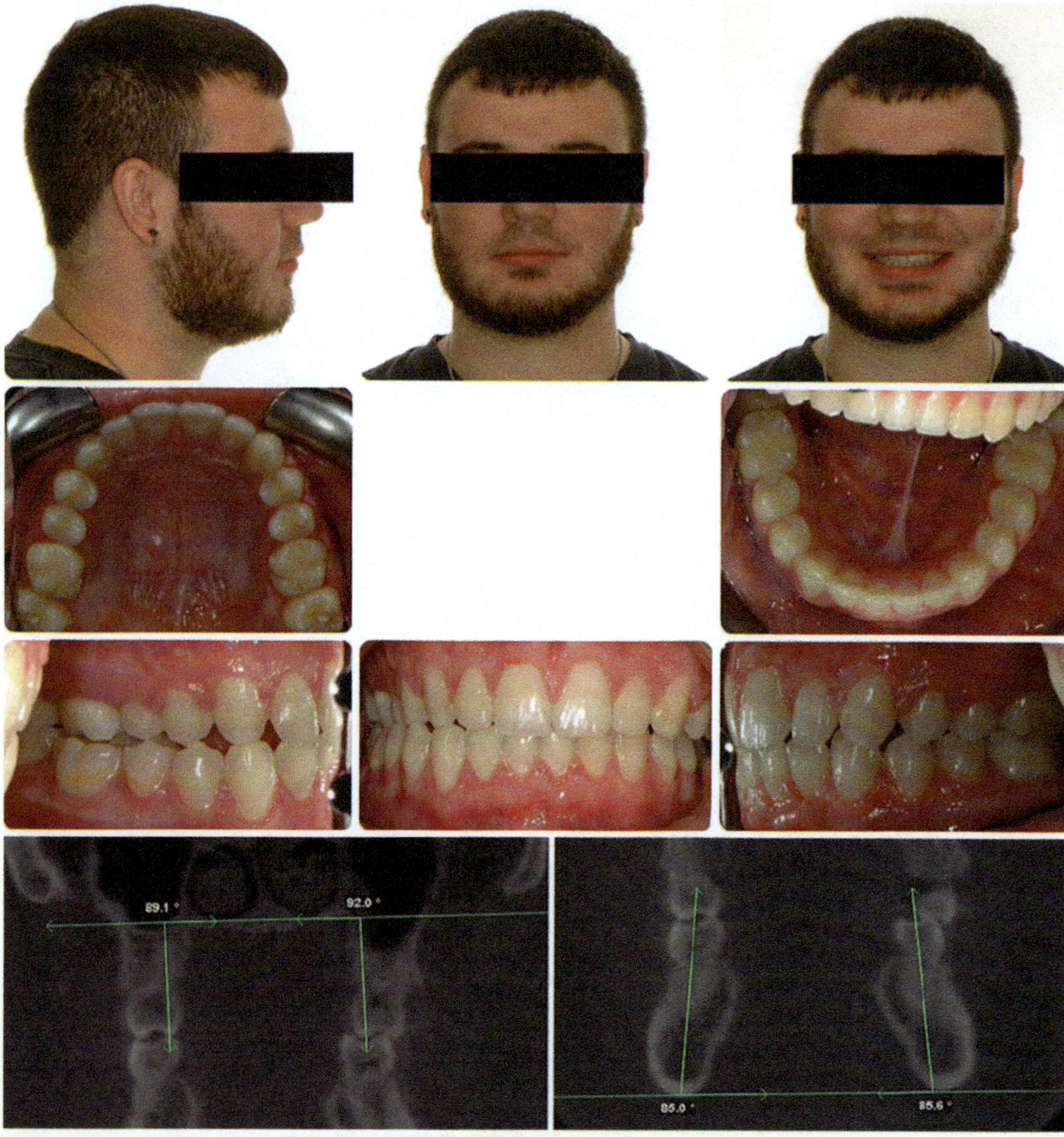

Fig. 8.10 Case B—posttreatment pictures and molar angulations

maxillary superimposition, and evaluation of dentoalveolar mandibular changes through local mandibular superimposition. However, cephalometrics was limited to viewing three-dimensional (3D) craniofacial structures in only two dimensions (2D).

The importance of the third dimension for orthodontic and surgical purposes has been emphasized for decades, and several attempts have been made to be able to view the third dimension [74–76]. Recently, advancement of CBCT devices and abundant development of software packages have led not only to better diagnosis and treatment planning but also in evaluation of treatment outcomes in 3D.

The resolution of CBCT imaging is determined by the individual volume elements (voxels) produced from the volumetric dataset [20]. The size of a voxel is defined by its height, width, and depth, and CBCT voxels are generally isotropic – equal in three dimensions [7]. The voxel size of a 3D image is equivalent to the pixel resolution in 2D images, and each voxel contains an intensity or density of the grayscale level. This grayscale level can be analyzed to identify areas of stability within an image when comparing two images.

The superimposition of CBCT volumes in 3D space after changes of the craniofacial structures over time due to growth or orthodontic/surgical treatment requires understanding different types of 3D superimpositions. With the development of software packages, different methods have been proposed for superimposition of volume images from a CBCT scan.

8.6.1 Landmark-Based Superimposition

Landmark-based superimposition requires accuracy in anatomical landmarks identification [77]. Landmark superimposition works by calculating the difference between selected anatomical landmarks on two CBCT images, and accordingly the software overlays the two images. Grauer et al. discussed registration and superimposition of 3D images from DICOM files using different commercial software packages. Most software programs provide superimposition tools for 3D images using anatomical landmarks [78].

Steps for Landmark Superimposition

The initial (T1) and the final (T2) CBCT images are uploaded to the software, and landmarks are placed on anatomically stable structures to serve as registration references. Each software requires a different number of landmarks, varying between 3 and 7.

Landmark superimposition works by calculating the difference between selected anatomical landmarks on the two CBCT images, and the software overlays the two images accordingly. Once the two images are overlaid, a position-refining tool can be used to manually refine the registration of the two images to reach the best fit or match of the cranial bases. Changes between two images can be evaluated through the rendered superimposed volumes or through superimposed slices. There is also an option to adjust image segmentation for soft tissue evaluation.

Landmark superimposition methods using Dolphin Imaging software version 11.9 (Dolphin Imaging & Management Solutions, Chatsworth, CA) are shown in Fig. 8.11, and the one using InVivo version 5.1 (Anatomage, San Jose, CA) is shown in Fig. 8.12.

8.6.2 Surface-Based Superimposition

Surface-based registration implies separately selecting corresponding unchanged surfaces in two images. Once surfaces are selected, a manual approximation is performed by translating one of the two images to align the two surfaces. Finally, the software program performs a surface-to-surface registration to refine the initial manual registration.

8.6.3 Voxel-Based Superimposition

Cevidanes et al. introduced the voxel-based superimposition method to the dental field [79, 80]. It has been widely used previously in the medical field for superimposing CT, CBCT, and MRI images. Voxel-based registration method measures the unchanged grayscale intensity within each voxel in a defined volume of interest of two scans to register the images. This makes it a fully automated superimposition method, which can overcome the drawbacks of the previously described ones that mainly depend on accurate landmark identification.

It can be performed using the Slicer open-source software (www.slicer.org), for which video tutorials are available at https://www.youtube.com/user/DCBIA/playlists [81]. The image analysis steps include (1) 3D registration and construction of segmentations, (2) construction of surface models, and (3) quantification of changes.

The voxel-based superimposition is considered the most advanced one and has been used in the literature for superimposition in both growing and nongrowing patients for assessment of facial soft tissues, orthognathic surgery procedures, and

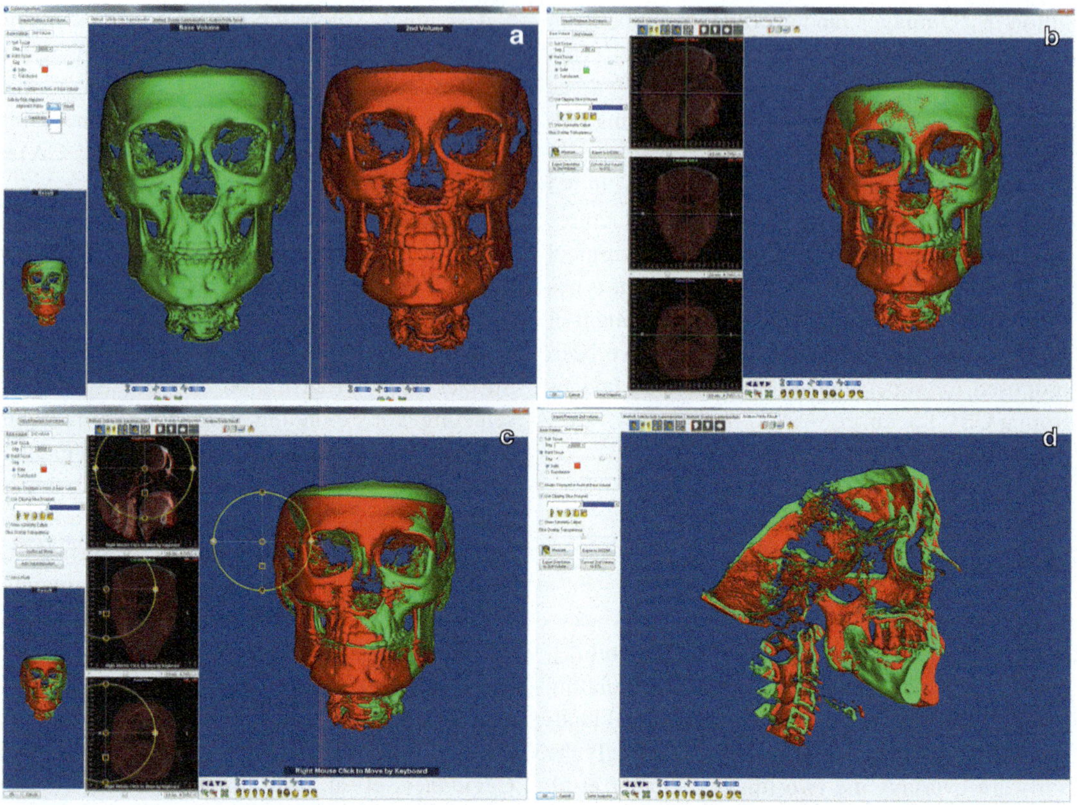

Fig. 8.11 Landmark-based superimposition using Dolphin 3D software. (**a**) Two CBCT images uploaded to the software. Same landmarks are placed on the two images as a registration reference. (**b**) The two images are overlaid together by the software after calculating the difference between the selected anatomical landmarks. (**c**) Manually refine the registration of the two images to reach the best fit or match of the cranial bases using the refining tool. (**d**) Final outcome after superimposition

temporomandibular joint in osteoarthritis patients [82–85]. However, it is not ideal. The major drawback is that it requires several steps performed using more than one software, which take about 1 h for a well-trained user. Recently, commercial software packages started to offer a new tool for cranial base superimposition that is also voxel based and does not require the construction of surface models prior to superimposition. This tool is user-friendly in most of the software packages, and superimposition can be performed in 30–40 s.

Several studies have compared fast 3D voxel-based superimposition using commercial software programs with the Cevidanes method, which is considered the gold standard for voxel-based superimposition. Bazina et al. [86] compared the fast 3D voxel superimposition on the cranial base using Dolphin 3D software (version 11.9, Dolphin Imaging & Management Solutions, Chatsworth, CA) to the Cevidanes method using a sample of nongrowing surgical patients to assess the accuracy of the software. No clinically significant differences were found between the two programs [86]. Ben Nasir et al. compared the same software packages in another study using a sample of growing patients and reached the same conclusion that the fast 3D voxel-based superimposition was an accurate method which can be used for clinical and research purposes [87]. Another study validated the method for fast 3D superimposition of CBCT images of growing patients and adults using commercial software (OnDemand3D;

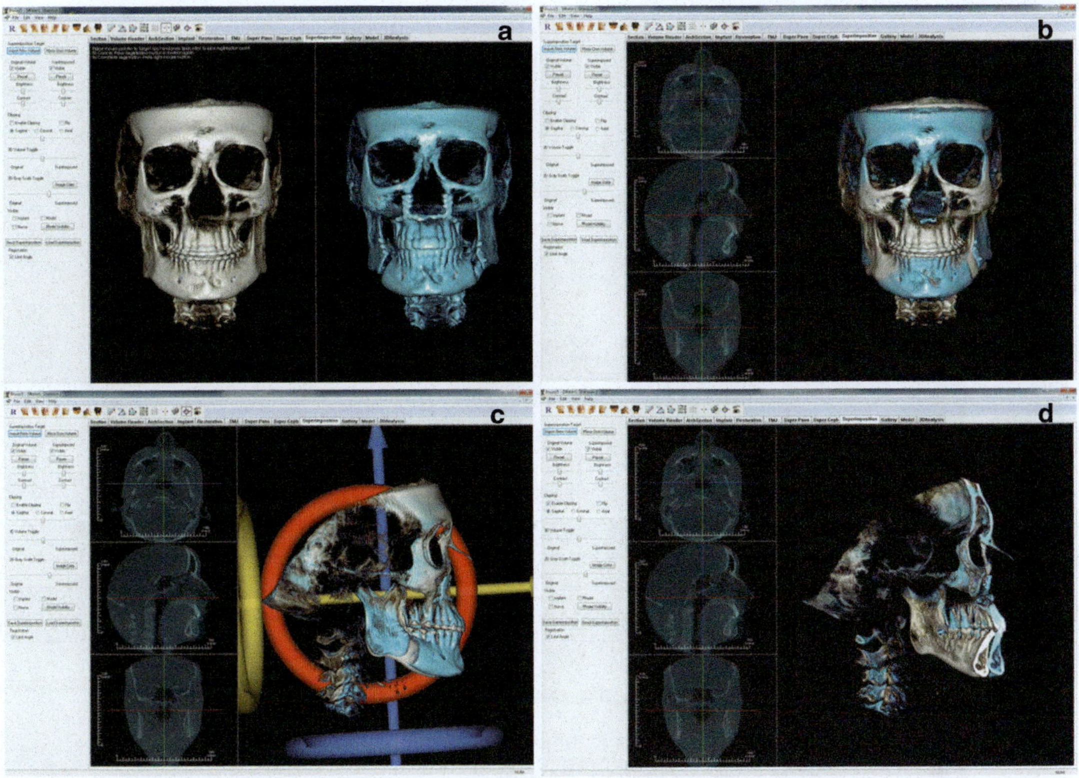

Fig. 8.12 Landmark-based superimposition using InVivo software. Same steps as Dolphin 3D

Cybermed, Seoul, South Korea) and concluded it was reproducible in different clinical conditions and applicable for research and clinical practice [88]. After validation of commercial software packages against the gold standard method, a new study compared the fast commercial software packages (Fig. 8.13). It was concluded that there were no clinically significant differences between the programs [89].

Steps for Fast 3D Voxel-Based Superimposition on the Cranial Base

The base volume and the second volume CBCT images are uploaded to the software and are approximated using at least three landmarks placed on each volume. Different software programs require different numbers of landmarks, usually a minimum of 3 and maximum of 7. After approximation, a position-refining tool is used to manually refine the registration of the two images to reach the best fit or match of the cranial bases. Anatomical structures of the anterior cranial base are then selected on different slice views of the used volumes by placing a size adjustable box on the area of interest. Next, the automated registration tool is performed to align the volumes using the unchanged voxels within the superimposition box of the two CBCT volumes. Figure 8.14 shows the voxel-based superimposition method using Dolphin software program.

8.6.4 Superimposition Assessment

Color maps: After complete registration of the two volumes, the outcome can be assessed based on the absolute value of the maximum distance between surfaces and then graphically displayed as color maps. Colored segments corresponding to the distance (mm) are used to highlight the differences between the two surfaces in the regions of interest (Fig. 8.15).

Another method to assess the superimposition is by visualizing of the semitransparent surface

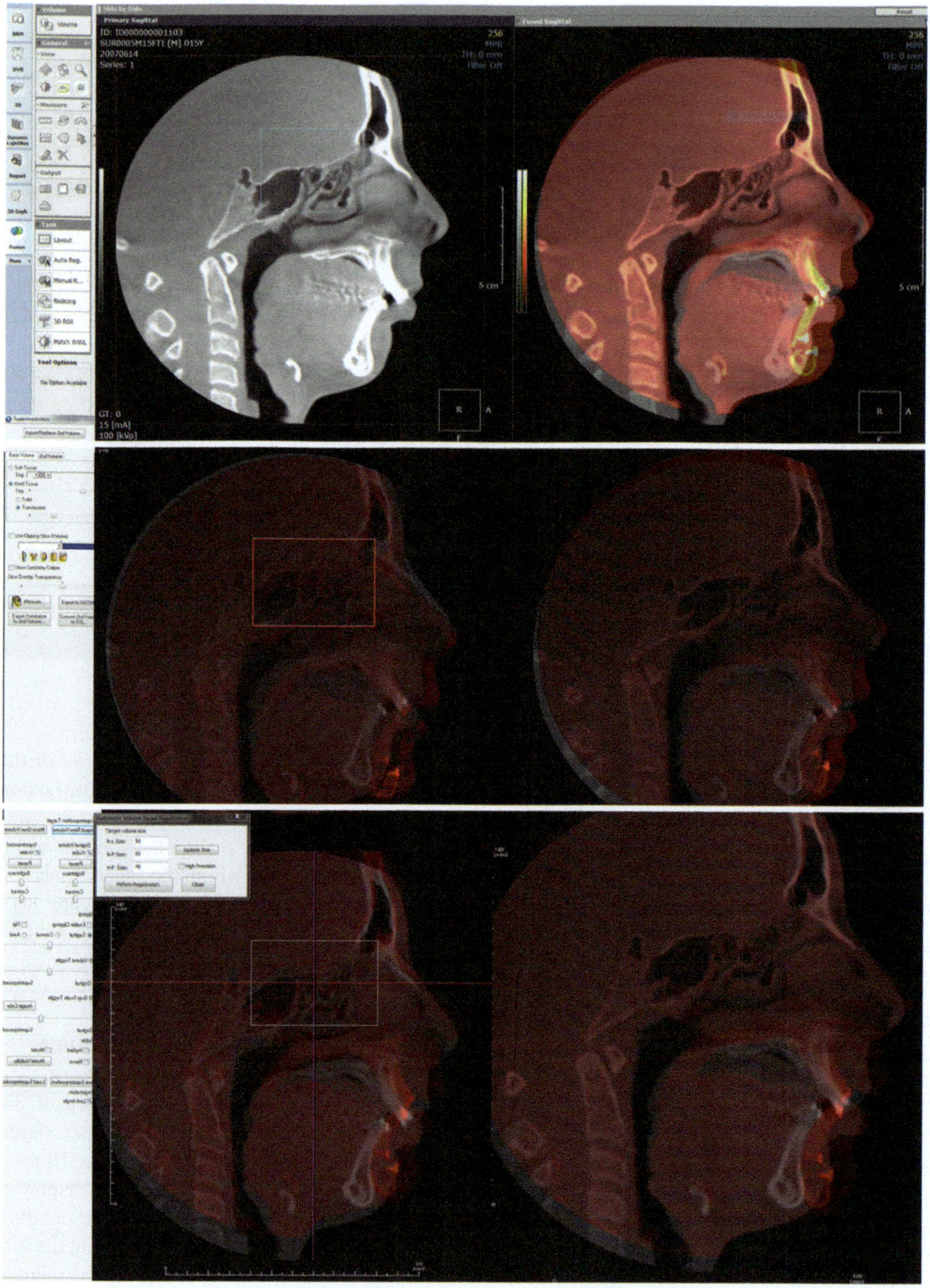

Fig. 8.13 Voxel-based superimposition on the cranial base for one patient using three different software programs (Ondemand 3D, Dolphin 3D, and InVivo)

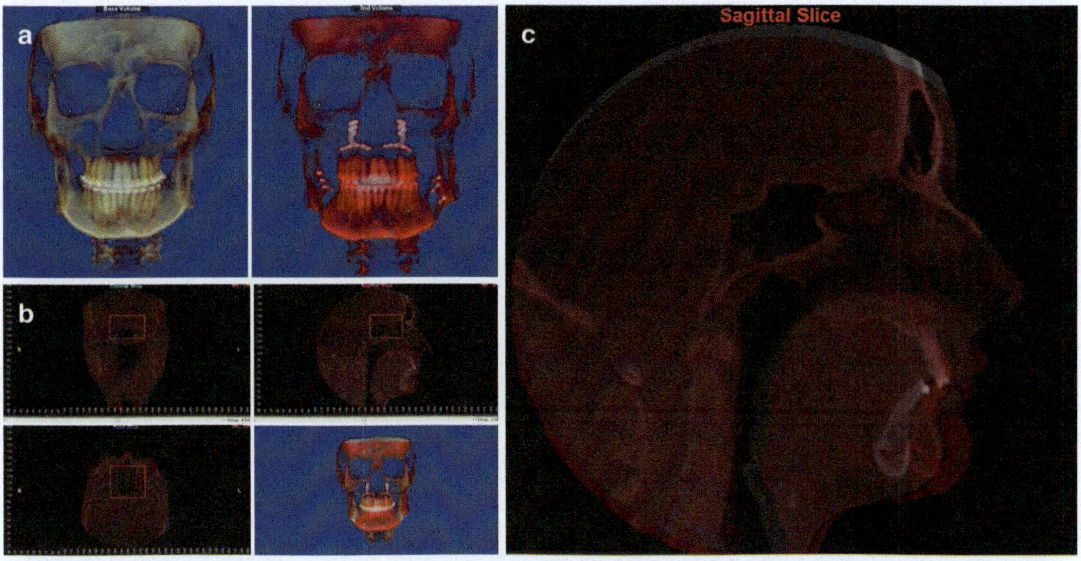

Fig. 8.14 Voxel-based superimposition. (**a**) Approximation of the two CBCT images using at landmarks placed on each volume. (**b**) Selection of the anterior cranial base on different slice views of the used volumes by placing a size adjustable box on the area of interest. (**c**) Final superimposition image after aligning the volumes using the unchanged voxels within superimposition box of the two CBCT volumes

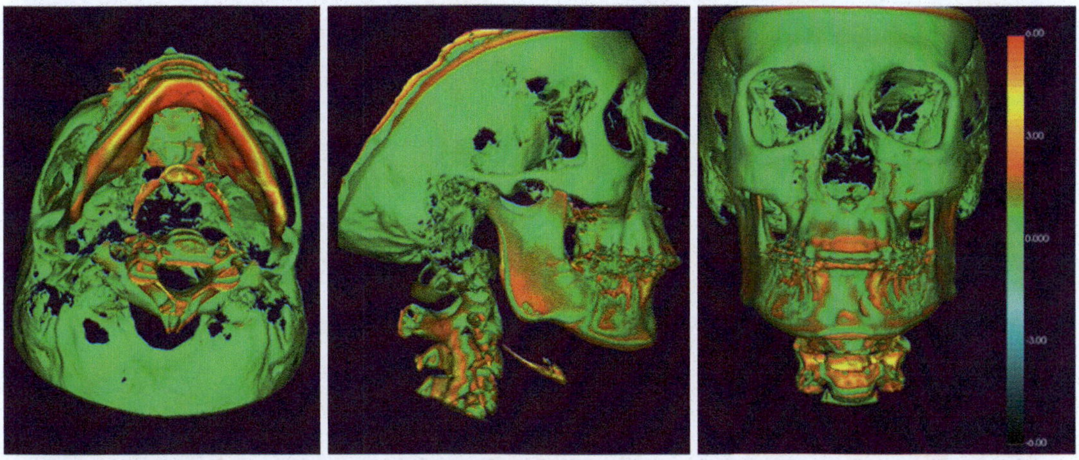

Fig. 8.15 CBCT superimposition of an adult patient had an orthognathic surgery. 3D models color maps showing the surface distances between presurgical and 6-month postsurgical. The areas in red and orange are changed due to surgical movements. The green color represents the areas with no changes

models, axial, sagittal, and coronal cross-sectional slices of the base and second volumes (Fig. 8.16).

Figures 8.17 and 8.18 show different views of 3D superimposition that can be assessed in volume rendering and different slices to allow clinicians to evaluate and assess their treatment outcome.

8.7 Chapter Conclusions

CBCT is not only useful as a static diagnostic tool. CBCT is changing the way an orthodontist treatment plans a case and provides the orthodontist a more comprehensive outcome assessment tool. With all this additional information, we

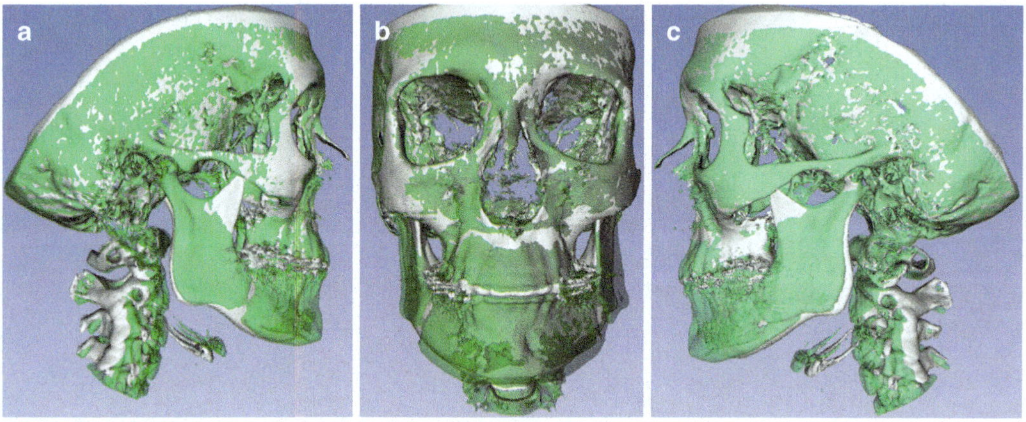

Fig. 8.16 CBCT superimposition of the same patient showing the presurgical surface in white and the postsurgical surface in semitransparent green color, showing the changes after mandibular advancement and maxillary rotation

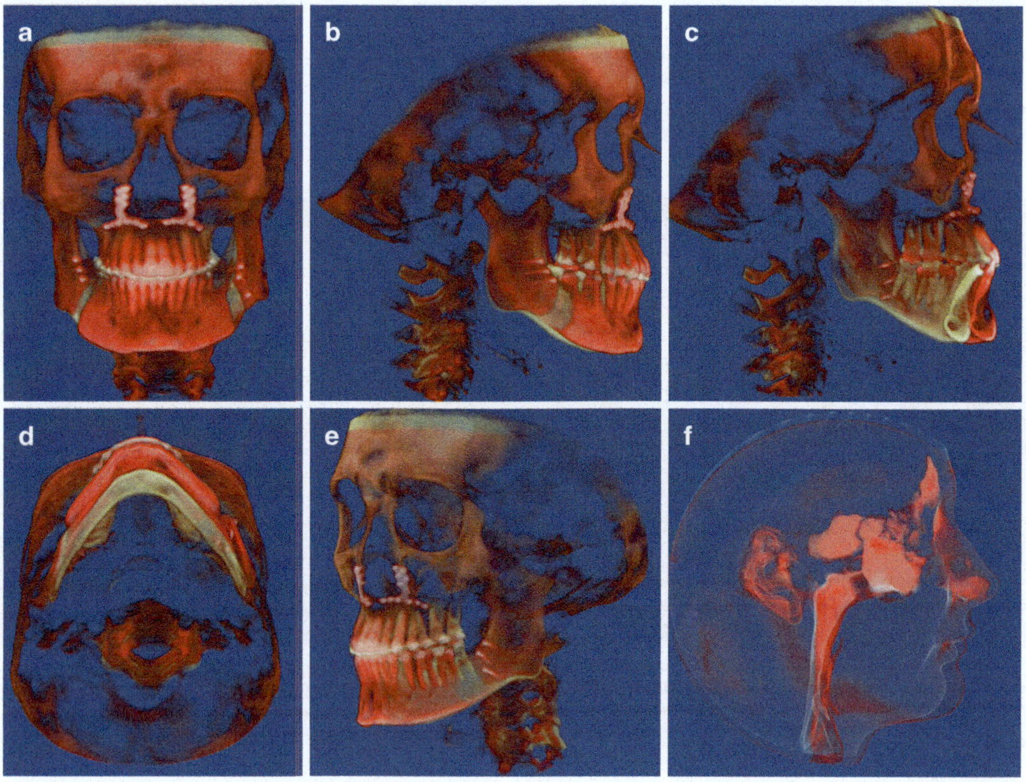

Fig. 8.17 3D superimposition allows clinicians to assess the craniofacial structures in different views. This figure shows 3D superimposed volume render of an orthognathic case (the red color represents the postsurgical volume) in (**a**) frontal view, (**b**) complete sagittal view, (**c**) clipped sagittal view, (**d**) submentovertex view, (**e**) oblique view, and (**f**) view presenting only soft tissue changes

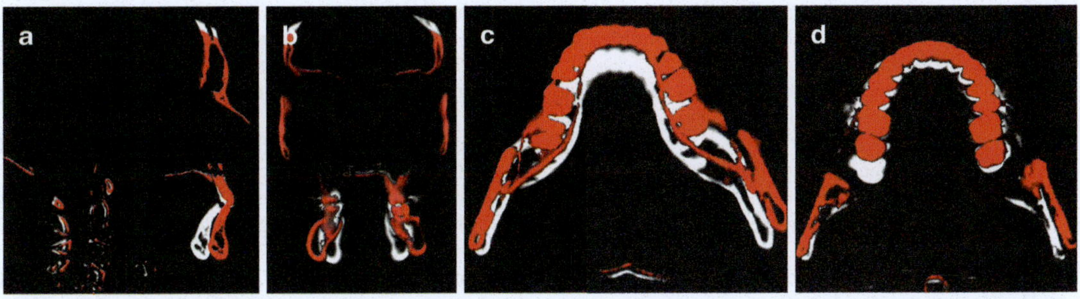

Fig. 8.18 3D superimposition of an orthognathic case (the red color represents the postsurgical volume) showing the changes after mandibular advancement and maxillary rotation in (**a**) sagittal view, (**b**) mandibular changes in coronal view, (**c**) asymmetric mandibular advancement in an axial view, and (**d**) maxillary changes in an axial view

hope that advanced imaging can help the specialty of orthodontics to further evidence-based treatment.

References

1. van Vlijmen OJ, Kuijpers MA, Berge SJ, Schols JG, Maal TJ, Breuning H, et al. Evidence supporting the use of cone-beam computed tomography in orthodontics. J Am Dent Assoc. 2012;143:241–52.
2. Palomo JM, El H, Palomo LB, Strohl KP. Upper airway, cranial morphology, and sleep apnea. In: Graber TM, Vanarsdall RL, KWL V, GJH H, editors. Orthodontics: current principles and techniques. St. Louis, MO: Elsevier; 2017. p. x, 1016.
3. Meyer W. On adenoid vegetations in the nasopharyngeal cavity: their pathology, diagnosis, and treatment. Med Chir Trans. 1870;53:191–216 191.
4. Angle EH. Malocclusion of the teeth. Philadelphia, PA: S.S. White Dental Mfg. Co.; 1907.
5. Aboudara C, Nielsen I, Huang JC, Maki K, Miller AJ, Hatcher D. Comparison of airway space with conventional lateral headfilms and 3-dimensional reconstruction from cone-beam computed tomography. Am J Orthod Dentofac Orthop. 2009;135:468–79.
6. Avrahami E, Englender M. Relation between CT axial cross-sectional area of the oropharynx and obstructive sleep apnea syndrome in adults. AJNR Am J Neuroradiol. 1995;16:135–40.
7. Ogawa T, Enciso R, Shintaku WH, Clark GT. Evaluation of cross-section airway configuration of obstructive sleep apnea. Oral Surg Oral Med Oral Pathol Oral Radiol Endod. 2007;103:102–8.
8. Mull RT. Mass estimates by computed tomography: physical density from CT numbers. AJR Am J Roentgenol. 1984;143:1101–4.
9. Mah P, Reeves TE, McDavid WD. Deriving Hounsfield units using grey levels in cone beam computed tomography. Dentomaxillofac Radiol. 2010;39:323–35.
10. Pauwels R, Araki K, Siewerdsen JH, Thongvigitmanee SS. Technical aspects of dental CBCT: state of the art. Dentomaxillofac Radiol. 2015;44:20140224.
11. El H, Palomo JM. Measuring the airway in 3 dimensions: a reliability and accuracy study. Am J Orthod Dentofac Orthop. 2010;137:S50 e51–9. discussion S50–2
12. El H, Palomo JM. Airway volume for different dentofacial skeletal patterns. Am J Orthod Dentofac Orthop. 2011;139:e511–21.
13. Sutera SP, Skalak R. The history of Poiseuille's law. Annu Rev Fluid Mech. 1993;25:1–19.
14. Hatcher DC. Cone beam computed tomography: craniofacial and airway analysis. Dent Clin N Am. 2012;56:343–57.
15. Schwab RJ, Gefter WB, Hoffman EA, Gupta KB, Pack AI. Dynamic upper airway imaging during awake respiration in normal subjects and patients with sleep disordered breathing. Am Rev Respir Dis. 1993;148:1385–400.
16. Schwab RJ, Gefter WB, Pack AI, Hoffman EA. Dynamic imaging of the upper airway during respiration in normal subjects. J Appl Physiol (1985). 1993;74:1504–14.
17. Battagel JM, Johal A, Smith AM, Kotecha B. Postural variation in oropharyngeal dimensions in subjects with sleep disordered breathing: a cephalometric study. Eur J Orthod. 2002;24:263–76.
18. Malhotra A, Pillar G, Fogel R, Beauregard J, Edwards J, White DP. Upper-airway collapsibility: measurements and sleep effects. Chest. 2001;120:156–61.
19. Bell WH. Modern practice in orthognathic and reconstructive surgery. St. Louis: W B Saunders Co; 1992.
20. Ballrick JW, Palomo JM, Ruch E, Amberman BD, Hans MG. Image distortion and spatial resolution of a commercially available cone-beam computed tomography machine. Am J Orthod Dentofac Orthop. 2008;134:573–82.
21. Brown AA, Scarfe WC, Scheetz JP, Silveira AM, Farman AG. Linear accuracy of cone beam CT derived 3D images. Angle Orthod. 2009;79:150–7.
22. Ludlow JB, Laster WS, See M, Bailey LJ, Hershey HG. Accuracy of measurements of mandibular

anatomy in cone beam computed tomography images. Oral Surg Oral Med Oral Pathol Oral Radiol Endod. 2007;103:534–42.
23. Kau CH. Creation of the virtual patient for the study of facial morphology. Facial Plast Surg Clin North Am. 2011;19:615–22. viii
24. Tanner JM, Weiner JS. The reliability of the photogrammetric method of anthropometry, with a description of a miniature camera technique. Am J Phys Anthropol. 1949;7:145–86.
25. Kau CH, Richmond S, Incrapera A, English J, Xia JJ. Three-dimensional surface acquisition systems for the study of facial morphology and their application to maxillofacial surgery. Int J Med Robot. 2007;3:97–110.
26. Oy P. Planmeca ProFace. Helsinki: Planmeca; 2017.
27. Plooij JM, Maal TJ, Haers P, Borstlap WA, Kuijpers-Jagtman AM, Berge SJ. Digital three-dimensional image fusion processes for planning and evaluating orthodontics and orthognathic surgery. A systematic review. Int J Oral Maxillofac Surg. 2011;40:341–52.
28. Cope JB. Temporary anchorage devices in orthodontics: a paradigm shift. Seminars in orthodontics. Amsterdam: Elsevier; 2005. p. 3–9.
29. Kanomi R. Mini-implant for orthodontic anchorage. J Clin Orthod. 1997;31:763–7.
30. Melsen B. Mini-implants: where are we? J Clin Orthod. 2005;39:539–47. quiz 531–2
31. Papadopoulos MA, Tarawneh F. The use of miniscrew implants for temporary skeletal anchorage in orthodontics: a comprehensive review. Oral Surg Oral Med Oral Pathol Oral Radiol Endod. 2007;103:e6–15.
32. Nucera R, Lo Giudice A, Bellocchio AM, Spinuzza P, Caprioglio A, Perillo L, et al. Bone and cortical bone thickness of mandibular buccal shelf for mini-screw insertion in adults. Angle Orthod. 2017;87(5):745–51.
33. Papadopoulos MA, Papageorgiou SN, Zogakis IP. Clinical effectiveness of orthodontic miniscrew implants: a meta-analysis. J Dent Res. 2011;90:969–76.
34. Papageorgiou SN, Zogakis IP, Papadopoulos MA. Failure rates and associated risk factors of orthodontic miniscrew implants: a meta-analysis. Am J Orthod Dentofac Orthop. 2012;142:577–95. e577
35. Landin M, Jadhav A, Yadav S, Tadinada A. A comparative study between currently used methods and small volume-cone beam tomography for surgical placement of mini implants. Angle Orthod. 2015;85:446–53.
36. Brisceno CE, Rossouw PE, Carrillo R, Spears R, Buschang PH. Healing of the roots and surrounding structures after intentional damage with miniscrew implants. Am J Orthod Dentofac Orthop. 2009;135:292–301.
37. Hembree M, Buschang PH, Carrillo R, Spears R, Rossouw PE. Effects of intentional damage of the roots and surrounding structures with miniscrew implants. Am J Orthod Dentofac Orthop. 2009;135:280–9. discussion 280–1.
38. Abbassy MA, Sabban HM, Hassan AH, Zawawi KH. Evaluation of mini-implant sites in the posterior maxilla using traditional radiographs and cone-beam computed tomography. Saudi Med J. 2015;36:1336–41.
39. Jung BA, Wehrbein H, Heuser L, Kunkel M. Vertical palatal bone dimensions on lateral cephalometry and cone-beam computed tomography: implications for palatal implant placement. Clin Oral Implants Res. 2011;22:664–8.
40. Jung BA, Wehrbein H, Wagner W, Kunkel M. Preoperative diagnostic for palatal implants: is CT or CBCT necessary? Clin Implant Dent Relat Res. 2012;14:400–5.
41. Kim SH, Kang SM, Choi YS, Kook YA, Chung KR, Huang JC. Cone-beam computed tomography evaluation of mini-implants after placement: is root proximity a major risk factor for failure? Am J Orthod Dentofac Orthop. 2010;138:264–76.
42. Shinohara A, Motoyoshi M, Uchida Y, Shimizu N. Root proximity and inclination of orthodontic mini-implants after placement: cone-beam computed tomography evaluation. Am J Orthod Dentofac Orthop. 2013;144:50–6.
43. Marquezan M, Osório A, Sant'Anna E, Souza MM, Maia L. Does bone mineral density influence the primary stability of dental implants? A systematic review. Clin Oral Implants Res. 2012;23:767–74.
44. Kapila SD, Nervina JM. CBCT in orthodontics: assessment of treatment outcomes and indications for its use. Dentomaxillofac Radiol. 2015;44:20140282.
45. Alsamak S, Psomiadis S, Gkantidis N. Positional guidelines for orthodontic mini-implant placement in the anterior alveolar region: a systematic review. Int J Oral Maxillofac Implants. 2013;28:470–9.
46. Baumgaertel S. Cortical bone thickness and bone depth of the posterior palatal alveolar process for mini-implant insertion in adults. Am J Orthod Dentofac Orthop. 2011;140:806–11.
47. Baumgaertel S, Hans MG. Assessment of infrazygomatic bone depth for mini-screw insertion. Clin Oral Implants Res. 2009;20:638–42.
48. de Rezende Barbosa GL, Ramirez-Sotelo LR, Tavora DM, Almeida SM. Comparison of median and paramedian regions for planning palatal mini-implants: a study in vivo using cone beam computed tomography. Int J Oral Maxillofac Surg. 2014;43:1265–8.
49. Fayed MM, Pazera P, Katsaros C. Optimal sites for orthodontic mini-implant placement assessed by cone beam computed tomography. Angle Orthod. 2010;80:939–51.
50. Kalra S, Tripathi T, Rai P, Kanase A. Evaluation of orthodontic mini-implant placement: a CBCT study. Prog Orthod. 2014;15:61.
51. Moslemzadeh SH, Sohrabi A, Rafighi A, Kananizadeh Y, Nourizadeh A. Evaluation of interdental spaces of the mandibular posterior area for orthodontic mini-implants with cone-beam computed tomography. J Clin Diagn Res. 2017;11:Zc09–12.

52. Pan F, Kau CH, Zhou H, Souccar N. The anatomical evaluation of the dental arches using cone beam computed tomography—an investigation of the availability of bone for placement of mini-screws. Head Face Med. 2013;9:13.
53. Watanabe H, Deguchi T, Hasegawa M, Ito M, Kim S, Takano-Yamamoto T. Orthodontic miniscrew failure rate and root proximity, insertion angle, bone contact length, and bone density. Orthod Craniofac Res. 2013;16:44–55.
54. Palomo JMVM, Hans MG. Cone beam computed tomography in orthodontics: indications, insights, and innovations. Ames, IA: John Wiley & Sons; 2014.
55. Casko JS, Vaden JL, Kokich VG, Damone J, James RD, Cangialosi TJ, et al. Objective grading system for dental casts and panoramic radiographs. Am J Orthod Dentofacial Orthop. 1998;114:589–99.
56. Sawchuk D, Currie K, Vich ML, Palomo JM, Flores-Mir C. Diagnostic methods for assessing maxillary skeletal and dental transverse deficiencies: a systematic review. Korean J Orthod. 2016;46:331–42.
57. Proffit WRSD, Fields HW. Contemporary orthodontics. St. Louis, MO: Mosby; 2007.
58. Betts NJ, Vanarsdall RL, Barber HD, Higgins-Barber K, Fonseca RJ. Diagnosis and treatment of transverse maxillary deficiency. Int J Adult Orthodon Orthognath Surg. 1995;10:75–96.
59. Enlow DHHM. Essentials of facial growth. Ann Arbor, MI: Needham Press; 2008.
60. Solow B. The dentoalveolar compensatory mechanism: background and clinical implications. Br J Orthod. 1980;7:145–61.
61. Miner RM, Al Qabandi S, Rigali PH, Will LA. Cone-beam computed tomography transverse analysis. Part I: normative data. Am J Orthod Dentofac Orthop. 2012;142:300–7.
62. Cheung G, Goonewardene MS, Islam SM, Murray K, Koong B. The validity of transverse intermaxillary analysis by traditional PA cephalometry compared with cone-beam computed tomography. Aust Orthod J. 2013;29:86–95.
63. Streit LM. CWRU's transverse analysis developing norms orthodontics. Cleveland, OH: Case Western Reserve University; 2012.
64. Shewinvanakitkul W, Hans MG, Narendran S, Palomo JM. Measuring buccolingual inclination of mandibular canines and first molars using CBCT. Orthod Craniofac Res. 2011;14:168–74.
65. Alkhatib R, Chung CH. Buccolingual inclination of first molars in untreated adults: a CBCT study. Angle Orthod. 2017;87:598–602.
66. Evangelinakis N. Changes in buccolingual inclination of mandibular canines and first molars after orthodontic treatment using CBCT. Orthodontics. Cleveland, OH: Case Western Reserve University; 2010.
67. Karamitsou E. Pretreatment buccolingual inclination of maxillary canines and first molars. Orthodontics. Cleveland, OH: Case Western Reserve University; 2011.
68. Miyamoto MJ. Changes in buccolingual inclination of maxillary canines and first molars after orthodontic treatment using CBCT. Orthodontics. Cleveland, OH: Case Western Reserve University; 2011.
69. Yehya Mostafa R, Bous RM, Hans MG, Valiathan M, Copeland GE, Palomo JM. Effects of Case Western Reserve University's transverse analysis on the quality of orthodontic treatment. Am J Orthod Dentofac Orthop. 2017;152:178–92.
70. Vanarsdall RL Jr. Transverse dimension and long-term stability. Semin Orthod. 1999;5:171–80.
71. Holly Broadbent B. A new x-ray technique and its application to orthodontia. Angle Orthod. 1931:45–66.
72. Tweed CH. Evolutionary trends in orthodontics, past, present, and future. Am J Orthod Dentofacial Orthop. 1953;39:81–108.
73. Steiner CC. Cephalometrics in orthodontics. Angle Orthod. 1959;29:8–29.
74. El H, Palomo JM. Measuring the airway in 3 dimensions: a reliability and accuracy study. Am J Orthod Dentofac Orthop. 2010;137:S50.e51–9.
75. Osorio F, Perilla M, Doyle DJ, Palomo JM. Cone beam computed tomography: an innovative tool for airway assessment. Anesth Analg. 2008;106:1803–7.
76. Palomo JM, Rao PS, Hans MG. Influence of CBCT exposure conditions on radiation dose. Oral Surg Oral Med Oral Pathol Oral Radiol Endod. 2008;105:773–82.
77. van Vlijmen OJ, Rangel FA, Berge SJ, Bronkhorst EM, Becking AG, Kuijpers-Jagtman AM. Measurements on 3D models of human skulls derived from two different cone beam CT scanners. Clin Oral Investig. 2011;15:721–7.
78. Grauer D, Cevidanes LS, Proffit WR. Working with DICOM craniofacial images. Am J Orthod Dentofac Orthop. 2009;136:460–70.
79. Cevidanes LH, Bailey LJ, Tucker GR Jr, Styner MA, Mol A, Phillips CL, et al. Superimposition of 3D cone-beam CT models of orthognathic surgery patients. Dentomaxillofac Radiol. 2005;34:369–75.
80. Cevidanes LH, Styner MA, Proffit WR. Image analysis and superimposition of 3-dimensional cone-beam computed tomography models. Am J Orthod Dentofac Orthop. 2006;129:611–8.
81. Cevidanes LH, Ruellas AC, Jomier J, Nguyen T, Pieper S, Budin F, et al. Incorporating 3-dimensional models in online articles. Am J Orthod Dentofac Orthop. 2015;147:S195–204.
82. Cevidanes LH, Heymann G, Cornelis MA, DeClerck HJ, Tulloch JF. Superimposition of 3-dimensional cone-beam computed tomography models of growing patients. Am J Orthod Dentofac Orthop. 2009;136:94–9.
83. Cevidanes LH, Bailey LJ, Tucker SF, Styner MA, Mol A, Phillips CL, et al. Three-dimensional cone-beam computed tomography for assessment of mandibular changes after orthognathic surgery. Am J Orthod Dentofac Orthop. 2007;131:44–50.
84. Heymann GC, Cevidanes L, Cornelis M, De Clerck HJ, Tulloch JF. Three-dimensional analysis of maxillary

protraction with intermaxillary elastics to miniplates. Am J Orthod Dentofac Orthop. 2010;137:274–84.
85. Cevidanes LH, Hajati AK, Paniagua B, Lim PF, Walker DG, Palconet G, et al. Quantification of condylar resorption in temporomandibular joint osteoarthritis. Oral Surg Oral Med Oral Pathol Oral Radiol Endod. 2010;110:110–7.
86. Bazina M, Cevidanes L, Ruellas A, Valiathan M, Quereshy F, Syed A, Wu R, Palomo JM. Precision and reliability of Dolphin 3-dimensional voxel-based superimposition. Am J Orthod Dentofac Orthop. 2018 Apr;153(4):599–606.
87. BenNasir E, Bazina M, Cevidanes L, Amberman BD, Palomo JM. Accuracy and Reliability of Dolphin 3D voxel based superimposition in growing patients. Cleveland, OH: Case Western Reserve University; 2017.
88. Weissheimer A, Menezes LM, Koerich L, Pham J, Cevidanes LH. Fast three-dimensional superimposition of cone beam computed tomography for orthopaedics and orthognathic surgery evaluation. Int J Oral Maxillofac Surg. 2015;44:1188–96.
89. Elshebiny T, Bennasir E, Palomo JM. Comparison of two fast three dimensional voxel based superimposition software programs. Cleveland, OH: Case Western Reserve University; 2017.

Part III

Clinical Applications in Oral Surgery and Orthodontics

Temporomandibular Joint Morphology and Orthognathic Surgery

9

Ryan Patrick O'Sullivan, Onur Kadioglu, G. Fräns Currier, Steven M. Sullivan, Tara Beard, and Dee Wu

Abstract

Orthognathic surgery is one of the most important treatment options available to an orthodontist, especially in cases of significant skeletal malocclusions. In order to accurately measure any changes that may have occurred following orthognathic surgery, data regarding shape changes occurring in the temporomandibular joint is necessary. CBCT imaging can provide more complete three-dimensional information related to these changes than has been possible in the past. The use of three-dimensional CBCT data has allowed for a more complete analysis of changes that have taken place in the TMJ with orthognathic surgery, with changes primarily taking place bilaterally in posterior, superior, and lateral aspects of the condylar heads and posterior regions of the fossae.

R. P. O'Sullivan
Private Practice, Fayetteville, AR, USA

O. Kadioglu (✉) · G. F. Currier
Division of Orthodontics, Department of Developmental Sciences, University of Oklahoma Health Sciences Center College of Dentistry, Oklahoma City, OK, USA
e-mail: onur-kadioglu@ouhsc.edu; frans-currier@ouhsc.edu

S. M. Sullivan
Deparment of Surgical Sciences, University of Oklahoma Health Sciences Center College of Dentistry, Oklahoma City, OK, USA
e-mail: Steven-Sullivan@ouhsc.edu

T. Beard
University of Oklahoma Health Sciences Center College of Dentistry, Oklahoma City, OK, USA
e-mail: tara-beard@ouhsc.edu

D. Wu
Department of Radiological Sciences, University of Oklahoma Health Sciences Center College of Medicine, Oklahoma City, OK, USA
e-mail: dee-wu@ouhsc.edu

9.1 Introduction

Orthognathic surgery is one of the most important treatment options available to an orthodontist, especially in cases of severe skeletal malocclusion where conventional orthodontics will not be able to achieve an acceptable result. In these cases, the patient's occlusion, function, and esthetics are dependent on a combination of orthodontics and orthognathic surgery, but there is still uncertainty about relapse and changes that may occur in the temporomandibular joint following surgery.

Orthognathic surgery originated in the United States in 1849 when Hullihen's procedure was reported to be the first successful operation to correct a malocclusion [1]. However, the combination of orthodontics and orthognathic surgery was pioneered in St. Louis by orthodontist Edward Angle and surgeon Vilray Blair, who reported the first successful mandibular subcon-

O. Kadioglu, G. F. Currier (eds.), *Craniofacial 3D Imaging*,
https://doi.org/10.1007/978-3-030-00722-5_9

dylar vertical ramus osteotomy for mandibular prognathia in 1898 [1]. From these early pioneers, the field has continuously advanced with new procedures to correct a variety of different malocclusions and adverse facial patterns.

Today, orthognathic surgery remains an acceptable and effective way to treat a spectrum of skeletal and dental irregularities to improve speaking, breathing, and occlusion [2, 3]. Although techniques have improved over the years, postsurgical relapse is still a major concern and varies greatly depending on the type of surgery performed [2, 4]. Two-jaw orthognathic surgery has been shown to be a stable and reliable means for the correction of skeletal discrepancies occurring in both arches. However, successful two-jaw orthognathic surgery is dependent on the movements being stable, the use of rigid fixation, and completion of skeletal growth of the patient [2].

Clinically, two-jaw surgeries to correct Class II or III malocclusion have shown excellent results in nearly 90% of patients when rigid fixation was used [5]. After the first year, however, a different pattern appeared. For example, approximately 20% of patients who had undergone mandibular advancement showed decreases in mandibular length 1–5 years after surgery [3]. Initial changes in mandibular stability have been attributed to the musculature, while longer-term stability problems have been attributed to remodeling of the condyles [3]. There are multiple risk factors for relapse and lack of stability following orthognathic surgery, including the amount of the advancement or the setback, counterclockwise rotation of the mandible, the presence of a high mandibular plane angle, fixation devices, preexisting pathologic conditions of the TMJ, gender of the patient, and skill of the surgeon [6–9].

Orthognathic surgery may lead to changes in condylar position during surgery, which may lead to adaptive changes in the temporomandibular joint, including remodeling of both the condyle and the fossa [6]. In order to accurately measure any changes that may have occurred following orthognathic surgery, accurate information regarding the positional and shape changes occurring in the temporomandibular joint is needed. CBCT imaging can provide more complete three-dimensional data related to these changes than has previously been possible. It is important that both orthodontists and oral surgeons fully understand the changes that may occur following orthognathic surgery so that patients can be fully informed of the risks and benefits of such a significant procedure before agreeing to undergo orthognathic surgery.

9.2 Effects of Orthognathic Surgery on the Temporomandibular Joint

9.2.1 Condylar Position

When repositioning the mandible during a mandibular setback or advancement, it initially needs to be segmented. Currently, the most commonly used procedure is the bilateral sagittal split osteotomy, or BSSO [10]. In this procedure, the mandible is segmented into an anterior portion and two posterior portions which are then repositioned and fixed according to whether a mandibular advancement or a setback is needed. This procedure is often performed in conjunction with a maxillary osteotomy [2]. Mandibular advancement procedures are considered to be predictable and very stable, while mandibular setbacks are less predictable and stable [3]. Regardless of the procedure, some positional changes of the condyle are expected to occur, leading to concerns that these positional changes may predispose the patient to relapse or TMJ issues following surgery.

Numerous studies have evaluated positional changes of the condyles after performing BSSO procedures. These studies have shown that condylar displacement may occur during or immediately after surgery [11, 12].

One of the most common positional changes that occurs after surgery is postoperative condylar displacement or sag [10, 13]. Noncontact condylar sag occurs when condyles are displaced inferiorly or anterior-interiorly, which does not allow B-point to be held, leading to relapse [6].

It has also been shown that when displacements occur, they most commonly occur after the sagittal split and are a combination of displacement and rotation or tilting of the axes of the condyles [14]. During sagittal split osteotomy procedures using rigid fixation, the condyle may be torqued if interferences are not relieved, which may lead to later instability [8, 13, 15, 16].

Freihofer and Petrešević [17] in 1975 followed 38 patients for at least 2 years following BSSO procedures and found that the most common displacement was anterior in the fossa. Marmulla and Mühling in 2007 found that the median malposition of the condyles during surgery was 2.4 mm without any computer-guided assistance to the surgeon. Due to this displacement, TMJ pain and dysfunction can develop following surgery, along with short-term relapse that could occur because of the intrinsic tendency of the condyle to return to its original position in the fossa [16].

9.2.2 Remodeling

Remodeling has also long been identified as a reason for instability and relapse after orthognathic surgery. Condylar resorption is the primary form of remodeling that has been examined in the literature. There are a variety of risk factors for condylar resorption including mandibular deficiency with high mandibular plane angles, counterclockwise rotation of the mandible during surgery, the gender of the patient, fixation type, compression of the condyles, and presence of an anterior open bite [7, 9, 18–24]. Condylar resorption and its risk factors should be taken into account prior to planning orthognathic surgery [25].

The shape of the condylar head has also been identified as a cause for relapse due to resorption. Hwang et al. in 2004 found that patients with preexisting mandibular hypoplasia, posteriorly inclined condylar necks, high mandibular plane angles, and short posterior face heights were at the highest risk for condylar resorption. Moore et al. in 1991 also found that condylar resorption occurred more in young women with condyles that tended to be slender and have a posterior incline, which may give them less ability to adapt to the additional load that often occurs during oral surgery [7].

Mobarak et al. in 2001 explained that counterclockwise rotation of the proximal segment leads to instability and relapse due to altered muscle orientation. If compression of the condyles and retrodiscal tissues occurs during surgery, remodeling will occur. If the compression occurs against the posterior wall of the glenoid fossa, the condyles will seat superiorly as the remodeling occurs, which will cause horizontal relapse. If the compression occurs against the medial or lateral walls of the glenoid fossae, similar reseating of the condyles can occur [20].

Arnett in 1990 explained that condylar sag can also occur if the condyles are displaced posteriorly, medially, or laterally. In such cases, the condyles can be seated inferiorly while still contacting the fossae, thus supporting B-point. However, this compression of the condyles and retrodiscal tissues can cause remodeling to occur beginning from 9 to 18 months following surgery. As the condyles remodel, they will begin to seat more superiorly, leading to additional relapse [6].

Kerstens et al. in 1990 found that the most common factor contributing to condylar atrophy following orthognathic surgery was a high mandibular plane along with a retrognathic mandible. They recommended avoiding rotational movements in these cases to prevent atrophy. Similarly, Bouwman et al. in 1994 evaluated condylar resorption following orthognathic surgery and found that increased loading occurred in cases with larger advancements or when counterclockwise rotations of the mandibular plane occurred [18].

Hoppenreijs et al. in 1998 found similar risk factors for condylar resorption as previously mentioned. Their study found that resorption primarily occurred at the anterior portion of the condyles. The remodeling continued beyond the first postoperative year, particularly in the superior and anterior regions of the condyles. Additionally, they found that rigid internal fixation led to a lower incidence of progressive condylar resorption but a higher incidence of remodeling [9].

Studies by Merkx and Van Damme in 1994 and Huang et al. in 1997 both found that condylar resorption was much more likely to be found following mandibular surgery as opposed to maxillary surgery alone [26].

Joss and Vassalli in 2009 found that condylar distraction occurred if the condyles were positioned inferiorly or anteriorly in the glenoid fossae and was unable to support the new mandibular position set by the surgeon, leading to relapse. They found that condylar resorption could occur if the condyles were pushed posteriorly into the fossae or torqued during fixation. These localized forces could lead to condylar resorption [8].

9.3 Methods of Evaluating TMJ Changes

9.3.1 Conventional Imaging Modalities

The vast majority of previous studies evaluating the effects of orthognathic surgery on the TMJ have done so using conventional radiographs, computed tomography, and magnetic resonance imaging. Conventional radiographs have included linear tomograms, submentovertex radiographs, lateral oblique radiographs, lateral cephalograms, and PA cephalograms [11, 14]. Most of these studies have used linear or angular measurements to evaluate TMJ changes. While these radiographs were the best technology available at the time, they are limited due to their technique sensitivity, magnification error, and distortion error [27].

Additionally, Moyers and Bookstein in 1979 argued that cephalometric conventions had little basis in biology or biometrics. They are based on landmarks and straight lines, so they cannot truly capture form. They argued that future conventions need to include tangents, curvatures, and biorthogonal grids. Many of these limitations have since been overcome with the advent of medical CT imaging, MRI, and cone beam computed tomography, although each of these still have limitations [28].

Kundert and Hadjianghelou in 1980 used posterior-anterior cephalograms, linear tomograms, and lateral oblique radiographs to analyze condylar position and found that condylar displacements were frequently found following sagittal split osteotomies. They reached the conclusion that the majority of the displacements were due to rotation and tilting of the condylar neck during surgery [14]. Spitzer et al. in 1984 reached similar conclusions in their study using computed tomography, finding that rotational movements of the condyles were the most common malpositions following surgery [29].

Sund et al. in 1983 used axial, frontal, and lateral radiographs to identify changes in the TMJ after oblique sliding osteotomies. They determined that the condyles were displaced in an anteroinferior direction, that a superior rotation of the lateral portion of the condyle occurred, and that the condyles rotated in both anterior and posterior directions. However, they also found that after 18 months, the TMJ often normalized [30].

Will et al. in 1984 used submentovertex radiographs, lateral cephalograms, and left and right TMJ tomograms to look at condylar position. During surgery, they found no significant rotational changes or changes in the anteroposterior position of the condyles. Counterclockwise inclination and inferior movement of the condyles were seen but were not statistically significant. During fixation, both the condyles moved superiorly and the left condyle having additional posterior movement. After fixation and removal of the splint, no significant movement was noted. Overall, the right condyle showed significant superior movement, and the left condyle showed significant counterclockwise rotation and posterior displacement. They noted no significant overall differences in displacements of the condyles with regard to the amount of movement taking place during surgery and found very little movement had been observed overall. Differences between the left and right condyles were attributed to the positioning of the surgeon during the procedure or due to more edema occurring in whichever side was split first [11].

Woodside et al. in 1987 looked into remodeling changes of the condyle and glenoid fossa in primates during Herbst treatment using cephalometric analysis and histologic assessment of the TMJ. They found that a large volume of new bone was formed along the anterior border of the postglenoid spine and resorption along the posterior border of the postglenoid spine, indicating an overall anterior remodeling of the fossa [31].

Hackney et al. in 1989 looked at intercondylar width and angular changes after BSSO with rigid fixation and found no significant differences before and after the surgery. They also found that symptoms of TMD did not increase following surgery, indicating that the TMJ was able to adapt to any displacement that did occur during surgery [32].

Rotskoff et al. in 1991 found displacement immediately after surgery in the sagittal, transverse, and long axes of the condyles when evaluating linear full-head tomograms and submentovertex radiographs. They also compared lateral cephalograms to assess superoinferior and anteroposterior condylar displacements [16].

Stroster and Pangrazio-Kulbersh in 1993 studied condylar position in patients undergoing BSSO for mandibular advancement using submentovertex and transcranial radiographs. They found no correlation between the amount of advancement and condylar displacement and concluded that rigid internal fixation results in a greater degree of condylar displacement than wire fixation [33].

Cutbirth et al. in 1998 evaluated panoramic radiographs and cephalometric tracings to determine correlation between amount of mandibular advancement and condylar changes. They found that condylar resorption usually occurred unilaterally on patients with previous TMJ symptoms as well as those who had underwent large mandibular advancements [34].

Kawamata et al. in 1998 used computed tomography to evaluate condylar displacement after BSSO and found inward rotation of the condylar long axis. Alder et al. in 1999 also used computed tomography to evaluate changes in condylar position associated with rigid fixation. They found a variety of changes in condylar position including displacements in all directions, condyle angle changes, and rotations of the proximal segments. The most common displacements noted were posterior and superior in 67% and 60% of the patients, respectively. Overall, the most frequently seen condylar positional changes following surgery were more lateral, with increased angle, higher coronoid processes, and more superior and posterior positions of the condyles in the fossae [35].

Ruf and Pancherz in 1999 analyzed the adaptive mechanisms of the TMJ using magnetic resonance imaging in patients being treated with the Herbst appliance. They took MRIs before treatment when the appliance was placed, 6–12 weeks after the appliance delivery, and at the end of treatment, with remodeling analyzed visually. They found that remodeling of the condyles and glenoid fossae during Herbst treatment contributed to the increase in mandibular prognathism. They also concluded that MRI was an excellent method to visualize the growth and remodeling of the TMJ [36].

Hu et al. in 2000 studied changes in TMJ function and condylar position following mandibular setback surgeries with rigid fixation. They used lateral oblique radiographs 6 months before and after surgery and found an anteroinferior displacement of the condyle after oblique ramus osteotomy. The displacement was attributed to intracapsular edema, manipulation of the segments, and the alteration to the direction of pull of the lateral pterygoid muscles and pterygomasseteric sling [37].

Voudouris et al. in 2003 looked at condyle-fossa changes that occurred in primates during Herbst treatment using EMG, computerized histomorphometry, tetracycline staining, and cephalometry using the Björk implant method. They determined that the glenoid fossa normally grows downward and backward in untreated primates. They found that Herbst treatment caused significant 1.2 mm average bone formation in the fossa over 12 weeks in a downward and forward direction [38].

Katsumata et al. in 2006 compared computed tomography, conventional radiographs, and MR imaging as a method for evaluation of condylar remodeling after mandibular setback surgery.

They found that after setback, a bone layer formed on the posterior medial aspect of the condyle, usually after 6 months or more post-surgery. New bone formation usually occurred in cases where the condylar head underwent an anterior-inferior displacement. Also noted was a limitation of the CT scan, in that it uses a fan-shaped beam to acquire axial slices, so structures parallel to the beam tended to be deficient. Therefore, remodeling on the superior aspect of the condylar heads would be more difficult to detect. Overall, all imaging modalities showed high agreement on remodeling, and a correlation was found between the degree of rotation of the condylar long axes and remodeling [39].

Cortez and Passeri in 2007 analyzed condylar position after Le Fort I osteotomies using submentovertex radiographs and tomographic images and found that the procedure did not cause any significant changes to condylar position following surgery [40].

Ueki et al. in 2012 evaluated changes to the TMJ and ramus after sagittal split ramus osteotomies in patients with and without asymmetrical development using magnetic resonance imaging and CT scans. Their study found that the anterior joint space was significantly larger postoperatively in all groups, indicating posterior displacement during surgery [41].

Han and Hwang [15] looked at condylar displacement using computed tomography scans and found that immediately after surgery, the condyles were displaced laterally but did not find any significant anterior-posterior or superior-inferior movement. Additionally, they found that in the postoperative period, the condyles tended to move in a medial and superior direction back toward its original position.

9.3.2 Evaluating TMJ Changes with CBCT

History of CBCT

Cone beam computed tomography was commercially introduced in 1998 in Europe and in the United States in 2001 with the NewTom 9000 [42–44]. In 2004, there were four primary CBCT machines in use; by 2008, there were 16 companies making 23 different machines; and by 2013, 20 companies were producing 47 different machines [44]. It has since been used throughout the dental field for implant placement, pain diagnosis, visualization of impacted teeth, fractures, TMD, and orthognathic surgery [44]. During a scan, the x-ray source emitting a cone-shaped beam and sensor rotate 360 degrees around the object of interest, capturing multiple images. Scan times are variable depending on the machine but generally range from 5 to 40 seconds [42, 44]. After image acquisition, an algorithm generates a viewable three-dimensional image.

CBCT imaging has a variety of advantages compared to conventional radiographs as well as medical CT and MRI. Lateral cephalograms have limitations such as the overlap of structures, magnification differences between the left and right sides, and distortion, all of which are mostly mitigated through the use of the CBCT [45, 46]. Additionally, CBCT allows for rapid volumetric data acquisition while needing a significantly smaller radiation dose than fan-beam CT to obtain a high-quality image [47, 48].

While CBCT imaging has significant advantages over many other imaging modalities, it also has important drawbacks. The machines are costly with a relatively high amount of radiation as problematic [42, 44, 46]. While the radiation doses have decreased over the years, the clinician must still find a balance between diagnostic quality and radiation dose [42, 44, 48]. CBCT images also tend to present with scatter, noise, truncated view artifacting, and artifacts from beam hardening [43, 48, 49]. Additionally, any motion distortion will affect the entire image due to the nature of the beam [49]. Finally, CBCT imaging cannot be used to estimate bone density due to distortion of Hounsfield units that are normally used to determine bone density [49].

Accuracy of CBCT

Multiple studies have examined the accuracy of CBCT imaging and have come to the consensus that it is highly accurate. Mozzo et al. in 1998 found good accuracy and a 1:1 representation of the volume scanned [43]. Similarly, Kobayashi et al. in 2004 found CBCT images to be accurate and useful due to their high resolution and field size [50].

Lascala et al. in 2004 examined the accuracy of linear measurements in CBCT imaging. To do this, CBCT images of dry skulls were obtained and measurements compared to actual measurements taken on the skulls using high-precision digital calipers. They found that the CBCT images underestimated the real distances but that the differences were only significant at the base of the skull. Therefore, they found that CBCT images were reliable for measuring structures commonly associated with dental and maxillofacial imaging [47].

Hilgers et al. in 2005 looked at the accuracy of linear TMJ measurements made on CBCT images as compared to conventional cephalometric radiographs. Dry skulls were used in order to take accurate measurements with digital calipers. Their findings showed that CBCT images were highly accurate when compared to direct measurements, whereas conventional radiographs, on the other hand, were much more inconsistent even when calibrated. The CBCT images allowed measurements to be taken on a 2D plane without the superimposition of structures and noise that often accompany lateral cephalograms [51].

Honey et al. in 2007 compared the accuracy of observers viewing images from CBCT, panoramic radiographs, and linear tomography to detect erosion of the condylar head. Greater intraobserver reliability was found with the CBCT images as compared to the other radiographs. Additionally, they concluded that CBCT imaging was more accurate in detecting condylar defects than conventional radiographs and that the CBCT images were more accurate and reliable in detecting condylar erosion than TMJ panoramic projections or angle linear tomography [52].

Zhang et al. in 2012 used dry skulls to measure joint spaces using CBCT imaging and found that CBCT images were accurate in measuring temporomandibular joint spaces [53].

Findings from CBCT Studies

Studies using CBCT images to evaluate various aspects of the TMJ have been undertaken since the early 2000s. Due to CBCT imaging being a relatively new technology, various studies have used different methods for examining changes in the TMJ. For example, Cevidanes et al. in 2005 evaluated CBCT images before and after maxillary orthognathic surgery to assess mandibular anatomy and position. In order to visualize changes in condylar position, they used 3D color-coded mapping [54].

Cevidanes et al. again used similar color mapping techniques in a 2007 study looking at condylar changes after orthognathic surgery. Only small condylar displacements were found for both one- and two-jaw surgeries. Small posterior and lateral displacements were noted following surgery. None of the changes noted were found to be significant [55].

Ikeda and Kawamura [56] sought to determine optimal condylar position using CBCT. They used linear measurements to determine condylar position. No gender differences were noted in optimal condylar position, and they found that optimal condylar position had an anterior to superior to posterior joint space ratio of 1.0 to 1.9 to 1.6.

Kim et al. in 2010 studied short- and long-term changes in condylar position following orthognathic surgery using CBCT imaging. They found that the condyles started in a more anterior position in the glenoid fossae before surgery, moved to a concentric position immediately after surgery, and then tended to return to their original positions after surgery [57].

Kim et al. in a later study in 2011 used CBCT imaging to evaluate changes in condylar position after two-jaw orthognathic surgery. They found no significant skeletal changes but significant differences in axial condylar angles and anteroposterior condylar position. Furthermore the condylar axes were rotated inward after surgery [58].

Motta et al. in 2011 evaluated positional changes to the condyles, mandibular rami, and chin 1 year after surgery using CBCT image superimposition and 3D color mapping to visualize changes. Their findings included lateral displacement of the condyles in 35 of 54 patients. Patients with larger displacements after surgery had significant condylar adaptations to the displacement. They also found an association between the position of the chin and the postoperative adaptation of the condyles and rami. Overall, 17% of patients had condylar displacements greater than 2 mm remaining 1 year after surgery [59].

De Clerck et al. in 2012 evaluated changes to the mandible and glenoid fossa in skeletal Class III patients treated with bone-anchored intermaxillary traction. CBCT scans were superimposed on the anterior cranial base along with color mapping to visualize changes. They found that the use of CBCT imaging allowed better visualization of changes to the condyles and glenoid fossae and found a high correlation between modeling of the anterior and posterior eminences of the glenoid fossae and displacement of the condyles [60].

Park et al. in 2012 used CBCT volumetric superimposition to evaluate remodeling of condylar heads following bimaxillary orthognathic surgery and found that condylar heights decreased after surgery. Resorption occurred primarily in the anterior, posterior, superior, and lateral portions of the condylar heads. They also found bone formation occurring primarily on the anteromedial portion of the condylar heads [61].

Chen et al. in 2013 evaluated short- and long-term changes in condylar positions using CBCT imaging. No significant differences were detected between the left and right condyles. Immediately after surgery, the condyles tended to move inferoposteriorly, which then changed to an anterosuperior movement 3 months later. The final condylar positions after surgery were posterosuperior as compared to the presurgical position [46].

LeCornu et al. in 2013 aimed to analyze skeletal changes using CBCT imaging in subjects with treated with the Herbst appliance versus control patients treated with elastics. All images were taken before treatment, then after Herbst removal or after treatment in the control group. The CBCT images were superimposed on the anterior cranial base, and changes were visualized using 3D color mapping. The study found anterior displacement of the glenoid fossae and condyles in the Herbst patients. The control patients, however, showed posterior displacement of the condyles and fossae [62].

Chen et al. in 2015 used CBCT superimposition to evaluate remodeling of the condyles before and after mandibular advancement. Their study found that remodeling of the condylar heads primarily occurred as resorption on the posterior of the condyles and apposition on the anterior surfaces of the condyles [63].

Xi et al. in 2015 analyzed volumetric changes of the condyles using CBCT imaging following orthognathic surgery. They used C-point, which is the most caudal point of the sigmoid notch as a landmark to define condylar volume. They found a correlation between skeletal relapse and decreased condylar volume, indicating that resorption had occurred following surgery and may have contributed to skeletal relapse following orthognathic surgery [64].

Beginning in 2015, in an effort to more effectively measure changes to the condyles and fossae, studies evaluating the TMJ were published using a new method of condylar mapping using Stratovan Checkpoint software (Stratovan Corporation, Davis, CA). The software allows for the condyles and fossae to be evaluated individually for morphological changes such as resorption and apposition or study the TMJ as a unit to evaluate positional changes.

Ikeda, in her 2014 dissertation, and Ikeda et al. in a 2016 article based on the dissertation established the use of checkpoint to evaluate the TMJ. Their study determined the optimal patch density used to accurately map condylar morphology was 11 × 11. Lower density did not show minor changes in morphology, as well as higher densities not adding significant information. They also found that Checkpoint was a reliable method of evaluating TMJ form and joint space [65, 66].

Contro in 2015 used Stratovan Checkpoint to study condylar head morphology using CBCT images, with the goal of seeing if there were morphological differences of condylar form based on the skeletal pattern of the subjects (dolichofacial, mesofacial, and brachyfacial). He found

Checkpoint to be a reliable method of mapping condylar morphology with a coefficient of variation of 1.81%, as well as finding morphological differences between skeletal classes [67].

None of these previous studies compared data collected with Stratovan Checkpoint software to evaluate condylar head and fossae remodeling before and after orthognathic surgery. The software has been shown to be accurate and reliable for mapping the TMJ, so comparing the data collected within, the software can provide valuable information to orthodontists and surgeons on some of the effects of orthognathic surgery on condyles and fossae.

9.4 Example Study

9.4.1 Materials and Methods

This study began with a selection of 69 patients treated at the University of Oklahoma Department of Oral and Maxillofacial Surgery with the following inclusion criteria: (1) initial skeletal Class II or Class III malocclusion; (2) orthognathic jaw surgery in conjunction with orthodontic treatment; (3) CBCTs at time points prior to surgery, immediately after surgery, and at least 3 months after surgery; (4) CBCTs of adequate size and quality to fully and accurately visualize entire TMJ in Stratovan Checkpoint software; and (5) CBCTs taken on the same machine with the same settings and technique. Subjects were excluded from the study if they met any of the following exclusion criteria: (1) history of TMD or joint dysfunction, (2) craniofacial deformities, (3) CBCTs with subject not in maximum intercuspation, (4) patients undergoing treatment with a Class II functional appliance, (5) poor quality CBCTs, and (6) inadequate time between CBCTs. Following exclusion of subjects that did not meet the criteria for the study, 31 subjects remained, with 22 female and 9 male subjects. There were 13 Class II (11 female, 2 male) and 18 Class III (14 female, 4 male) individuals.

Stratovan Checkpoint was used in order to plot the morphology of the TMJ; and a modified protocol was developed based on Ikeda [65].

Step 1: Import DICOM file and Crop Image
The CBCT DICOM file was imported into Checkpoint software, and the window and level settings were adjusted to best visualize the TMJ, which was then cropped to include the most caudal point of the sigmoid notch as the inferior border, just anterior to the most caudal point of the sigmoid notch as the anterior border, the posterior border of the external auditory meatus, and the entire roof of the glenoid fossa as the superior border (Fig. 9.1).

The TMJ file was then exported as a NIfTI file, which allowed for a smaller file size and eliminated unnecessary data. With less data included in the image file, less computer resources were needed, allowing for faster processing of the image.

Step 2: Import TMJ, Adjust, and Orient
The TMJ file was loaded back into Checkpoint, and the level and width settings were again adjusted to best visualize the bony outline of the TMJ while minimizing noise. The isosurface was then adjusted using the histogram on the right side of the screen to find the best representation of the cortical outlines of the articular surface (Fig. 9.2).

The TMJ was then oriented following the protocol developed by Ikeda et al. [66].

The axes were then adjusted beginning with moving the avatar to the widest part of the condylar head mediolaterally in the coronal slice. The sagittal axis was then rotated so that it passed through the long axis of the condyle (Fig. 9.3).

Step 3: Determine Boundaries of the Condylar Head
Next, it was necessary to determine the boundaries of the condylar head. Here, the protocol from Ikeda et al. was modified in order to include the entirety of the condylar head rather that only the superior portion. First, a landmark was added at the most caudal point of the sigmoid notch or C-point [64]. A second landmark was then added and moved to the most posterior point of the angle of the condyle using the axial window in order to keep it in the same

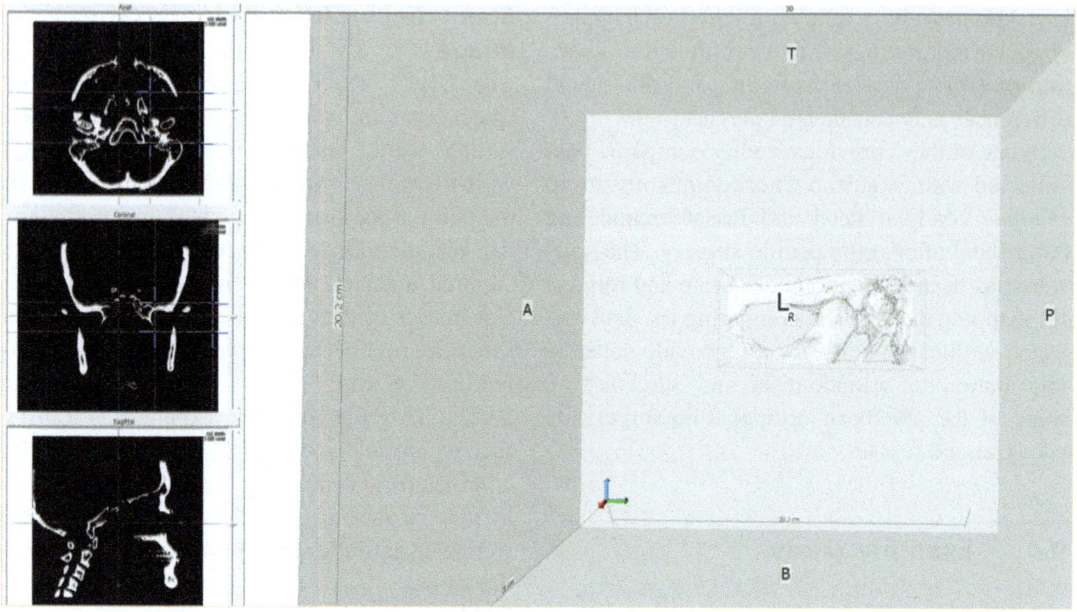

Fig. 9.1 Import DICOM file and Crop image

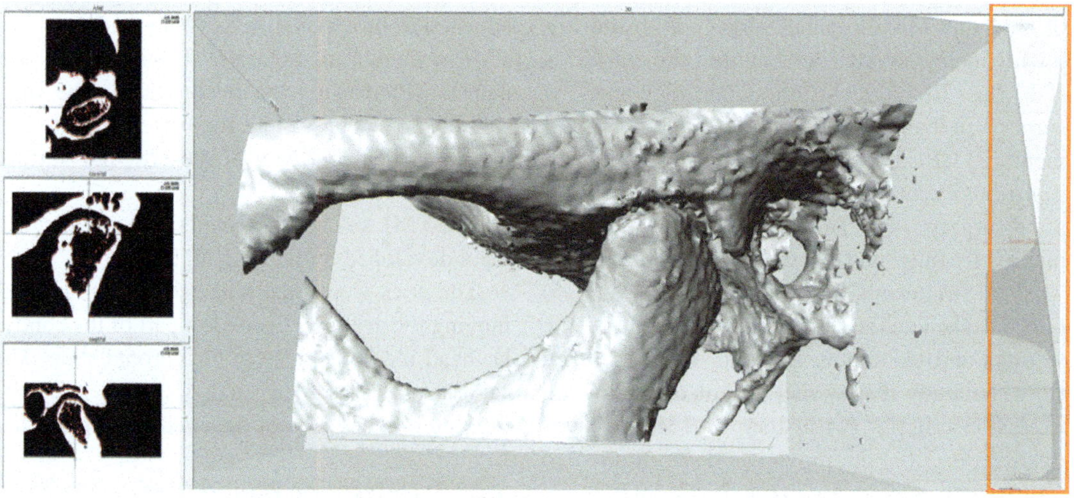

Fig. 9.2 Note histogram at right side outlined in red

plane as the first landmark. Then, a third landmark was added and moved to the most superior point of the condylar head using the sagittal window to again maintain the points' orientation in the same plane. From there, the angle measurement tool was used to move the third landmark in the sagittal window until a 90° angle was formed. Using the linear measurement tool, the distance between points two and three was determined, and the third landmark was moved inferiorly to one half the distance found, while maintaining the 90° angle. This point was then moved perpendicularly to the lateral aspect of the condylar neck using the axial window, which determined the inferior border of the joint overlay (Fig. 9.4).

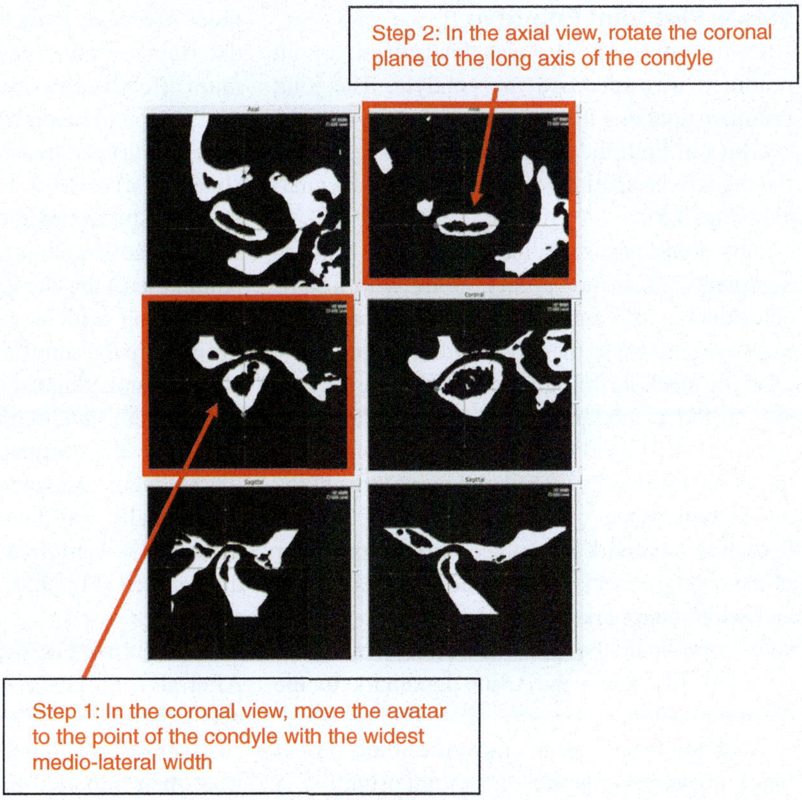

Fig. 9.3 Orientation of the TMJ

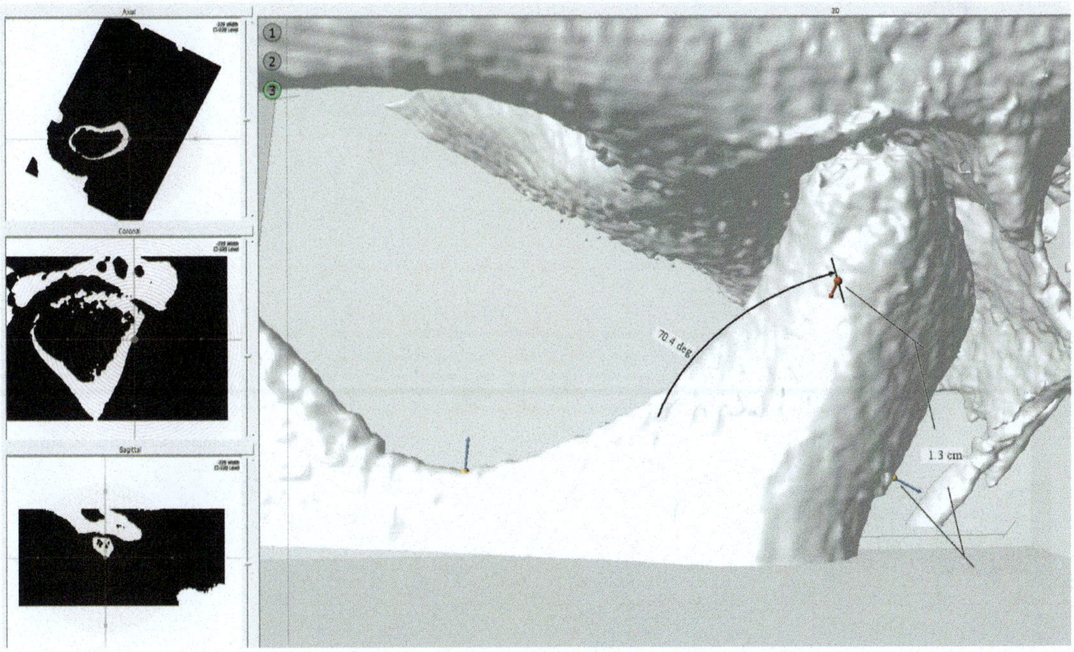

Fig. 9.4 Example of condyle after inferior boundary of the condyle was determined

Step 4: Plot Joint Primitive

After the inferior border was determined, a joint primitive was added to the condyle. The joint primitive contains the data points for the 13 × 13 overlays of both the condyles and fossae, allowing them to be spatially related in later data analysis (Fig. 9.5).

The joint primitive includes three primary landmarks, or anchor points, along with a variable number of semi-landmarks. The red landmark was moved to the most medial aspect of the condylar neck in the same plane as inferior border of the condylar head that was previously determined. The yellow landmark was placed on top of the previously mentioned point three at the most lateral aspect of the condylar neck. Finally, the white landmark was placed on the posterior of the condylar neck in the same plane as the red and white landmarks using the "snap to sagittal slice" option in the joint primitive window to automatically move the white landmark to the desired location.

With the anchor points moved to their proper place, it was possible to plot the joint primitive. A field size of at least 9 × 9 semi-landmarks was determined in the previous study to be the minimum effective density, but 11 × 11 field size was determined to be optimal [66]. However, because a larger surface area was being measured in this study, a larger 13 × 13 semi-landmark field size was used to ensure adequate semi-landmark density. The software automatically placed the semi-landmarks at the chosen density over the condylar head along with a matching set for the fossa. From there the semi-landmarks were individually checked and adjusted as needed (Fig. 9.6).

Any semi-landmarks that were not accurately placed on the cortical outline of the condyle or fossa were moved by hand as needed. Additionally, any landmarks that fell outside of the fossa or condyle were marked as missing and disregarded (Fig. 9.7).

Step 5: Export and Statistical Analysis

After all semi-landmarks were adjusted, the data was exported in CSV format for each TMJ for each subject at all three timepoints. These datasets were then imported into MATLAB (Mathworks,

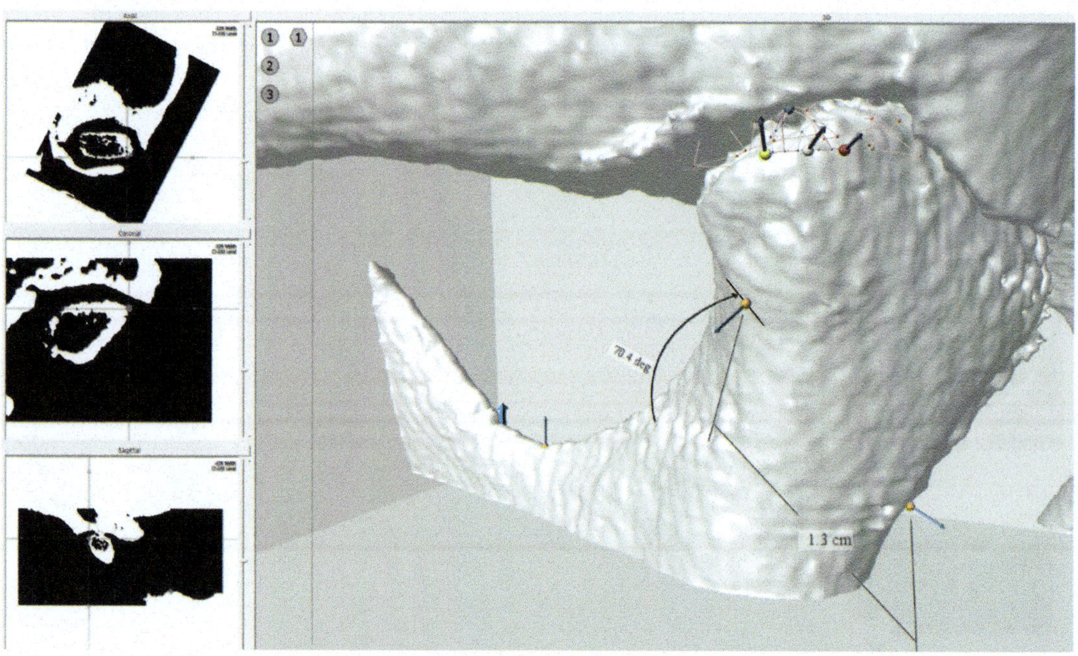

Fig. 9.5 A joint primitive with three anchor points was added. The yellow point is the most lateral point of the condylar neck, the red point is the most medial point of the condylar neck, and the white point is placed on the posterior of the condylar neck equidistant to the yellow and red points

Inc., Natick, MA), where they were able to be visualized and checked for completeness (Fig. 9.8). A generalized Procrustes analysis was then used to compare the condyle and the fossa separately. The Procrustes analysis removes orientation and positional data so that the condyles can be superimposed based on their shape [65]. The condyles from T_1 were compared to T_2 and T_1 compared to T_3. The same process was completed for the fossa. In order to evaluate regional differences that may or may not exist, the condyles and fossae were divided into regions based on the positioning of the semi-landmarks. For the condyle the regions were anterior, posterior, superior, lateral, and medial. For the fossa, the regions were anterior, posterior, superior, and medial. Due to the lack of a bony lateral wall of the fossa, no lateral region was defined.

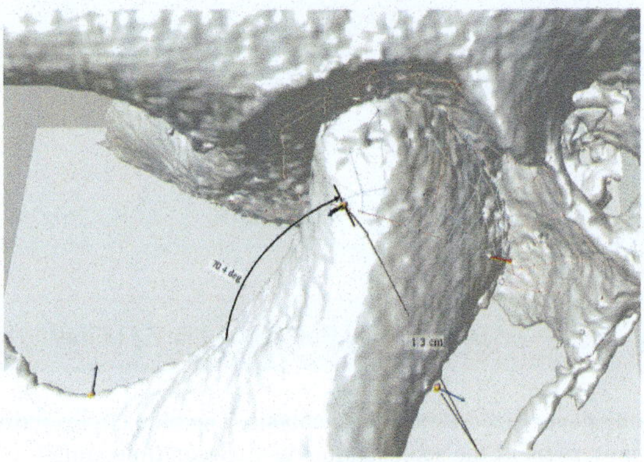

Fig. 9.6 Example of joint primitive including anchor points and semi-landmarks for the condyle and fossa

9.4.2 Statistical Analysis

For the statistical analysis, comparisons were planned to determine if there were differences between left and right sides in order to maximize final sample size (Fig. 9.9). If no differences were found between left and right condyles and fossae, they could be treated as separate samples for our analysis. After it was determined that no side differences were present, the main analysis was performed to see if any regional differences in remodeling were detected. Class, gender, and time effects were also analyzed using a repeated measures design. The within-subject factors were region and poststate (time), and the between-

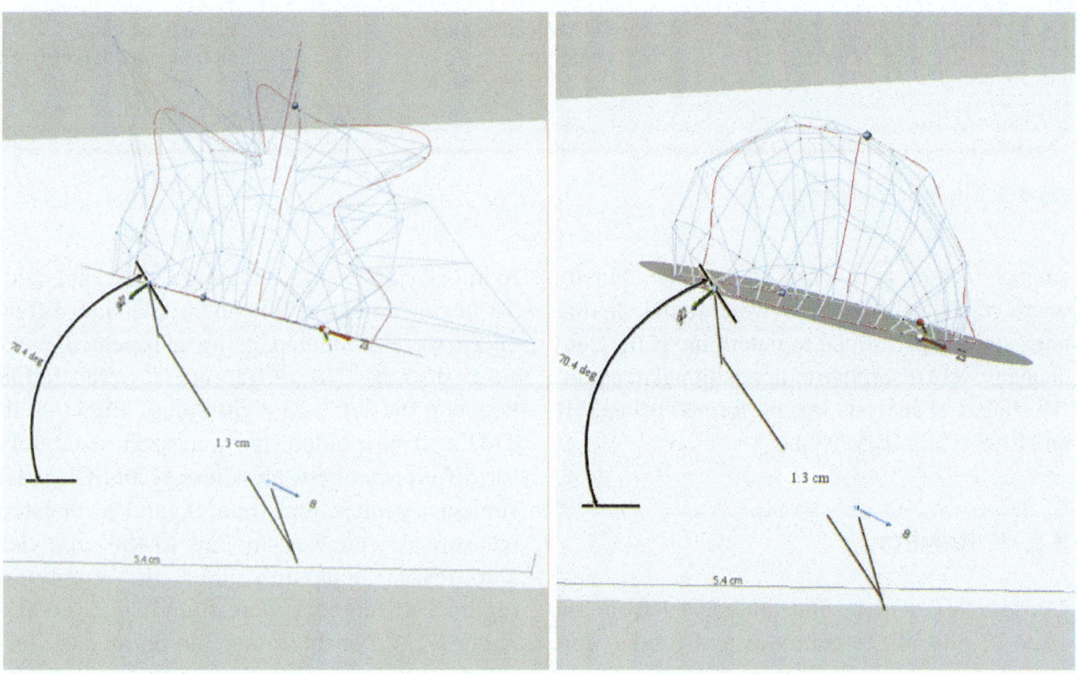

Fig. 9.7 Example of an overlay of a condylar head before (left) and after (right) adjustment of semi-landmarks

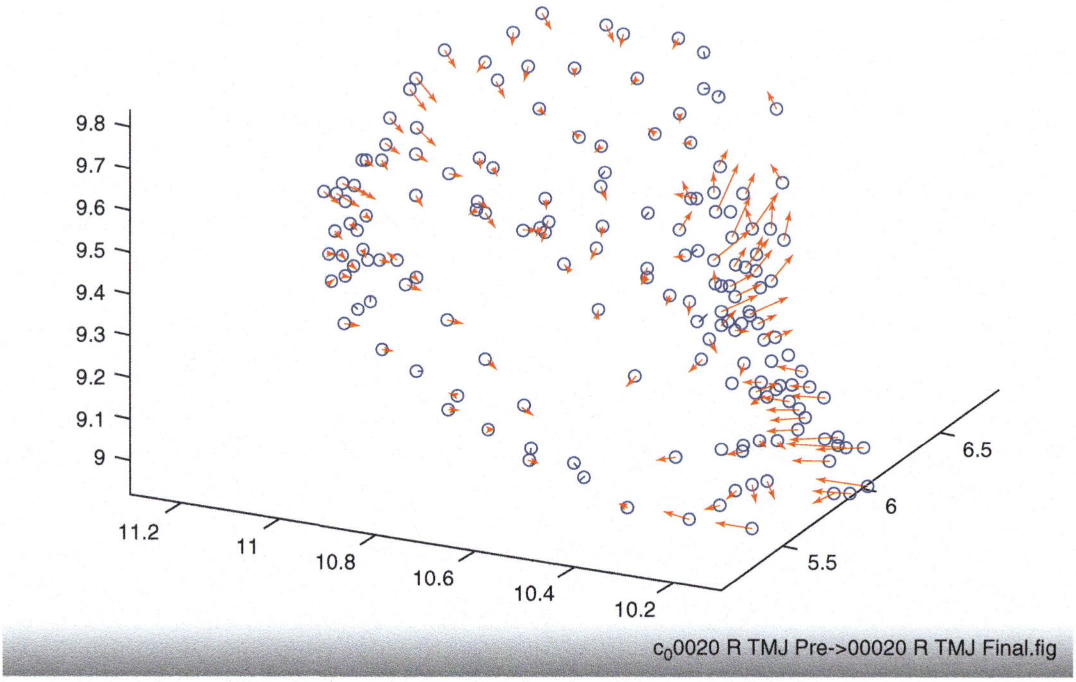

Fig. 9.8 Example of condylar head comparison of T_1–T_3 using generalized Procrustes analysis. The blue circles indicate the T_1 position, and the red arrows indicate direction of movement of the semi-landmarks between timepoints

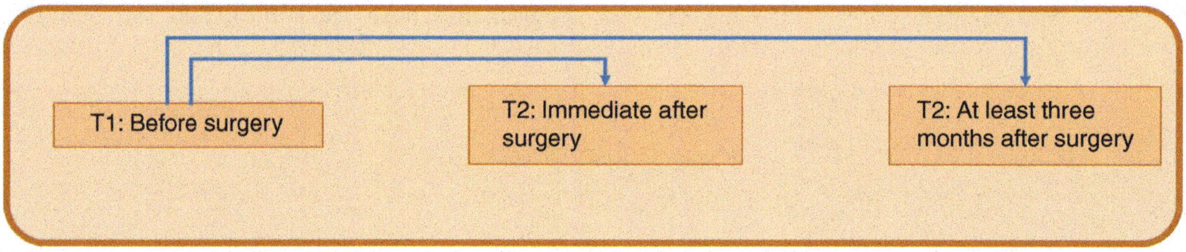

Fig. 9.9 Timepoint comparisons

subject factors were class and gender. Finally, when regional differences were found, a final analysis was performed to determine if the bony changes were resorptive or appositional in nature. All statistical analysis was performed using JMP and SAS (JMP, Cary, NC).

9.5 Results

Coordinates of the semi-landmarks from the 13 × 13 plot of the condyles and fossae were recorded for each TMJ at each timepoint for all 31 subjects included in the study. Overall, a total of 62,868 semi-landmarks were analyzed. To begin, potential left and right side differences were evaluated using a repeated measure design. No differences were found between the left and right sides, allowing all TMJ's to be pooled and analyzed separately for differences between Class II and Class III subjects, gender, and time. Again, a repeated measure design was applied to the condyles and fossae separately. For the condyles, regional differences were found ($p < 0.0001$; Table 9.1a). For the fossae, no regional differences were found, but there were time differences (Table 9.1c).

After it was determined that regional differences were found in the condyles, the regions were further analyzed to determine where the changes occurred. It was found that significant changes occurred in the lateral, posterior, and superior regions. Additionally, a slight interaction ($p = 0.0487$) was found between T_3 and the superior region (Table 9.2a).

In order to determine if the regional changes were resorptive or appositional, a repeated measures design was applied with the data separated by the direction of the movements of each semi-landmark to determine what types of remodeling were occurring. When remodeling took place in the condyles, there were regional differences (Table 9.1b), with the resorption occurring in the posterior and superior regions (Table 9.2b). When appositional condylar remodeling took place, it occurred in the lateral, posterior, and superior regions (Table 9.2c).

For the fossae, when resorption was detected, there were both regional differences and time differences (Table 9.1d). Additionally, when resorption was detected in the fossae, it occurred in the posterior regions, and there were time differences between T_1 and T_3 (Table 9.2d). Finally, in subjects where apposition took place in the fossae, only a time difference was noted with a slight interaction between time and region (Table 9.1e).

Table 9.1 The statistically significant differences (*) for the condyles and fossae by region and time

Source	Permutations	DF	DFDen	*F* ratio	Prob > *F*
A: Repeated measures for condylar regional differences					
Region	4	4	141.4	12.028	<0.0001*
B: Repeated measures for condyle resorption regional differences					
Region	4	4	138.5	6.2199	0.0001*
C: Repeated measures for fossa time differences					
Time	1	1	34.91	7.4371	0.0099*
D: Repeated measures for fossa resorption differences					
Time	1	1	34.66	4.6119	0.0388*
Region	3	3	97.77	7.1388	0.0002*
E: Repeated measures for fossa apposition differences					
Time	1	1	35.29	7.4204	0.0100*
Time × region	3	3	96.22	4.0215	0.0096*

Table 9.2 The statistically significant differences (*) for the specific regions of the condyles and fossae with pertinent time interactions

Source	Estimate	Std error	DFDen	*T* ratio	Prob > |*t*|
A: Repeated measures for condylar specific regions					
Lateral region	−0.00615	0.002339	152.4	−2.63	0.0094*
Posterior region	−0.0083	0.002286	139.1	−3.63	0.0004*
Superior region	0.014371	0.002279	137.3	6.31	<0.0001*
T3 × Superior region interaction	0.002797	0.001406	133.4	1.99	0.0487*
B: Repeated measures for condylar resorption-specific regions					
Posterior region	−0.00887	0.003636	137.5	−2.44	0.0160*
Superior region	0.016877	0.003619	134.9	4.66	<0.0001*
C: Repeated measures for condylar apposition-specific regions					
Lateral region	−0.00566	0.002547	149.1	−2.22	0.0279*
Posterior region	−0.00667	0.002462	130.9	−2.71	0.0076*
Superior region	0.008063	0.002465	132	3.27	0.0014*
D: Repeated measures for fossa resorption region (posterior)					
Posterior region	−0.00805	0.002284	101.4	−3.53	0.0006*
T_3	−0.00403	0.001875	34.66	−2.15	0.0388*
T3 × Posterior region interaction	0.004519	0.001678	105.6	2.69	0.0082*

After determining that statistically significant amounts of remodeling were taking place, the mean amounts of apposition and resorption over the sample were evaluated. An ANOVA analysis was completed with the effects contrasted to the medial region. It was found that in the condyles, posterior resorption averaged 0.56 mm, superior resorption averaged 0.80 mm, lateral apposition averaged 0.48 mm, posterior apposition averaged 0.44 mm, and superior apposition averaged 0.63 mm. In the fossae, the only statistically significant regional change occurred with posterior resorption with an average of 0.53 mm (Table 9.3).

9.6 Discussion

9.6.1 Technique and Methodology Assessment

Using Stratovan Checkpoint to plot the TMJ has been used in previous studies by [66, 67]. The majority of the protocol used in this study was adapted from the protocol developed by Ikeda et al. in 2016. Unlike previous CBCT studies which have evaluated TMJ changes using three-dimensional color mapping [2, 54, 55, 59, 60, 62] or volumetric superimposition [61, 63], Checkpoint allows for both the accurate analysis of condylar positions relative to the fossae in addition to any morphological changes of the condylar head and fossa individually. Modifications to the protocol from Ikeda et al. [66] were made in order to capture the entirety of the condylar head, rather than only the superior portion. While that study had determined that the ideal field size to accurately detect morphological changes was 11 × 11, because a larger portion of the condylar head was being mapped in this study, a field size of 13 × 13 was used to compensate for the great area.

Table 9.3 Mean resorption and apposition by region

Region	Mean resorption (mm)	Mean apposition (mm)
Condyle anterior	0.61	0.56
Condyle lateral	0.58	0.48
Condyle posterior	0.56	0.44
Condyle superior	0.80	0.63
Condyle medial	0.62	0.51
Fossa anterior	0.57	0.64
Fossa posterior	0.53	0.70
Fossa superior	0.61	0.63
Fossa medial	0.68	0.66

This modification led to difficulties in attempting to standardize the region of the condyle being measured. The most caudal point of the sigmoid notch was used as the primary landmark to attempt to ensure that the portion of the condyle being measured remained consistent across time-points. This was determined to be the best method available considering the limited size of the CBCT image after cropping. Ideally, it would be possible to map both joints simultaneously along with the placement of points on non-changing structures within the skull that could be used for accurate superimposition.

The software is designed to be semiautomated with the placement of the semi-landmarks. The three main anchor points are meant to be manually adjusted to their correct placement, at which point the semi-landmarks would be automatically placed over the cortical surface of the condyle along with the fossa. In practice, however, a high percentage of the semi-landmarks would be placed a significant distance away from the cortical surface of the condyle. For the fossa, the lack of a lateral wall led to many of the landmarks being projected into space. These misplaced landmarks had to be either manually adjusted to the cortical surface or marked as missing when there was no bony surface to map. This not only significantly increased the time it took to map the TMJ's, but it also increased error as the placement was based on the best visualization. Much of this will likely improve over time as the image quality from the CBCT machines improves, noise reduced, and the software further refined. If the semi-landmarks did not have to be manually adjusted, it would be much more feasible to greatly increase sample size.

9.6.2 Evaluation of Condylar Changes

Condylar changes have been evaluated in a multitude of previous studies. The majority of these studies primarily examined positional changes of the condyle, likely due to difficulty in accurately measuring morphological changes with two-dimensional radiographs, whereas positional changes can be more easily evaluated using linear and angular measurements on conventional radiographs. With the advent of three-dimensional imaging, studies have begun analyzing morphological changes to the condylar head after orthognathic surgery.

The findings of this study show that there are significant differences in the amount of change between the fossa and the condyle. No previous studies have been found evaluating the difference in these types of changes between the condyles and fossae. Further examination found regional differences in the condyles in the lateral, posterior, and superior regions. Unlike previous studies, no significant differences were found with regard to gender, classification, or left and right sides. Previous studies have found that females were more likely to undergo resorption than males [7, 9] and that large mandibular advancements tended to most commonly cause resorption [7, 18, 19, 34]. Previous studies have found that often resorption will occur unilaterally [34].

When condylar resorption was detected, it occurred in the posterior and superior regions of the condyle. This is contradictory to the study by Hoppenreijs et al. [9], where resorption was found to primarily occur at the anterior site of the condyle. They also found that morphologic changes continue beyond 1 year following surgery, which is supported by this study. Mobarak et al. [20] found that medial or lateral resorption of the condyle can occur if medial or lateral compression occurs during fixation. Park et al. [61] used CBCT superimposition to evaluate remodeling after mandibular advancement and found a statistically significant amount of remodeling 1 year after surgery. They found that resorption occurred primarily in the anterior, posterior, superior, and lateral portions of the condylar head. However, Chen et al. [63] found that resorption primarily occurred on the posterior of the condyle, with the overall differences in condylar head dimensions before and 1 year after surgery at 0.37 ± 0.11 mm. Finally, Xi et al. [64] examined volumetric changes of the condyles using CBCT imaging and found that condylar resorption occurred following surgery and contributed to skeletal relapse. While there have been a number of discordant results, most studies seem to agree that posterior resorption is a common finding. This makes sense because, as previously discussed, the condyle is often displaced in a posterior and superior direction during surgery, with compression leading to resorption. In this study, posterior condylar resorption averaged 0.56 mm, and superior condylar resorption averaged 0.80 mm.

When apposition was detected on the condyle, it primarily occurred in the lateral, posterior, and superior regions. Yamada et al. [23] found condylar bone changes in 35.7% of subjects, which most frequently presented at osteophyte formation. Katsumata et al. [39] found that after mandibular setback, a bone layer formed on the posterior medial aspect of the condyle and that condylar remodeling was found in 51.1% of intraoral vertical ramus osteotomy (IVRO) patients and 10.3% of BSSO patients. Park et al. [61] found that condylar apposition primarily occurred on the anteromedial region of the condyle, which is opposite of the findings of this study. Chen et al. [63] also found that apposition tended to occur on the anterior aspect of the condyle, which was not found in this study. In this study, condylar apposition was 0.48 mm in the lateral region, 0.44 mm in the posterior region, and 0.63 mm in the superior region.

Overall, the resorptive condylar changes observed in this study seem to mostly agree with previous studies, while the appositional changes conflict with previous studies.

9.6.3 Evaluation of Fossae Changes

The fossae has not been as heavily evaluated as the condyles in previous studies, likely due to the difficulty in accurately identifying it with traditional radiographs and also the difficulty in accurately measuring it with three-dimensional imaging. The primary findings here are that there are time differences in the fossa between timepoints. It was also found that some cases exhibited significant resorption of the fossa; and when this resorption occurred, there were both regional differences and time differences. It was determined that the resorption was only significant in the posterior region between the initial and the final CBCT's. In that case, the posterior resorption was averaged 0.53 mm. In other cases, apposition occurred within the fossa. However, in these cases, there were only generalized differences over time. No significant regional differences were able to be detected.

Studies on morphological changes of the glenoid fossa after orthognathic surgery have been very limited. Sanromán et al. [68] found no significant changes in the condyle of glenoid fossa using both CT and MR imaging at any timepoint up to 1 year following orthognathic surgery. De Clerk et al. [60] used CBCT imaging and three-dimensional color mapping to visualize changes to the condyles and glenoid fossae in Class III nonsurgical cases and found evidence of remodeling changes on both the anterior and posterior eminences of the glenoid fossae. Woodside et al. in 1987 evaluated remodeling changes of the condyle and glenoid fossa in primates during Herbst treatment and found that a large volume of new bone was formed along the anterior border of the postglenoid spine and resorption along the posterior border of the postglenoid spine, indicating an overall anterior remodeling of the fossa. Finally, LeCornu et al. [62] found anterior displacement of the glenoid fossae and condyles in Class II patients undergoing treatment with Herbst appliances.

The findings of this study regarding fossa remodeling are interesting, because, unlike the Sanromán et al. [68] study, remodeling was found after orthognathic surgery. Additionally, both resorption and appositional changes were detected within the glenoid fossae.

9.6.4 Study Limitations

This study presented a number of challenges that needed to be overcome. One of the biggest challenges in the study was the quality of the CBCT images. Many images were noisy in the TMJ region, leading to issues in accurately identifying the cortical outlines of the condyles and fossae. This issue alone was the greatest contributor to the reduction in our initial sample size of 69 down to the final sample size of 31. A large part of this was the fact that three timepoints were analyzed. If any one of the three timepoints was noisier or distorted, the entire subject had to be removed from the study. Therefore, it would have been beneficial to begin with a larger initial sample, knowing that over 50% would be lost due to issues with the CBCT images.

Additionally, the manual adjustment of semi-landmarks may lead to error. Ideally, the software would automatically place all of the semi-landmarks accurately, and minimal manual adjustment would be needed. In actuality, however, a high number of the semi-landmarks had to be manually adjusted, which could have led to higher error in the measurements. Again, much of this was due to noise and difficulty identifying cortical outlines of the condyle and fossa in the CBCT images. Although the investigators were calibrated and completed the placement of semi-landmarks for the entirety of each subject at all timepoints, the fact that manual adjustment was needed could have led to higher error. It should also be noted that the window, level, and isosurface settings had to be changed not only for each subject but also for each timepoint within the study, as the settings that led to the best representation of the cortical outline of the condyle and fossa varied greatly between each CBCT. Ideally, standardized settings could have been determined and used across all subjects, but in this study, the settings were adjusted based on the best visual reference of the cortical outline.

The use of Checkpoint also led to difficulty in the superimposition of the CBCT data to allow for an analysis of any positional changes of the condyle that may have occurred after surgery.

Ideally, the data would have been accurately superimposed on the fossae so that condylar movement could be evaluated. However, this was not possible in this study. Additional anchor points would need to be placed to better allow for the superimposition of the fossae, which would allow for an extremely accurate analysis of any positional changes that may have occurred with surgery. Then, the positional data could have been compared to the large number of previously completed studies evaluating positional changes.

Similarly, for statistical analysis, the generalized Procrustes analysis was adequate for showing that remodeling changes did occur in the condyles and fossae, but again, a more ideal solution would have been to superimpose both the condyles and fossae based on an unchanging structure elsewhere in the mandible or cranial base. By including more of the condylar head and using a larger 13 × 13 field size, we believe that we were able to still gain valuable information using the generalized Procrustes analysis, but this could be improved upon by identifying unchanging structures that could be used for more accurate superimposition in future studies.

9.7 Summary and Conclusions

This was a retrospective study of 31 subjects treated by the University of Oklahoma Department of Oral and Maxillofacial Surgery from July 2013 to July 2015. All of the subjects presented with skeletal Class II or Class III malocclusions. Of these 31, 22 were female and nine were male with 13 Class II and 18 Class III malocclusions. All subjects had CBCT images taken at three timepoints which were prior to surgery (T_1), immediately after surgery (T_2), and at least 3 months after surgery (T_3). The TMJ's of all subjects were plotted using Stratovan Checkpoint software with a 13 × 13 semi-landmark field size to obtain the three-dimensional morphology of the condyles and the fossae. The data was then imported into MatLab where it was able to be visualized and checked for accuracy. Statistical analysis was performed using SAS and JMP software to analyze if changes occurred over time and in what regions the changes occurred in both the condyles and the fossae. The use of three-dimensional CBCT data allowed for a more complete analysis of changes taking place in the TMJ. Stratovan Checkpoint software is useful in evaluating morphological changes to the TMJ and holds a great deal of potential for future studies regarding morphological and positional changes.

The conclusions of this study were:

1. There were no statistically significant differences between left and right TMJ structures in both the condyles and fossae over the three timepoints.
2. Remodeling of the condylar head primarily occurred on the posterior, superior, and lateral regions similar to previous studies.
3. Resorption of the condylar head occurred in areas of the posterior and superior regions, while apposition occurred in areas of the superior, posterior, and lateral regions.
4. Resorption in the glenoid fossae was primarily seen in the posterior regions with a significant difference seen over time. When apposition occurred in the fossa, only time differences were detected with no specific regional differences.

References

1. Steinhäuser EW. Historical development of orthognathic surgery. J Craniomaxillofac Surg. 1996;24(4):195–204.
2. Bailey LTJ, Cevidanes LH, Proffit WR. Stability and predictability of orthognathic surgery. Am J Orthod Dentofac Orthop. 2004;126(3):273.
3. Proffit WR, Turvey TA, Phillips C. The hierarchy of stability and predictability in orthognathic surgery with rigid fixation: an update and extension. Head Face Med. 2007;3(1):21.
4. Kretschmer WB, et al. Transverse stability of 3-piece Le Fort I osteotomies. J Oral Maxillofac Surg. 2011;69(3):861–9.
5. Proffit WR, Turvey TA, Phillips C. Orthognathic surgery: a hierarchy of stability. Int J Adult Orthodon Orthognath Surg. 1996;11(3):191–204.

6. Arnett GW, Tamborello JA. Progressive class II development: female idiopathic condylar resorption. Oral Maxillofac Surg Clin North Am. 1990;2:699–716.
7. Moore KE, Gooris PJJ, Stoelinga PJW. The contributing role of condylar resorption to skeletal relapse following mandibular advancement surgery: report of five cases. J Oral Maxillofac Surg. 1991;49(5):448–60.
8. Joss CU, Vassalli IM. Stability after bilateral sagittal split osteotomy advancement surgery with rigid internal fixation: a systematic review. J Oral Maxillofac Surg. 2009;67(2):301–13.
9. Hoppenreijs TJM, et al. Condylar remodelling and resorption after Le Fort I and bimaxillary osteotomies in patients with anterior open bite: a clinical and radiological study aesthetic and reconstructive surgery. Int J Oral Maxillofac Surg. 1998;27(2):81–91.
10. Alder ME, et al. Short-term changes of condylar position after sagittal split osteotomy for mandibular advancement. Oral Surg Oral Med Oral Pathol Oral Radiol Endod. 1999;87(2):159–65.
11. Will LA, et al. Condylar position following mandibular advancement: its relationship to relapse. J Oral Maxillofac Surg. 1984;42(9):578–88.
12. Marmulla R, Mühling J. Computer-assisted condyle positioning in orthognathic surgery. J Oral Maxillofac Surg. 2007;65(10):1963–8.
13. Arnett GW. A redefinition of bilateral sagittal osteotomy (BSO) advancement relapse. Am J Orthod Dentofac Orthop. 1993;104(5):506–15.
14. Kundert M, Hadjianghelou O. Condylar displacement after sagittal splitting of the mandibular rami: a short-term radiographic study. J Maxillofac Surg. 1980;8:278–87.
15. Han JJ, Hwang SJ. Three-dimensional analysis of postoperative returning movement of perioperative condylar displacement after bilateral sagittal split ramus osteotomy for mandibular setback with different fixation methods. J Craniomaxillofac Surg. 2015;43(9):1918–25.
16. Rotskoff KS, Herbosa EG, Villa P. Maintenance of condyle-proximal segment position in orthognathic surgery. J Oral Maxillofac Surg. 1991;49(1):2–7.
17. Freihofer HPM, Petreśević D. Late results after advancing the mandible by sagittal splitting of the rami. J Maxillofac Surg. 1975;3:250–7.
18. Bouwman JPB, Kerstens HC, Tuinzing DB. Condylar resorption in orthognathic surgery: the role of intermaxillary fixation. Oral Surg Oral Med Oral Pathol. 1994;78(2):138–41.
19. Kerstens HCJ, et al. Condylar atrophy and osteoarthrosis after bimaxillary surgery. Oral Surg Oral Med Oral Pathol. 1990;69(3):274–80.
20. Mobarak KA, et al. Mandibular advancement surgery in high-angle and low-angle class II patients: different long-term skeletal responses. Am J Orthod Dentofac Orthop. 2001;119(4):368–81.
21. Huang YL, Pogrel MA, Kaban LB. Diagnosis and management of condylar resorption. J Oral Maxillofac Surg. 1997;55(2):114–9.
22. Hwang S-J, et al. Surgical risk factors for condylar resorption after orthognathic surgery. Oral Surg Oral Med Oral Pathol Oral Radiol Endod. 2000;89(5):542–52.
23. Yamada K, et al. Condylar bony change, disk displacement, and signs and symptoms of TMJ disorders in orthognathic surgery patients. Oral Surg Oral Med Oral Pathol Oral Radiol Endod. 2001;91(5):603–10.
24. Hwang S-J, et al. Non-surgical risk factors for condylar resorption after orthognathic surgery. J Craniomaxillofac Surg. 2004;32(2):103–11.
25. Mousoulea S, et al. Condylar resorption in orthognathic patients after mandibular bilateral sagittal split osteotomy: a systematic review. Eur J Orthod. 2016;39(3):294–309.
26. Merkx MAW, Van Damme PA. Condylar resorption after orthognathic surgery: evaluation of treatment in 8 patients. J Craniomaxillofac Surg. 1994;22(1):53–8.
27. Athanasiou AE, Mavreas D. Tomographic assessment of alterations of the temporomandibular joint after surgical correction of mandibular prognathism. Int J Adult Orthodon Orthognath Surg. 1991;6(2):105–12.
28. Moyers RE, Bookstein FL. The inappropriateness of conventional cephalometrics. Am J Orthod Dentofac Orthop. 1979;75(6):599–617.
29. Spitzer W, Rettinger G, Sitzmann F. Computerized tomography examination for the detection of positional changes in the temporomandibular joint after ramus osteotomies with screw fixation. J Maxillofac Surg. 1984;12:139–42.
30. Sund G, Eckerdal O, Åstrand P. Changes in the temporomandibular joint after oblique sliding osteotomy of the mandibular rami: a longitudinal radiological study. J Maxillofac Surg. 1983;11:87–91.
31. Woodside DG, Metaxas A, Altuna G. The influence of functional appliance therapy on glenoid fossa remodeling. Am J Orthod Dentofac Orthop. 1987;92(3):181–98.
32. Hackney FL, Van Sickels JE, Nummikoski PV. Condylar displacement and temporomandibular joint dysfunction following bilateral sagittal split osteotomy and rigid fixation. J Oral Maxillofac Surg. 1989;47(3):223–7.
33. Stroster TG, Pangrazio-Kulbersh V. Assessment of condylar position following bilateral sagittal split ramus osteotomy with wire fixation or rigid fixation. Int J Adult Orthodon Orthognath Surg. 1993;9(1):55–63.
34. Cutbirth M, Van Sickels JE, Thrash WJ. Condylar resorption after bicortical screw fixation of mandibular advancement. J Oral Maxillofac Surg. 1998;56(2):178–82.
35. Kawamata A, et al. Three-dimensional computed tomography evaluation of postsurgical condylar displacement after mandibular osteotomy. Oral Surg Oral Med Oral Pathol Oral Radiol Endod. 1998;85(4):371–6.
36. Ruf S, Pancherz H. Temporomandibular joint remodeling in adolescents and young adults during Herbst treatment: a prospective longitudinal mag-

netic resonance imaging and cephalometric radiographic investigation. Am J Orthod Dentofac Orthop. 1999;115(6):607–18.
37. Hu J, Wang D, Zou S. Effects of mandibular setback on the temporomandibular joint: a comparison of oblique and sagittal split ramus osteotomy. J Oral Maxillofac Surg. 2000;58(4):375–80.
38. Voudouris JC, et al. Condyle-fossa modifications and muscle interactions during Herbst treatment, part 1. New technological methods. Am J Orthod Dentofac Orthop. 2003;123(6):604–13.
39. Katsumata A, et al. Condylar head remodeling following mandibular setback osteotomy for prognathism: a comparative study of different imaging modalities. Oral Surg Oral Med Oral Pathol Oral Radiol Endod. 2006;101(4):505–14.
40. Cortez ALV, Passeri LA. Radiographic assessment of the condylar position after Le Fort I osteotomy in patients with asymptomatic temporomandibular joints: a prospective study. J Oral Maxillofac Surg. 2007;65(2):237–41.
41. Ueki K, et al. Changes in temporomandibular joint and ramus after sagittal split ramus osteotomy in mandibular prognathism patients with and without asymmetry. J Craniomaxillofac Surg. 2012;40(8):821–7.
42. Hatcher DC. Operational principles for cone-beam computed tomography. J Am Dent Assoc. 2010;141:3S–6S.
43. Mozzo P, et al. A new volumetric CT machine for dental imaging based on the cone-beam technique: preliminary results. Eur Radiol. 1998;8(9):1558–64.
44. Nemtoi A, et al. Cone beam CT: a current overview of devices. Dentomaxillofac Radiol. 2013;42(8):20120443.
45. Vandenberghe B, Jacobs R, Bosmans H. Modern dental imaging: a review of the current technology and clinical applications in dental practice. Eur Radiol. 2010;20(11):2637–55.
46. Chen S, et al. Short-and long-term changes of condylar position after bilateral sagittal split ramus osteotomy for mandibular advancement in combination with Le Fort I osteotomy evaluated by cone-beam computed tomography. J Oral Maxillofac Surg. 2013;71(11):1956–66.
47. Lascala CA, Panella J, Marques MM. Analysis of the accuracy of linear measurements obtained by cone beam computed tomography (CBCT-NewTom). Dentomaxillofac Radiol. 2004;33(5):291–4.
48. Loubele M, et al. A comparison of jaw dimensional and quality assessments of bone characteristics with cone-beam CT, spiral tomography, and multi-slice spiral CT. Int J Oral Maxillofac Implants. 2007;22(3):446–54.
49. De Vos W, Casselman J, Swennen G. Cone-beam computerized tomography (CBCT) imaging of the oral and maxillofacial region: a systematic review of the literature. Int J Oral Maxillofac Surg. 2009;38(6):609–25.
50. Kobayashi K, et al. Accuracy in measurement of distance using limited cone-beam computerized tomography. Int J Oral Maxillofac Implants. 2004;19(2):228–31.
51. Hilgers ML, et al. Accuracy of linear temporomandibular joint measurements with cone beam computed tomography and digital cephalometric radiography. Am J Orthod Dentofac Orthop. 2005;128(6):803–11.
52. Honey OB, et al. Accuracy of cone-beam computed tomography imaging of the temporomandibular joint: comparisons with panoramic radiology and linear tomography. Am J Orthod Dentofac Orthop. 2007;132(4):429–38.
53. Zhang Z-l, et al. Measurement accuracy of temporomandibular joint space in Promax 3-dimensional cone-beam computerized tomography images. Oral Surg Oral Med Oral Pathol Oral Radiol Endod. 2012;114(1):112–7.
54. Cevidanes LHS, et al. Superimposition of 3D cone-beam CT models of orthognathic surgery patients. Dentomaxillofac Radiol. 2005;34(6):369–75.
55. Cevidanes LHS, et al. Three-dimensional cone-beam computed tomography for assessment of mandibular changes after orthognathic surgery. Am J Orthod Dentofac Orthop. 2007;131(1):44–50.
56. Ikeda K, Kawamura A. Assessment of optimal condylar position with limited cone-beam computed tomography. Am J Orthod Dentofac Orthop. 2009;135(4):495–501.
57. Kim YI, et al. The assessment of the short-and long-term changes in the condylar position following sagittal split ramus osteotomy (SSRO) with rigid fixation. J Oral Rehabil. 2010;37(4):262–70.
58. Kim Y-I, et al. Cone-beam computerized tomography evaluation of condylar changes and stability following two-jaw surgery: Le Fort I osteotomy and mandibular setback surgery with rigid fixation. Oral Surg Oral Med Oral Pathol Oral Radiol Endod. 2011;111(6):681–7.
59. Motta AT, et al. Three-dimensional regional displacements after mandibular advancement surgery: one year of follow-up. J Oral Maxillofac Surg. 2011;69(5):1447–57.
60. De Clerck H, et al. Three-dimensional assessment of mandibular and glenoid fossa changes after bone-anchored class III intermaxillary traction. Am J Orthod Dentofac Orthop. 2012;142(1):25–31.
61. Park S-B, et al. Effect of bimaxillary surgery on adaptive condylar head remodeling: metric analysis and image interpretation using cone-beam computed tomography volume superimposition. J Oral Maxillofac Surg. 2012;70(8):1951–9.
62. LeCornu M, et al. Three-dimensional treatment outcomes in class II patients treated with the Herbst appliance: a pilot study. Am J Orthod Dentofac Orthop. 2013;144(6):818–30.
63. Chen S, et al. Three-dimensional evaluation of condylar morphology remodeling after orthognathic surgery in mandibular retrognathism by cone-beam computed tomography. J Peking Univ. 2015;47(4):703–7.

64. Xi T, et al. 3D analysis of condylar remodelling and skeletal relapse following bilateral sagittal split advancement osteotomies. J Craniomaxillofac Surg. 2015;43(4):462–8.
65. Ikeda R. Clinical application of a novel three-dimensional analysis to evaluate temporomandibular joint space changes after orthognathic surgery. San Francisco, CA: University of California; 2014. p. 107.
66. Ikeda R, et al. Novel 3-dimensional analysis to evaluate temporomandibular joint space and shape. Am J Orthod Dentofac Orthop. 2016;149(3):416–28.
67. Contro C. Evaluating condylar head morphology as it relates to the skeletal vertical facial dimension: a three-dimensional semi-automated landmark study. San Francisco, CA: University of California; 2015. p. 50.
68. Sanromán JF, et al. Morphometric and morphological changes in the temporomandibular joint after orthognathic surgery: a magnetic resonance imaging and computed tomography prospective study. J Cranio-Maxillofac Surg. 1997;25(3):139–48.

Anterior Limit of the Mandibular Dentition as Evaluated by Cone-Beam CT

10

Sercan Akyalcin, Jeremy Scarpate, and Jeryl English

Abstract

One of the most common limiting factors in correcting dental and skeletal deformities is the limited ability to orthodontically move the teeth within the buccal and lingual cortical plates of the mandibular symphysis. In this chapter, we will define the anatomic characteristics of the mandibular symphysis area and review the relationships between mandibular incisors and their bony support. Since cephalometric radiographs overestimate the width of the buccal bone due to superimposition errors, a cone-beam computed tomography (CBCT) evaluation of the anterior mandibular dentition was presented in a group of individuals with varying growth patterns (hypodivergent, normodivergent, and hyperdivergent) and incisor inclinations (retroclined, upright, and proclined). Using the CBCT data, the relationship between mandibular incisors and their structural support was presented for each group of the mandibular incisors. Normative data and imaging methods presented in this chapter can be utilized to enhance the clinicians' treatment planning strategies by providing a template on the limits of mandibular anterior teeth.

10.1 Background

Since the advent of lateral cephalometric radiography, the mandibular symphysis area has been a focus of investigation in search of critical clues that may improve orthodontic diagnosis and treatment planning. Anatomy of the area has also served as a predictive tool for the evaluation of the facial growth pattern. From a clinical point of view, the buccal and lingual cortical plates of the mandibular symphysis define the anatomic limits of orthodontic tooth movement that can rationally be expected. Since the cortical plates provide a physical boundary for tooth movement that cannot be exceeded without expecting deleterious effects, they can be considered "orthodontic walls" [1].

Traditionally, clinicians have relied upon the use of cephalometric analyses to determine the most stable position of the mandibular incisors as part of their treatment plans. Tweed [2] was an early outspoken proponent of establishing a definitive position and angulation of the incisors within the mandibular symphysis, as well as emphasiz-

S. Akyalcin (✉)
Department of Orthodontics, Tufts School of Dental Medicine, Boston, MA, USA
e-mail: sercan.akyalcin@tufts.edu

J. Scarpate
Royal Air Force, Lakenheath, UK

J. English
Department of Orthodontics, School of Dentistry, The University of Texas Health Sciences Center, Houston, TX, USA
e-mail: Jeryl.D.English@uth.tmc.edu

O. Kadioglu, G. F. Currier (eds.), *Craniofacial 3D Imaging*,
https://doi.org/10.1007/978-3-030-00722-5_10

ing the importance the mandibular incisors have in long-term stability and esthetics. In his experience, he found that patients with excessive crowding often suffered iatrogenic periodontal issues or significant relapse after the teeth were aligned orthodontically, without the presence or creation of space. He stated the necessity of positioning the incisors over basal bone and subsequently emphasized that incisors within 5° of 90° provided optimum esthetics. He included this measurement as a critical tenant to his diagnostic triangle. Likewise, other early pioneers in cephalometrics, such as Steiner [3] and Ricketts [4], also held that the mandibular incisors were a key component to both stability of the dentition and to the patient's overall esthetics. While modern orthodontic treatment philosophies often utilize the maxillary incisor position to dictate the profile and smile esthetics, the mandibular arch and final position of the mandibular incisors remain of primary diagnostic importance.

In essence, conventional cephalometric studies evaluating the ideal position for mandibular incisors are hampered by the inherent limitations of this radiographic method. The lateral cephalogram utilizes a divergent x-ray beam that enlarges the anatomical structures; thus, minute measurements are rendered relatively inaccurate. Additionally, the natural superimposition of anatomic landmarks formed by the lateral cephalogram creates an inherently flawed image and prevents accurate assessment of the symphysis [5–8].

With the introduction of cone-beam computed tomography (CBCT), the inherent shortcomings of the lateral cephalogram have been overcome. Through software reconstruction, the CBCT three-dimensional properties obtained are true to form. Investigators can utilize the CBCT images to measure localized areas in all dimensions with high accuracy that was not available with the lateral cephalogram or any previous radiographic method, except medical-grade computed tomography (CT) [9, 10]. However, the CT exposes the patient to a high amount of radiation, and the risk/benefit for the patient was only acceptable in fairly extreme cases, such as pathological identification. With the reduced radiation exposure that now approaches that of routinely taken orthodontic x-rays and its increased availability, the CBCT has become more acceptable as a diagnostic adjunct. The research that has emerged from the CBCT scans continues to expand our understanding of the anatomical features of the mandibular symphysis and incisors and how these features correlate with the patterns of facial growth in the individual patient.

In this chapter, we will define the anatomic characteristics of the mandibular symphysis area and review the relationships between mandibular incisors and their structural support in length. Additionally, effects of facial growth patterns on the osseous support of the mandibular anterior teeth will be presented as a guide to enhance clinicians' treatment planning strategies by providing guidelines on the limits of anterior movement of mandibular incisors.

10.2 Growth Effects on the Mandibular Symphysis and Incisors

The mandible does not simply grow. It remodels in the entire outline, except for the periosteal contour of the chin just below the pogonion. The chin area is one of the most variable areas in the entire mandible as seen among the different basic facial types and patterns. The anterior chin is a resorptive area, while the bone is added on the lingual side at a variable rate. As a result, there are correspondingly marked variations in the shape and the size of the chin among different individuals [11]. The mandibular symphysis undergoes morphological changes and modifications throughout growth as well. The only exception is the inner contour of the cortical plate at the lower border of the symphysis located at the most inferior aspect of the trabecular bone. This area is a primary landmark for adjusting mandibular superimpositions vertically [12].

Aki et al. [13] demonstrated that with age the symphysis increases in height and width with the height increasing at a greater intensity, thus increasing the height to width ratio. Similar findings were shown by Gutermann et al. [14] with less increase in the symphysis width and most of the vertical growth of the symphysis occurring superior to the B point. The depth of the symphysis (measured at B point) was shown to decrease over time and had a high correlation with the lower incisor to the mandibular plane angle [14]. Ricketts [4] showed that a thick, heavy symphysis was associated with the forward rotation of the mandible. Similarly, a thin symphysis was observed in individuals with a dolichocephalic growth pattern. In essence, the anterior growth direction of the mandible would cause a shorter height and larger depth in the mandibular symphysis resulting in a smaller height to depth ratio in both males and females. On the contrary, in cases with a posterior growth direction, the symphysis would exhibit a taller height, a smaller depth, and a larger ratio [13, 15]. The angular relationship of the symphysis to the mandibular plane is larger in the hypodivergent patient, whereas it is more acute in the hyperdivergent patient [15].

Cortical plate thickness increases in low-angle subjects and decreases in high-angle subjects [16, 17]. According to Swasty et al. [17], while keeping age constant, for every 1° increase in mandibular plane angle, the cortical bone would get thinner, accordingly. According to Yamada et al. [18], as the tooth becomes more upright, or lingually inclined, the supporting alveolar bone becomes thinner. When the relationship of teeth to the anterior wall of the symphysis is considered, central incisors seem to be closer to the cortical plates in hyperdivergent facial growth patterns compared to hypodivergent patterns [15]. There is a negative correlation between the incisor mandibular plane angle (IMPA°) with IMPA° lower in individuals with increased mandibular divergence [14, 19]. This is a direct result of compensatory growth in hyperdivergent individuals. The increase in the anterior facial height would require the incisors to erupt to maintain overbite. As a consequence the alveolus becomes attenuated with thinning of the width between labial and lingual walls [1].

Position of the mandibular incisors seems to be affected with the anterior-posterior (A-P) growth as well. In less than ideal A-P skeletal growth patterns, a natural dentoalveolar compensation often occurs to help improve the malrelationship between the basal bone of the maxilla and the mandible [20–27]. This compensation is typically expressed as proclined mandibular incisors when the skeletal relationship is Class II and retroclined mandibular incisors when in Class III [24, 27, 28]. Accordingly, significant differences in the bone width labial or lingual to the apices are reported as a result of osseous compensation to A-P discrepancies [1]. While the position of the incisors may change as a result of the differential A-P growth, the shape of the symphysis is more likely dependent on the degree of vertical divergency in any given malocclusion.

10.3 Prediction of Mandibular Growth Pattern by the Symphysis

Because mandibular symphysis remodeling is affected by the growth pattern, in an early investigation, Bjork [21] used the inclination of the mandibular symphysis in predicting mandibular rotation. He stated that in a vertical condylar growth pattern, the symphysis will swing forward in the face and with increased chin prominence, while in the horizontal type, it is swung back with a receding chin. Similarly, it was deemed as possible to retrospectively predict mandibular growth rotations by attributing 86% of the variance to changes in certain mandibular structures including the inclination of the mandibular symphysis [29]. These reports included individuals with extreme mandibular rotations. It was not until much later that these early observations were deemed as not applicable to mild to moderate rotations of the mandible [30, 31]. In fact, a study proved that clinicians could not predict mandibular rotation better than chance itself using cephalometric variables [32]. On the contrary, Gutermann et al. [14] observed a significant to highly significant negative correlation between the mandibular divergence and the angulation of the lower incisor. They noted that this correlation was best observed late in puberty, which might be the reason why other studies had not found any link in the prediction of growth rotation of the jaws.

10.4 Clinical Significance

A 2008 report demonstrated that only 18% of orthodontic cases presenting with arch-length deficiency are treated with extractions [33]. This number has been on a decline in the United States. The primary goal of orthodontic case management is to achieve proper articulation of teeth that are ideally aligned along a functional occlusal plane with adequate axial and bucco-lingual inclinations. Having an accurate understanding of the anatomic and/or physiologic distance in which teeth can be moved through the alveolar bone while ensuring long-term stability and minimal iatrogenic effects is critical for proper treatment planning in orthodontics. Proffit [34] argues that anterior movement of the mandibular incisors is limited to 2 mm from a stability point of view. In cases with moderate to significant space discrepancies, the fundamental decision that needs to be made by the clinician is whether there would be a need for tooth extractions in the individual's treatment planning.

Extraction decision may not be straightforward in all orthodontic cases. Even though the soft tissue profile following orthodontic treatment is primarily a result of the biomechanical management of the case, most orthodontists are still concerned about extractions. This is especially true for individuals with flattened profiles and retrusive lips. The 13-year-old female presented in Fig. 10.1a–d was cognizant of her soft tissue profile and smile. She presented with bilateral Class II molar and canine relationships. Both maxillary and mandibular incisors were proclined. She had an OJ of 6.2 mm and virtually no crowding in the mandibular arch. However, the maxillary right canine was partially blocked out, and there was 6.1 mm crowding in the maxillary arch. While she did not present with any significant A-P skeletal discrepancies, she possessed a horizontal growth pattern and a dental OB of 4 mm. The main problems of treating a case like this lie in the facial appearance. She had a slightly convex soft tissue profile with increased nasolabial angle (125°) and a relatively prominent nose with thin lips.

From a diagnostic point of view, the lower arch did not warrant any extractions. Because of the maxillary arch crowding and moderate increase in overjet, upper premolar extractions were considered. However, soft tissue examination, mainly the prominence of the nose and flat retrusive lip profile, made this treatment option a bit challenging. Maintaining the upper lip position by maintaining the maxillary incisor position seemed to be a better treatment objective for the facial esthetics. Expansion of the upper arch and interproximal enamel reduction would easily provide space for the resolution of crowding. The remaining problems were the correction of inter-arch A-P discrepancy and overjet. Considering that she did not have much remaining growth, using a fixed-type Class II corrector to advance the lower dentition was considered. This thought brought the attention to the mandibular incisors.

She was presenting with proclined mandibular incisors situated in a fairly thick and short mandibular symphysis, which are common in the brachyfacial growth type. Careful examination of the symphysis and bone support using radiological data supported with clinical evaluation led to the decision to allow for 1–2 mm advancement in the incisor position in the treatment plan. Class II mechanics could potentially cause more than 2 mm of incisor advancement. Therefore, her

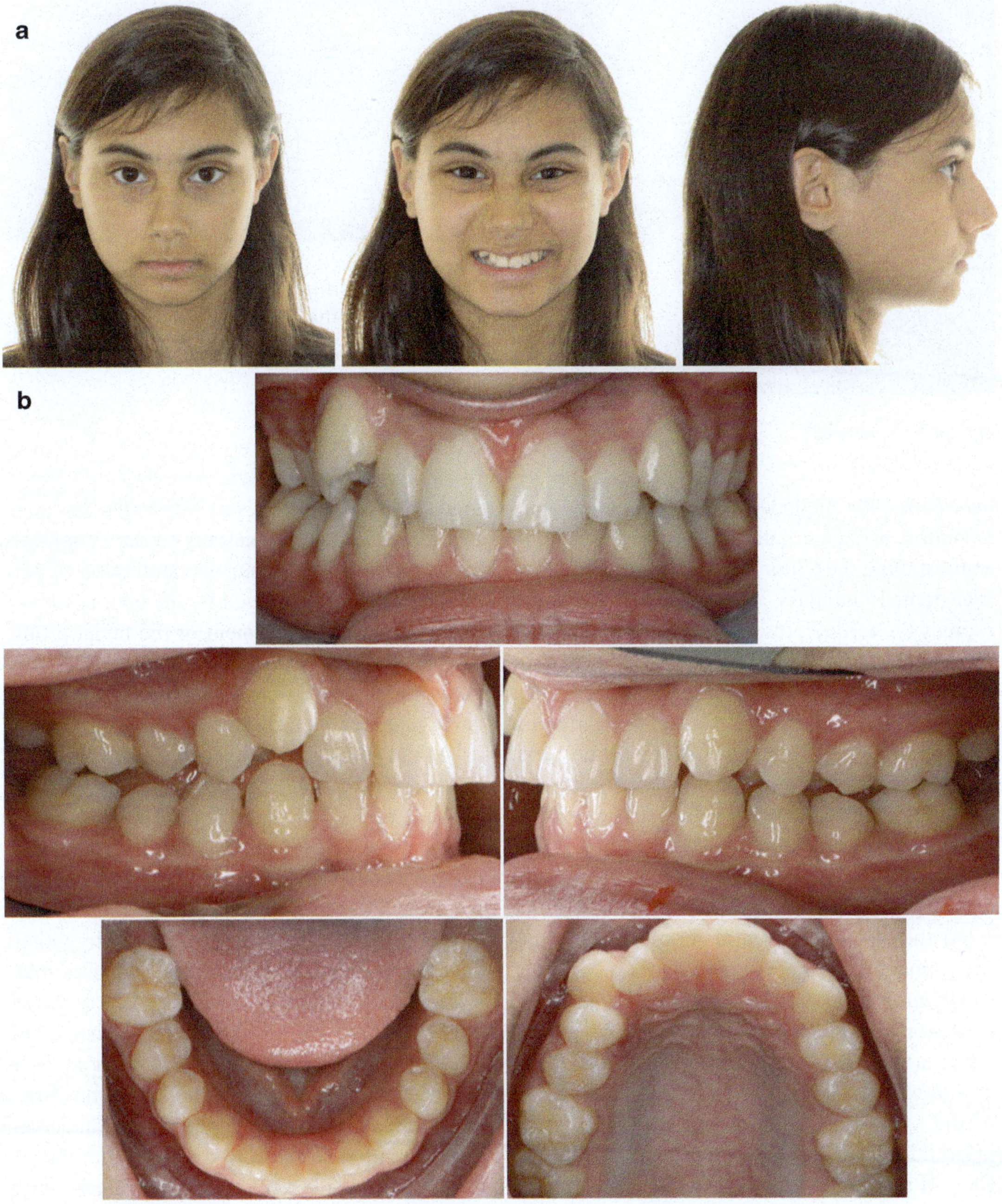

Fig. 10.1 (**a**) Case, pretreatment extraoral photographs. (**b**) Case, pretreatment intraoral photographs. (**c**) Case, pretreatment cephalometric analysis. (**d**) Case, pretreatment panoramic radiograph

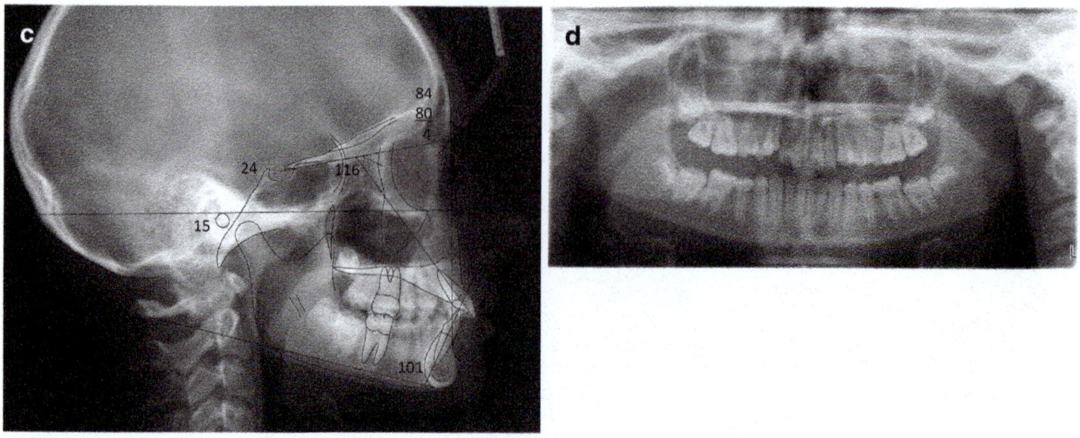

Fig. 10.1 (continued)

treatment plan included interproximal enamel reduction, using a negative 6° torque prescription and installing the Class II corrector on a full-size heat-treated stainless steel archwire to fully express the torque prescription. A fixed Class II corrector was used for a total of 4 months until A-P correction was achieved. Detailing of the case was achieved through bracket repositioning and heat-activated NiTi wires. At that time, seating elastics began to be used. Final settling of the case was achieved by segmenting the maxillary archwire distal to the incisors and 10 more days of multiple triangle seating elastics followed with a tooth positioner.

At the end of the treatment, all of the treatment objectives were met (Fig. 10.2a–e) without significantly flaring the mandibular incisors. Cephalometric evaluation of the mandibular incisor area revealed that there was adequate bone covering the incisors. The short and thick nature of the symphysis allowed for advancement of mandibular incisors (1 mm; 2°) as a result of Class II mechanics. Cephalometric evaluation of the mandibular incisors in the classical sense did not serve as a limitation to the treatment planning. While this case report serves an example of "treat according to the face" approach, the ultimate clinical decision was supported by many factors including the anatomy of the symphysis, growth pattern, and clinical examination of periodontal tissues. As judged by the quality of final occlusion and improvement of the facial profile, the treatment was deemed as successful. However, this type of approach may not be applicable in a random orthodontic patient.

As the facial growth becomes more divergent, the mandibular incisors become more upright. Yamada et al. [18] found that as the mandibular incisors become more upright, the supporting bone becomes thinner. Thus, in high-angle patients, incisors are typically observed to be closer to both cortical plates, which hypothetically reduces the distance an incisor can move without encountering detrimental sequelae. Understanding the position of the incisors, specifically the apices, in the mandibular alveolus provides context to what orthodontic movement may successfully be accomplished. Advancement of the mandibular incisors would require the apices to follow the crowns for excellent torque control as exemplified in the clinical case.

In the following study, we aimed to evaluate mandibular incisor position and morphological measurements of the mandibular symphysis in a

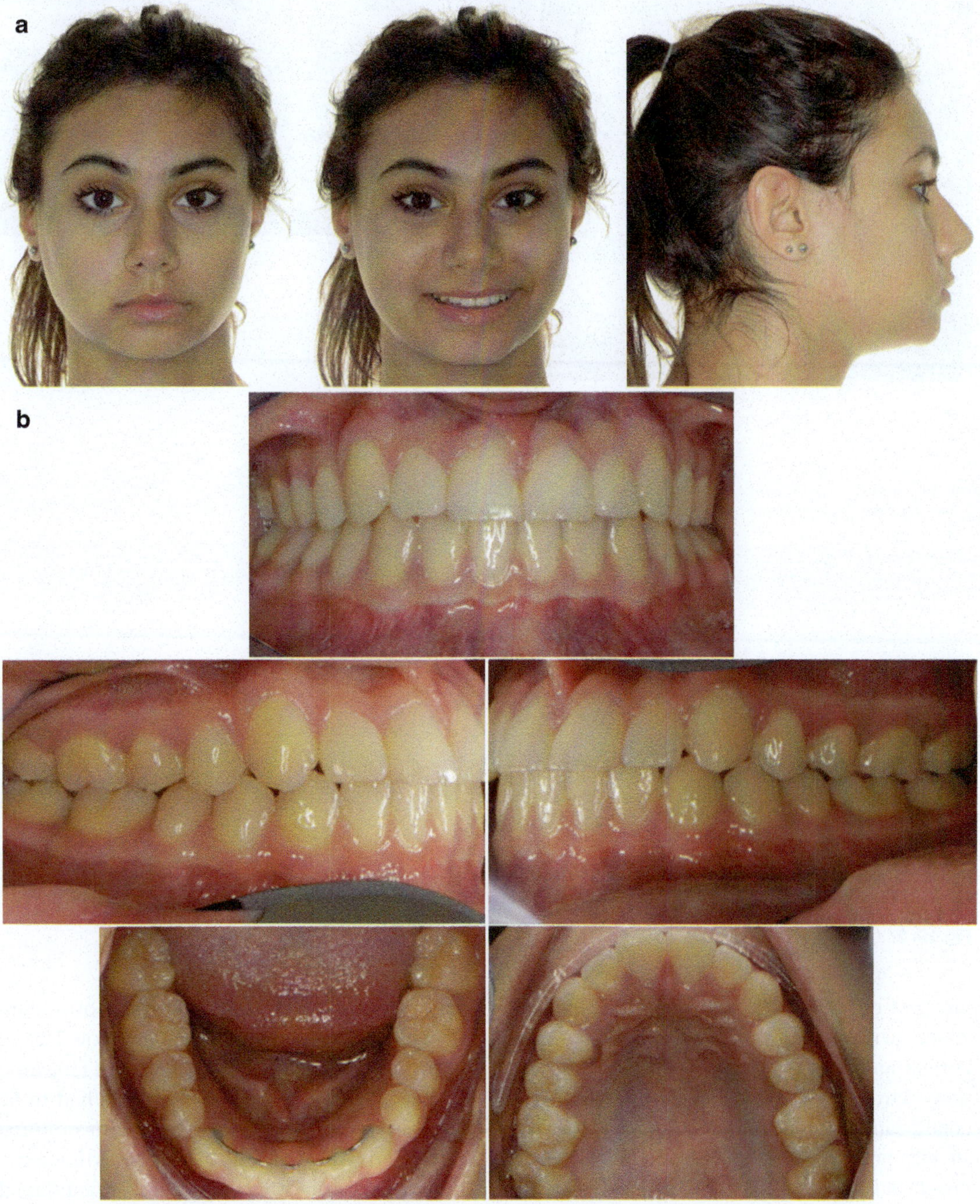

Fig. 10.2 (**a**) Case, posttreatment extraoral photographs. (**b**) Case, posttreatment intraoral photographs. (**c**) Case, posttreatment cephalometric analysis. (**d**) Case, posttreatment panoramic radiograph. (**e**) Cranial base and local superimpositions

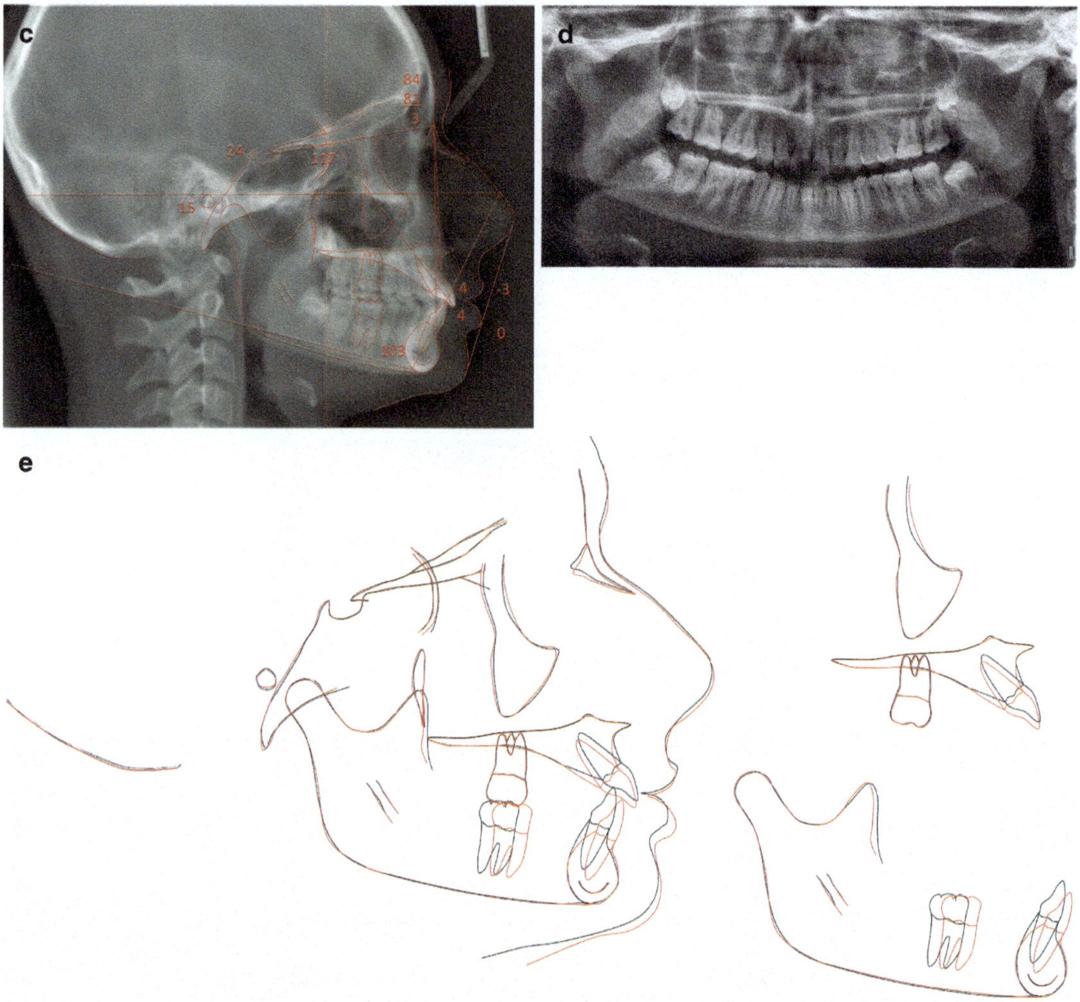

Fig. 10.2 (continued)

group of orthodontic patients classified in accordance with the growth pattern (high-angle, normal-angle, and low-angle) and incisor inclination (retroclined, normal, and proclined) using cone-beam computed tomography. Our data should provide the clinicians with additional insight adjunct to cephalometric analysis when planning their cases.

10.5 The Sample

The sample was selected from an existing database of patients who had received a lateral cephalometric x-ray and cone-beam computed tomography (CBCT) as part of their orthodontic evaluation at the University of Texas Health Science Center School of Dentistry at Houston. To be enrolled in the study, subjects had to be between 11 and 18 years of age with minimal incisor crowding and no previous orthodontic treatment, trauma, syndromes, craniofacial malformations, surgical intervention, or large dental restorations that would interfere with a CBCT evaluation of the dentition. A total of 510 patients were identified. Individuals with ANB values <0° and >6° were excluded from the study. Ninety subjects (48 boys, 42 girls) who met the inclusion and exclusion criteria were included in the study. The study sample

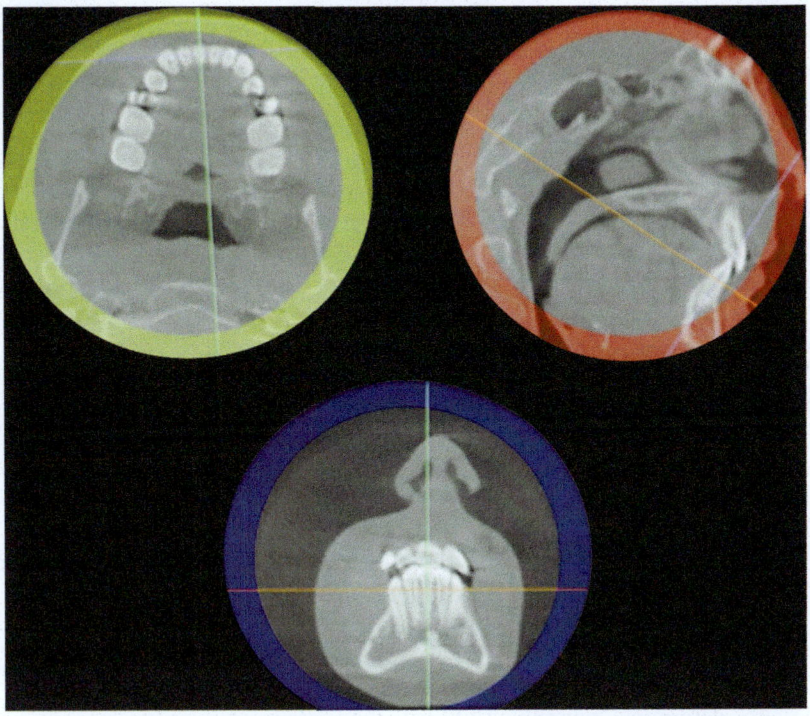

Fig. 10.3 Orientation of the symphysis area for image analysis: axial, sagittal, and coronal slices were arranged for each individual incisor to target the center along the long axis of the tooth to display the bony support and symphysis area

Table 10.1 Sample distribution

Groups	Retroclined	Upright	Proclined
Hypodivergent	$n = 8$	$n = 12$	$n = 10$
Normodivergent	$n = 10$	$n = 13$	$n = 7$
Hyperdivergent	$n = 11$	$n = 12$	$n = 7$

was categorized into three divergent facial growth patterns utilizing Frankfort horizontal to mandibular plane angle (FMA): hypodivergent (n = 30, FMA < 21°), normodivergent (n = 30, FMA 21°–29°), and hyperdivergent (n = 30, FMA > 29°). The subjects were then subcategorized within the growth pattern groups utilizing the lower incisor to mandibular plane angle: retroclined (n = 29 L1-MP < 86°), upright (n = 37, L1-MP = 87°–99°), and proclined (n = 24, L1-MP > 100°) (Table 10.1; Fig. 10.3).

Cone-beam computed tomography (CBCT) were obtained using the Galileos Comfort (Sirona Dental Systems GmbH, Bensheim, Germany) x-ray unit at the following exposure parameters: 85 KVp; 21 mA; exposure time, 14 s; 0.3 mm voxel size; and volume dimensions of 15 cm × 15 cm × 15 cm. CBCT scans were evaluated by the same examiner using Invivo 5 (version 5.1, Anatomage, San Jose, CA) for analysis. Each of the mandibular central and lateral incisors was analyzed individually. Images were oriented using sagittal, coronal, and axial slices. The sagittal alignment was through a straight line from the midpoint of incisor tip to the apex. The coronal and axial alignments were through a straight line that bisected the incisor in half along the long axis of the tooth (Fig. 10.3). The resulting sagittal slice (0.5 mm thickness) was used for analysis. On the sagittal slices, the center of resistance (C_{res}) was marked on half distance of the root from the apex to the marginal bone level. Then a full circle was drawn on the C_{res} to perform the measurements. Measurements (Table 10.2, Fig. 10.4) were repeated for each mandibular incisor.

Two-way analysis of variance (ANOVA) was utilized for comparing the measurements with the two main effects: growth pattern and incisor inclination. Multiple comparisons were made using the Bonferroni post hoc tests. Level of significance was established at

Table 10.2 Measurements

Alveolar width	Thickness of the alveolar bone at the apex as measured along the arc centered on the hypothetical center of rotation to the external cortical plates on a sagittal slice
Buccal medullary width	Thickness of the facial (*F*) medullary bone segment at the apex as measured along an arc centered on the hypothetical center of rotation from the apex to the internal cortical plate on a sagittal slice
Symphysis height	Height of the mandibular symphysis as measured from the most superior portion of the mandibular alveolar process to the most inferior border of the cortical plate on a sagittal slice parallel to the NB line (Sp-Ca)
W/*H* ratio	Ratio of the alveolar width to symphysis height

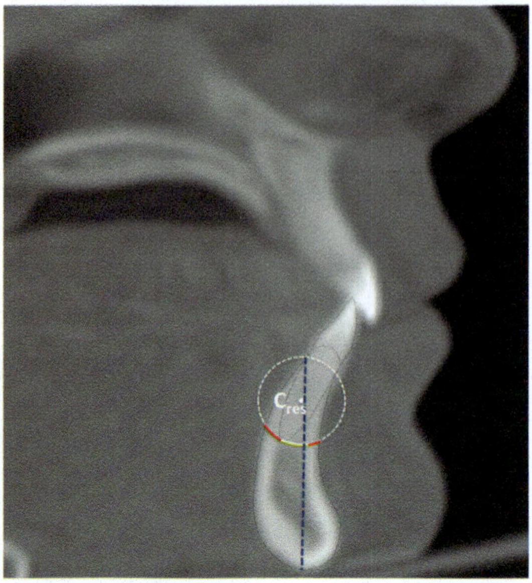

Fig. 10.4 Measurements used in the study: buccal medullary bone width (yellow), alveolar width (orange + yellow), symphysis height (blue)

$p < 0.05$. All statistical analyses were accomplished using the SPSS for Mac (version 21; IBM, Armonk, NY).

To determine reproducibility of the data, Dahlberg's formula was used [35]. The method error was calculated from the equation, $\mathrm{ME} = \sqrt{\frac{\Sigma d^2}{nx2}}$, where *d* is the difference between duplicated measurements and n is the number of replications. All measurements were repeated after 4 months for ten randomly selected individuals (Fig. 10.4).

Our results indicated that the error of method varied between 0.09 and 0.62 for the linear measurements used in this study.

10.6 Findings

Mean values, standard deviations, and comparisons between the groups were individually listed for mandibular right lateral (Table 10.3), right central (Table 10.4), left lateral (Table 10.5), and left central (Table 10.6) incisors. Alveolar width and buccal medullary width showed decreases and symphysis height increased in mean values as the mandibular divergence increased. Significant differences occurred mostly between low- and high-angle followed by low- and normal-angle individuals ($p < 0.05$). Alveolar width to symphysis height ratio also showed decreases as the downward mandibular plane rotation increased. Similarly, hypodivergent and hyperdivergent individuals demonstrated significant differences in their *W*/*H* ratio.

Incisor inclination, as a main effect, yielded variable findings. Symphysis height was not affected by incisor inclination. As a general trend, alveolar width was thinnest in individuals with retroclined incisors and got thicker as proclination increased. No significant growth pattern and

Table 10.3 Measurements for mandibular right lateral incisor

Measurements	Hypodivergent, mean (SD)	Normodivergent, mean (SD)	Hyperdivergent, mean (SD)	*p*	Retroclined, mean (SD)	Upright, mean (SD)	Proclined, mean (SD)	*p*
Alveolar width (mm)	10.77 (2.33)	9.69 (2.06)	8.86 (1.49)	<0.05[a]	9.00 (1.98)	9.79 (1.98)	10.68 (2.20)	<0.05[b]
Buccal medullary width (mm)	3.95 (1.50)	3.29 (0.90)	2.63 (1.04)	<0.05[a]	2.98 (0.81)	3.60 (1.56)	3.19 (1.21)	NS
Symphysis height (mm)	29.05 (2.35)	30.49 (2.18)	31.85 (2.75)	<0.05[a]	30.25 (2.95)	30.30 (2.70)	30.97 (2.28)	NS
W/*H* ratio	0.37 (0.10)	0.32 (0.07)	0.28 (0.05)	<0.05[a, c]	0.30 (0.09)	0.33 (0.08)	0.35 (0.08)	NS

$p < 0.05$
[a]Hypodivergent—hyperdivergent
[b]Retroclined—proclined
[c]Hypodivergent—normodivergent

Table 10.4 Measurements for mandibular right central incisor

Measurements	Hypodivergent, mean (SD)	Normodivergent, mean (SD)	Hyperdivergent, mean (SD)	*p*	Retroclined, mean (SD)	Upright, mean (SD)	Proclined, mean (SD)	*p*
Alveolar width (mm)	10.64 (2.11)	9.18 (2.18)	8.58 (1.69)	<0.05[a, b]	8.70 (1.96)	9.80 (2.20)	9.86 (2.20)	NS
Buccal medullary width (mm)	3.70 (1.30)	3.02 (0.96)	2.45 (1.00)	<0.05[a, b]	2.49 (0.91)	3.30 (1.31)	3.37 (1.12)	<0.05[c, d]
Symphysis height (mm)	28.96 (2.25)	30.22 (2.40)	31.95 (2.56)	<0.05[b, e]	30.20 (2.88)	30.18 (2.78)	30.88 (2.29)	NS
W/*H* ratio	0.37 (0.09)	0.31 (0.07)	0.27 (0.06)	<0.05[a, b]	0.29 (0.09)	0.33 (0.09)	0.32 (0.08)	NS

$p < 0.05$
[a]Hypodivergent—normodivergent
[b]Hypodivergent—hyperdivergent
[c]Retroclined—upright
[d]Retroclined—proclined
[e]Normodivergent—hyperdivergent

Table 10.5 Measurements for mandibular left lateral incisor

Measurements	Hypodivergent, mean (SD)	Normodivergent, mean (SD)	Hyperdivergent, mean (SD)	p	Retroclined, mean (SD)	Upright, mean (SD)	Proclined, mean (SD)	p
Alveolar width (mm)	10.44 (2.15)	9.18 (2.14)	9.23 (2.52)	NS	8.43 (1.74)	9.95 (2.05)	10.54 (2.80)	$<0.05^{a,b}$
Buccal medullary width (mm)	4.02 (1.44)	3.09 (1.03)	2.88 (1.20)	$<0.05^{c,d}$	2.84 (0.97)	3.70 (1.52)	3.35 (1.19)	$<0.05^{a}$
Symphysis height (mm)	29.09 (2.21)	30.21 (1.98)	31.95 (2.77)	$<0.05^{d,e}$	30.26 (2.75)	30.22 (2.70)	30.90 (2.30)	NS
W/*H* ratio	0.36 (0.08)	0.31 (0.07)	0.29 (0.08)	$<0.05^{c,d}$	0.28 (0.08)	0.33 (0.08)	0.34 (0.09)	$<0.05^{a,b}$

$p < 0.05$
[a]Retroclined—upright
[b]Retroclined—proclined
[c]Hypodivergent—normodivergent
[d]Hypodivergent—hyperdivergent
[e]Normodivergent—hyperdivergent

Table 10.6 Measurements for mandibular left central incisor

Measurements	Hypodivergent, mean (SD)	Normodivergent, mean (SD)	Hyperdivergent, mean (SD)	p	Retroclined, mean (SD)	Upright, mean (SD)	Proclined, mean (SD)	p
Alveolar width (mm)	10.38 (1.98)	9.41 (2.08)	8.46 (1.55)	$<0.05^{a}$	8.75 (2.00)	9.61 (1.93)	9.92 (2.06)	NS
Buccal medullary width (mm)	3.87 (1.33)	3.16 (0.95)	2.49 (0.92)	$<0.05^{a,b}$	2.76 (0.89)	3.39 (1.36)	3.36 (1.22)	NS
Symphysis height (mm)	28.74 (2.26)	30.04 (2.25)	32.04 (2.62)	$<0.05^{a,c}$	30.13 (2.87)	30.07 (2.80)	30.75 (2.45)	NS
W/*H* ratio	0.36 (0.09)	0.32 (0.08)	0.27 (0.05)	$<0.05^{a,b,c}$	0.30 (0.09)	0.32 (0.08)	0.33 (0.08)	NS

$p < 0.05$
[a]Hypodivergent—hyperdivergent
[b]Hypodivergent—normodivergent
[c]Normodivergent—hyperdivergent

incisor inclination interactions were found for any of the variables (Table 10.7). In general, these findings showed that varying degree of incisor inclination caused similar trends in the growth pattern groups.

When the averages from all incisors were combined (Table 10.8), the trends became more evident. As the incisor proclination increases, alveolar width increases in all growth pattern groups. In general, individuals with retroclined incisors have less buccal medullary bone and alveolar widths.

10.7 Clinical Interpretation and Discussion

In less-than-ideal growth patterns, natural dento-alveolar compensations often occur to help improve the relationship between the maxilla and mandible. The findings from our study showed that the low-angle group had a wider alveolar and medullary width compared to the normal-angle and high-angle groups. Corresponding to these measurements, the overall ratio of width to height was greatest in the low-angle group. This was in agreement with the findings of a previous CBCT study [15] that identified a shorter and thicker symphyseal pattern in a low-angle patient. When looking at the high-angle group, the height of the symphysis was significantly greater compared to the other two groups, which was also consistent with findings of Swasty et al. [17] and Molina-Berlanga et al. [26].

Overall, it appears that as the facial growth pattern becomes more divergent, there is a compensatory symphysis growth mechanism to lengthen and narrow, which masks the underlying discrepancy and improves the occlusion. Clinically, this mechanism can severely limit the abilities of the clinician to treat high-angle cases with sagittal discrepancies. In our study, buccal medullary width exhibited differences, with the low-angle group demonstrating a greater width than the other two groups. In analyzing the mandibular incisor inclination patterns, several findings were significant. The upright incisor groups demonstrated a fairly consistent pattern of having a narrower medullary width at the incisor apices compared to the other two groups. This is in agreement with the findings of Yamada et al. [18] and Gutermann et al. [14].

In our study, the subjects were divided into three incisor inclination groups (retroclined, upright, proclined) within the divergence facial growth pattern groups (hypodivergent, normodivergent, and hyperdivergent). It appeared that the dimensions of the symphysis were influenced by both the mandibular incisor inclination and growth pattern. Therefore, both of these factors should be accounted for in the final positioning of

Table 10.7 Growth pattern interaction with incisor inclination

	LR2	LR1	LL1	LL2
	p	*p*	*p*	*p*
Alveolar width-S	0.78	0.74	0.39	0.99
Buccal medullary width	0.07	0.71	0.57	0.55
Symphysis height	0.33	0.46	0.29	0.41
W/*H* ratio	0.65	0.77	0.21	0.92

Table 10.8 Means and standard deviations of measurements for all four incisors

		Alveolar width		Buccal medullary width		Symphysis height		*W*/*H* ratio	
Growth pattern	Incisors	Mean	SD	Mean	SD	Mean	SD	Mean	SD
Hypodivergent	Retroclined	9.94	2.46	3.13	0.94	27.99	2.39	0.36	0.12
	Upright	10.69	1.97	3.49	1.48	28.39	2.03	0.38	0.08
	Proclined	10.89	1.66	3.76	0.92	30.41	1.61	0.36	0.07
Normodivergent	Retroclined	8.24	1.17	2.81	0.57	29.87	2.75	0.28	0.06
	Upright	9.63	2.28	3.28	1.05	30.61	2.06	0.32	0.08
	Proclined	10.48	1.66	3.35	0.69	30.08	1.42	0.35	0.06
Hyperdivergent	Retroclined	8.27	1.30	2.47	0.78	32.13	1.81	0.26	0.04
	Upright	9.06	1.05	2.72	0.86	31.53	3.04	0.29	0.04
	Proclined	9.10	2.75	2.66	1.19	32.35	3.24	0.28	0.08

the mandibular incisors. In individuals with Angle Class I malocclusions and mild to moderate crowding, the distance between the cemento-enamel junction and the alveolar crest around the mandibular incisors was about 1.7 mm and increased to 2.2 mm at the end of the orthodontic treatment [36]. According to Garlock et al. [37], marginal alveolar bone loss following nonextraction therapy was 1.1 mm. These numbers are derived from study samples without accounting for growth pattern and initial incisor inclination. Correction of deep curve of Spee, mesially directed forces from Class II mechanics, and the addition of other discrepancies could possibly cause more problems as the anterior limit of the dentition would be compromised.

Historically, orthodontists relied on cephalometric variables to control for the incisor position and inclination. However, appreciation of bony support around the incisors before and after the orthodontic treatment might not be evident from the cephalogram since all four incisors superimpose on each other. Figure 10.5 demonstrates a digitally scanned dry mandible displaying uneven marginal bone around the mandibular incisors. It is clear that a two-dimensional cephalogram would not be able to reproduce the actual clinical scenario adequately. Cone-beam computed tomography (CBCT) and the associated rendering software are free of magnification and superimposition errors with the ability to isolate areas through custom slices. Methods described in this chapter could be used to determine whether or not required anterior dental movement and consequent uprighting would cause any periodontal sequelae in borderline cases using CBCT. It is demonstrated in Fig. 10.6a–d how the orthodontist could visualize the bone support around the anterior teeth step by step.

Using the sagittal slice, orthodontists can measure the buccal medullary width and compare it to the norms provided in Table 10.8 for the individual patient. One should be cognizant of the fact that these are averages from a normative sample. Additionally, individuals included in our study had minor mandibular incisor crowding. Therefore, if the initial buccal medullary width is smaller than the averages provided in Table 10.8 for the initial malocclusion, maintaining the incisor position at the very least would be indicated for a healthy and stable bony support at the end of the treatment. Even though the CBCT provides much better detail than the cephalogram, it is still limited by factors such as the voxel size and spatial resolution. Therefore, the findings reported in our study should be used

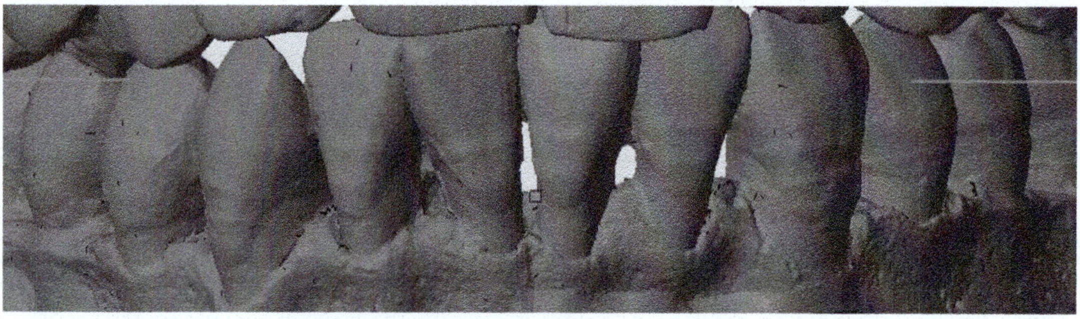

Fig. 10.5 Marginal bone levels around mandibular incisors shown on a dry skull

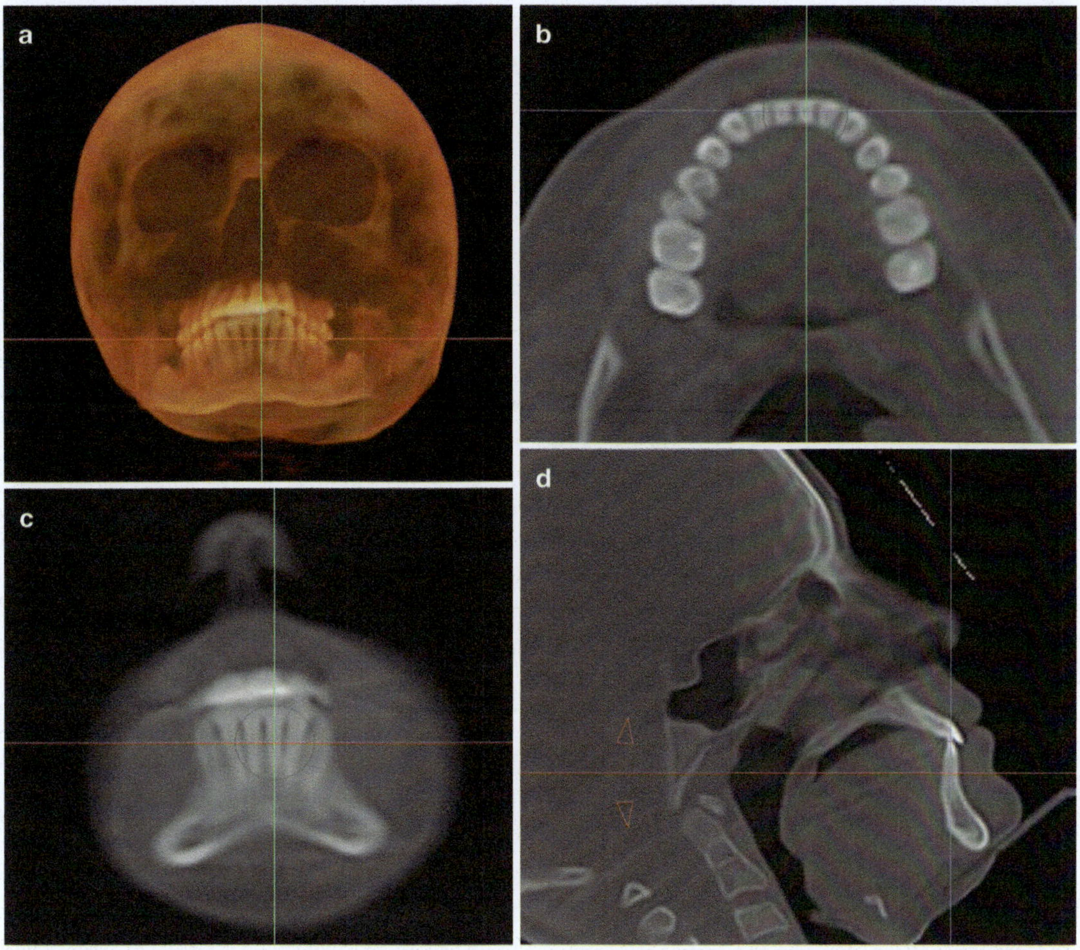

Fig. 10.6 (**a**) Start with the tooth in question. Mandibular left central incisor is viewed in this example. (**b**) On the axial slice, at about where the marginal bone starts, adjust the frontal (blue) and sagittal (green) planes to intersect on the center of the root as shown. (**c**) Frontal slice: confirm the level of axial plane (orange), and readjust according to the CEJ and marginal bone level. Confirm that the sagittal plane is along the long axis of the tooth. (**d**) Resulting sagittal slice, which was utilized for the measurements used in our study, would display the bone support around the tooth individually. This view was used for all the measurements in our study

cautiously. There is always a possibility of a thin layer of the bone covering the incisors that is not captured due to the voxel size of the scan. In some circumstances, scanning the entire buccal surface of the tooth by adjusting the sagittal plane (green) mesiodistally could also provide a thorough visual evaluation of the buccal bone. Mandibular right central incisor shown in Fig. 10.7 demonstrates the lack of bony support on the buccal aspect of the tooth toward the distal. These local dehiscences might never be

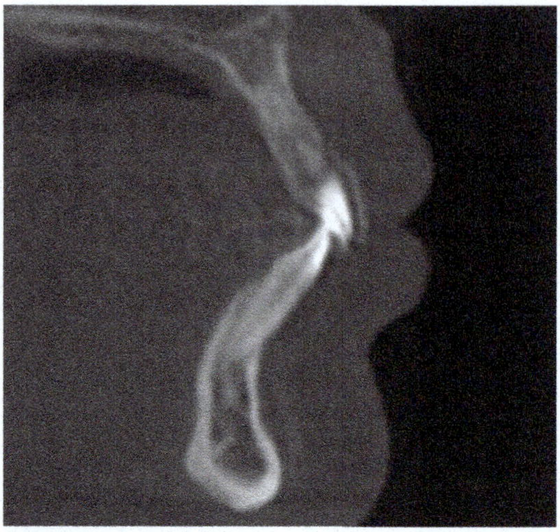

Fig. 10.7 Presence of dehiscence on the facial aspect and distal half of a central incisor

noticed by conventional radiography and could worsen with orthodontic treatment.

10.8 Conclusions

Anterior limit of the mandibular dentition is affected by both the initial incisor inclination and the growth pattern. During the initial diagnosis and treatment planning phase, cephalometric measurements can be supported by measurements obtained from CBCT images. Visual and analytic inspection of the mandibular symphysis and the bony support around the mandibular incisors can help with the decision-making process of the orthodontist so as to prevent violation of the anterior limit of the dentition.

References

1. Handelman CS. The anterior alveolus: its importance in limiting orthodontic treatment and its influence on the occurrence of iatrogenic sequelae. Angle Orthod. 1996;66:95–110.
2. Tweed C. The Frankfort-mandibular incisor angle in orthodontic diagnosis, treatment planning and prognosis. Angle Orthod. 1954;24:121–69.
3. Steiner CC. Cephalometrics in clinical practice. Angle Orthod. 1959;29:8–29.
4. Ricketts CM. Cephalometric synthesis. Am J Orthod. 1960;46:647–73.
5. Baumrind S, Frantz R. The reliability of head film measurements. 1. Landmark identification. Am J Orthod. 1971a;60:111–27.
6. Baumrind S, Frantz R. The reliability of head film measurement. 2. Conventional angular and linear measures. Am J Orthod. 1971b;60:505–17.
7. Fuhrmann R. Three-dimensional interpretation of labiolingual bone width of the lower incisors. J Orofac Orthop. 1996;57:168–85.
8. Nauert K, Berg R. Evaluation of labio-lingual bony support of lower incisors in orthodontically untreated adults with the help of computed tomography. J Orofac Orthop. 1999;60:321–34.
9. Baumgartel S, Palomo JM, Palomo L, Hans MG. Reliability and accuracy of cone-beam computed tomography dental measurements. Am J Orthod Dentofac Orthop. 2009;136:19–28.
10. Timock AM, Cook V, McDonald T, Leo MC, Crowe J, Benninger BL, Covell DA Jr. Accuracy and reliability of buccal bone height and thickness measurements from cone-beam computed tomography imaging. Am J Orthod Dentofac Orthop. 2011;140:734–44.
11. Enlow DH, Hans MG. Essentials of facial growth. In: W. B. Saunders. Philadelphia, PA: Pennsylvania; 1996.
12. Buschang PH, LaPalme L, Tanquay R, Demirjian A. The technical reliability of superimposition on cranial base and mandibular structures. Eur J Orthod. 1986;8:152–6.
13. Aki T, Nanda RS, Currier GF, Nanda SK. Assessment of symphysis morphology as a predictor of the direction of mandibular growth. Am J Orthod Dentofac Orthop. 1994;106:60–9.
14. Gutermann C, Peltomaki T, Markic G, Hanggi M, Schaltzle M, Signorelli L, Patcas R. The inclination of mandibular incisors revisited. Angle Orthod. 2014;84:109–19.
15. Gracco A, Luca L, Bongiorno MC, Siciliani G. Computed tomography evaluation of mandibular incisor bony support in untreated patients. Am J Orthod Dentofac Orthop. 2010;138:179–87.
16. Tsunori M, Mashita M, Kazutaka K. Relationship between facial types and tooth and bone characteristics of the mandible obtained by CT scanning. Angle Orthod. 1998;68:557–62.
17. Swasty D, Lee J, Huang JC, Maki K, Gansky SA, Hatcher D, Miller AJ. Cross-sectional human mandibular morphology as assessed in vivo by cone-beam computed tomography in patients with different vertical facial dimensions. Am J Orthod Dentofac Orthop. 2011;139:e377–89.
18. Yamada C, Kitai N, Kakimoto N, Murakami S, Furukawa S, Takada K. Spatial relationships between the mandibular central incisor and associated alveolar bone in adults with mandibular prognathism. Angle Orthod. 2007;77:766–72.
19. Eroz UB, Ceylan I, Aydemir S. An investigation of mandibular morphology in subjects with different vertical facial patterns. Aust Orthod J. 2000;16:16–22.

20. Bjork A. Variations in the growth pattern of the human mandible: longitudinal radiographic study by the implant method. J Dent Res. 1963;42:400–11.
21. Bjork A. Prediction of mandibular growth rotation. Am J Orthod. 1969;55:585–99.
22. Enlow DH, Kuroda T, Lewis AB. Intrinsic craniofacial compensations. Angle Orthod. 1971;41:271–85.
23. Corelius M, Linder-Aronson S. The relationship between incisor inclination and various reference lines. Angle Orthod. 1976;46:111–7.
24. Bibby RE. Incisor relationship in different skeletofacial patterns. Angle Orthod. 1980;50:41–4.
25. Nielsen IL. Vertical malocclusions: etiology, development, diagnosis and some aspects of treatment. Angle Orthod. 1991;61:247–60.
26. Molina-Berlanga N, Llopis-Perez j F-MC, Puigdollers A. Lower incisor compensation and symphysis dimensions among class I and III malocclusion patients with different facial vertical skeletal patterns. Angle Orthod. 2013;83:948–55.
27. Hernandez-Sayago E, Espinar-Escalona E, Barrerra-Mora JM, Ruiz-Navarro MB, Llamas-Carreras JM, Solano-Reina E. Lower incisor position in different malocclusions and facial patterns. Med Oral Patol Oral Cir Bucal. 2013;18:e343–50.
28. Baysal A, Ucar FI, Buyuk SK, Ozer T, Uysal T. Alveolar bone thickness and lower incisor position in skeletal class I and class II malocclusions assessed with cone-beam computed tomography. Korean J Orthod. 2013;43:134–40.
29. Skieller V, Bjork A, Linde-Hansen T. Prediction of mandibular growth rotation evaluated from a longitudinal implant sample. Am J Orthod. 1984;86:359–70.
30. Lee RS, Daniel FJ, Swartz M, Baumrind S, Korn EL. Assessment of a method for the prediction of mandibular rotation. Am J Orthod Dentofac Orthop. 1987;91:395–402.
31. Leslie LR, Southard TE, Southard KA, et al. Prediction of mandibular growth rotation: assessment of the Skieller, Bjork, and Linde-Hansen method. Am J Orthod Dentofac Orthop. 1998;114:659–67.
32. Baumrind S, Korn EL, West EL. Prediction of mandibular rotation: an empirical test of clinical performance. Am J Orthod Dentofac Orthop. 1984;86:371–85.
33. Keim RG, Gottlieb EL, Nelson AH, Vogels DS 3rd. 2008 JCO study of orthodontic diagnosis and treatment procedures, part 1: results and trends. J Clin Orthod. 2008;42:625–40.
34. Proffit WR, Fields HW Jr, Sarver DM. Contemporary Orthodontics. 5th ed. St. Louis, MO: Mosby; 2015.
35. Dahlberg G. Statistical methods for medical and biological students. London: George Allen and Unwin; 1940. p. 122–32.
36. Castro LO, Castro IO, de Alencar AH, Valladares-Neto J, Estrela C. Cone beam computed tomography evaluation of distance from cementoenamel junction to alveolar crest before and after non-extraction orthodontic treatment. Angle Orthod. 2016;86:543–9.
37. Garlock DT, Buschang PH, Araujo EA, Behrents RG, Kim KB. Evaluation of marginal alveolar bone in the anterior mandible with pretreatment and posttreatment computed tomography in non-extraction patients. Am J Orthod Dentofac Orthop. 2016;149:192–201.

11 The Alveolar Bone and Its Limits

Alejandro Romero-Delmastro, G. Fräns Currier, and Onur Kadioglu

Abstract

The alveolar bone has always been a factor in the decision-making process of the orthodontists, and there has recently been an increasing interest in the dental profession for evaluating the effects of orthodontic treatment on the alveolar bone. Both medical computed tomography (CT) and cone-beam computed tomography (CBCT) have made such evaluations possible under circumstances where direct observation was not practical or feasible.

CBCTs provide accurate imaging of the alveolar bone and other anatomical structures surrounding the teeth. Unlike on conventional 2D radiographs, both the facial and the lingual surfaces of the alveolar bone can be observed and measured on CBCT images. This yields much needed data for clinical in-vivo studies that intend to evaluate alveolar bone changes during and after orthodontic treatment. Several studies have been completed assessing bone changes both in the anterior and posterior segments, as well as in the presence or absence of expansion devices, and in the presence or absence of extractions. Along with these studies, methods have been developed for the purpose of measuring facial and lingual alveolar bone.

11.1 Introduction

Bonding agents and pre-adjusted appliances are probably ranked on the top of the list of things that have revolutionized the practice of orthodontics in the twentieth century, but the introduction of digital imaging technology in the 1990s has led, without a doubt, to many breakthroughs in orthodontics. The way patients are treated today, how diagnosis is approached, and what we are able to understand about the human body have remarkably evolved over the first part of the twenty-first century.

An evidence-based discipline is currently regarded as the standard of care delivery. Evidence-based density (EBD) is the application of an evidence-based approach to dental care. The American Dental Association defines EBD as "an approach to oral health care that requires the judicious integration of systematic assessments of clinically relevant scientific evidence, relating to the patient's oral and medical condition and history, with the dentist's clinical expertise and the patient's treatment needs and preferences" [1]. EBD requires gathering of data and available evidence, assessing its validity, and

A. Romero-Delamstro · G. F. Currier · O. Kadioglu (✉)
Division of Orthodontics, Department of Developmental Sciences, University of Oklahoma Health Sciences Center College of Dentistry, Oklahoma City, OK, USA
e-mail: alejandro-romerodelmastro@ouhsc.edu; frans-currier@ouhsc.edu; onur-kadioglu@ouhsc.edu

O. Kadioglu, G. F. Currier (eds.), *Craniofacial 3D Imaging*,
https://doi.org/10.1007/978-3-030-00722-5_11

using quality evidence to make informed decisions and provide care. According to Harrel [2], 3D imaging and its use over the fourth dimension—time—provide the true anatomical data necessary to expand clinical practice and researchers into EBD.

11.2 Cone-Beam Computed Tomography (CBCT)

CBCT is a relatively low-risk, noninvasive method for assessment of craniofacial structures in situations where direct measurements are not possible or the burden on the patient would be greater than the benefit they would receive from the procedure. CBCT scans currently offer the possibility of observing teeth and bone in ways never thought before.

Tomography was developed in 1967 by Sir Godfrey Hounsfield. The CT was developed in the 1970s as an evolutionary process resulting from the demand for 3D information obtained by conventional computerized tomography (CT) scans. Custom-built craniomaxillofacial CBCTs started appearing in the market place over the last decade of the twentieth century, and since then a variety of applications to the facial and dental environments have been established [3].

Medical CTs create images from multiple slices which are "stacked" to obtain a final complete image, making it time-consuming and less cost-efficient. With dental CBCTs, an entire region of interest is captured with a single rotation of the source as the radiation emitted falls onto a 2D detector. CBCT images are captured utilizing a relatively more focused beam, with lower exposure and acquisition times, considerably less scattered radiation, more X-ray utilization, and at a lower cost than medical CTs [4–11].

The 3D images from a scan are composed of a series of isotropic voxels equal in width, thickness, and height with sizes commonly ranging between 0.1 and 0.4 mm. The reconstruction of these images provides secondary perpendicular axial, coronal, and sagittal slices [9, 12–17]. The raw data obtained from a CBCT is stored in the computer as DICOM files (Digital Imaging and Communications in Medicine files are the standard for handling, storing, printing, and transmitting information in medical imaging). It is the reconstructions of those files what allows for different views to be created from one scan. All of the views are accurate and not subjected to geometric projection errors that are inherent with traditional 2D images as sophisticated computer algorithms correct for the geometric distortions that are inherent in 2D imaging.

Without a doubt, CBCT has rapidly gained popularity as a tool that allows for versatile visualization for diagnosis, comparison, and treatment simulation. It offers views that were not possible with conventional radiographic techniques. The errors related to patient positioning are not present as the data acquired represents a volume that can be oriented as needed.

11.2.1 Radiation and Dental CBCT

Radiation remains a general concern with CBCTs; although its dose is up to four times lower than that of a medical CT, it is still higher than digital 2D panoramic or lateral cephalometric radiographs [12, 18].

The American Association of Orthodontists (AAO) developed in 2010 a set of general guidelines for the optimization, justification, and referral criteria for users of CBCT. The use of this type of technology was still not required for routine orthodontic radiographs, but it is certainly increasing as the third dimension and volumetric measurements can be added to the diagnosis and treatment planning.

Background exposure comes from naturally occurring radiation. People are exposed to radiation from sources in the nature, with the average person in the United States receiving an effective dose of about 3000 μSv per year (+/− 8 μSv per day) from natural sources. The soil and cosmic radiation from outer space are the most common sources of naturally occurring radiation, and people living in high altitudes are more exposed to cosmic radiation than people living at sea level. The largest source of radiation comes from radon gas in homes, which constitutes about 2000 μSv per year. A round trip from Paris to Tokyo adds 139 μSv of effective dose to each passenger due to exposure to cosmic radiation. Most dental radiographs are equivalent to 1 day to 2 weeks of background radiation. The radiation dose for a complete volume of the maxillofacial area is usually more than a panoramic image but less than a periapical survey. The effective radiation dose from a CBCT machine is between 40 and 135 μSv which is about 5–17 days of equivalent natural background radiation (ENBR). One bitewing adds 1 μSv of effective radiation dose, while an occlusal film adds 5 μSv. A cephalometric radiograph adds 5–7 μSv which is less than one day of ENBR; a panoramic radiograph adds between 3 and 24 μSv which is between less than half to 3 days of ENBR. A TMJ series adds 20–30 μSv or 3–4 days of ENBR, and a full mouth series adds 30–170 μSv or 4–21 days of ENBR. A chest X-ray adds 100 μSv or about 10–12 days of ENBR, and a mammography adds 700 μSv or about 88 days of ENBR. A medical CT adds 8000 μSv or about 1000 days of ENBR [19–23].

The amount of radiation depends greatly on the specifications of the particular equipment. Until some years ago, the factors known to affect radiation dose were (1) scan time, (2) voxel size, (3) FOV size which determines the overall amount of ionizing radiation, (4) kVp and mA, (5) rotation as full 360° vs 180°, (6) pulsed vs continuous beam, and (7) radiation filter (type and shape) [24]. The use of lower mAs and collimation are some of the ways to reduce the amount of radiation the patient receives. The main concern with decreasing radiation dose has always been concomitant decrease in image quality. In this regard, some mathematical algorithms appear to be very promising, as they allow for the reconstructions of reliable 2D images at a relatively lower radiation dose and with fewer projections. This means that clinicians should be able to reduce the amount of radiation when obtaining diagnostic radiographs with CBCT [25].

Gamache et al. [26] demonstrated that the radiation exposure dose can be reduced while maintaining adequate image quality and diagnostic accuracy, by performing scans using low kV and moderate-to-high mA settings. In their study, low kV (60 kV) in combination with 4.0–15 mA (moderate-to-high mA settings) produced the best image quality. These settings resulted in a 56% reduction in total radiation amount, from 898 to 396 mGy cm^2. They also showed that even the low image quality settings were able to produce images that presented high levels of accuracy in the detection of root resorption volume as compared to actual physical volume. Manufacturers have recommended settings that do not necessarily need to be labeled as the best settings possible. In general, these are just recommendations or suggestions, giving the ability to control kVp and mA settings on the CBCT to the operator with slight differences from one manufacturer to another.

The ADA Council on Scientific Affairs recommends the use of techniques that would reduce the amount of radiation received during dental radiography. Known as the "as low as reasonably achievable" (ALARA) principle, this includes taking radiographs based on the patient's needs as determined by an examination, using the fastest film compatible with the diagnostic task, collimating the beam to a size as close to that of the film as feasible, and using leaded aprons and thyroid shields.

Currently, the duration of CBCT scans varies from roughly 5 to 75 s. The shorter the exposure, the lower the radiation dose will be. In general, the longer the duration of the scan, the higher the potential for distortion due to possible head movement. Short exposures decrease the total cumulative radiation dose for the patient and are considered beneficial for young patients or those who have a difficulty remaining stationary.

11.3 CBCT in Orthodontics

Custom dental CBCTs appeared in the market place over the first decade of the twenty-first century, and a variety of applications for dentistry have been established since then. In 2005 Kau et al. [3] reported four main system providers, but by 2011, 23 different providers and more than 40 different CBCT scanners were available representing an interest in developing digital technology for dental purposes.

The drastic increase in number of manufacturers and devices will hopefully translate into more options that will offer better image quality and, hopefully, lower prices. Three-dimensional imaging has become more accessible and popular during the second decade of the twenty-first century, which has helped close the gap between the educational environment and the private practice setting [27].

One of the most important features to consider when buying a CBCT scanner for orthodontic purposes is the scanner's field of view (FOV). This determines how much of the patient's anatomy may be visualized in one scan. If one is to replace all other forms of 2D images, then the largest FOV needs to be considered that would include the areas necessary for cephalometric analyses. Scanners with large FOV—height equal to or greater than 16 cm—are useful for traditional orthodontic surveys based on a typical adult male. Medium FOV scanners capture the middle of the orbits down to menton (vertically) and from one condyle to the contralateral condyle (horizontally). Scanners with a medium FOV are useful for panoramic radiographs and implant surveys, but are not suitable for cephalometric analysis. Scanners with a small FOV capture a user-defined region, usually symmetrical in shape. These scanners are useful for implant placement, TMJ surveys, and the localization of impacted teeth [27].

CBCT technology has helped in providing new discoveries in orthodontic research. As the images obtained can show density variations on bone surfaces, it appears to be the tool that was for so long needed to clinically study in live patients the changes associated with growth and clinical orthodontic care.

11.3.1 CBCT Accuracy and Image Quality

Advances in CBCT technology have produced the capability of rendering high-quality 3D images at a submillimeter level that allow 1:1 views along any plane. Therefore, morphologic changes can be evaluated and quantified over time using a relatively low dose of radiation [16, 19, 21, 28].

Still, there is some discrepancy in the literature between the agreements of CBCT measurements and direct measurements. For the most part, it is widely accepted that CBCT can be used to assess bone heights with a high level of accuracy and reliability, but the thickness evaluation is a source for a wider discrepancy.

When comparing CBCT and multi-slice spiral CT (MSCT) measurements, Loubele et al. [29] found that CBCT provided a better visualization of small bony structures, whereas MSCT provided better visualization of cortical bone and the gingival tissues. They found no statistical differences between the accuracy of alveolar bone measurements using CBCT and MSCT, with both showing accuracy to submillimeter levels.

Timock et al. [7] investigated the accuracy and reliability of buccal alveolar bone heights and thicknesses derived from CBCT images. The author used 12 embalmed human cadaver heads (five female and seven male with a mean age of 77 years) that were scanned with an i-CAT® 17–19 unit (Imaging Sciences International, Hatfield, Pennsylvania) at 0.3 mm voxel size. Buccal bone height and thickness measurements of 65 teeth were made in standardized radiographic slices and compared to direct measurements made by dissection. Agreement between the direct and CBCT measurement methods was higher for measurements of heights than thicknesses, but the results showed that CBCT measurements did not differ significantly from direct measurements, with no identifiable patterns of under- or overestimation. The authors concluded that the CBCT could be used with high accuracy and reliability to quantitatively assess buccal bone heights and thicknesses.

Creed et al. [30] evaluated the accuracy of linear measurements obtained from CBCT images and compared them to images acquired from digital models to determine if cone-beam digital models were as accurate as OrthoCAD digital models. Their results showed that linear measurements obtained from CBCT images had a good level of accuracy when compared with OrtoCAD models, with this accuracy adequate for diagnosis and treatment planning.

CBCTs are accurate over long distances, but they may not be when measuring an object in close proximity such as buccal bone width [31]. While CBCTs are capable of producing views of alveolar bone to the submillimetric levels with high levels of reproducibility, the use of CBCT to view buccal bone is not infallible as there might be a small amount of error when comparing direct measurements to CBCT measurements [32, 33].

CBCT measurements are very accurate for short linear distances. They have been shown to be reliable when compared to direct measurements as these two do not differ significantly. The accuracy is higher for measurements of buccal bone heights than thickness and with the differences between physical measurements and CBCT ones being as low as 0.1 mm. There is robust literature supporting the fact that distances can be accurately measured using CBCT and it is adequate for observation of mandibular bony structures, including interproximal, facial, and lingual bone levels [34–37].

Some authors have reported that CBCTs are detailed enough to show dehiscences and fenestrations and have accuracy compatible to periodontal probing for detecting bony defects and identifying root fenestrations and dehiscences. The CBCTs have the highest sensitivity and diagnostic accuracy for detection of multiple periodontal defects among multiple radiographic modalities [38, 39]. On the other hand, others have reported a discrepancy in the identification of dehiscences and fenestrations on CBCTs as compared to direct observations [40].

Currently, there is no difference in accuracy of measurements among scanners with deferring arcs of rotation (360° vs 180°) and number of projection images. Additionally, Kusnoto et al. [25] showed that landmark identification on 2D cephalometric radiographs derived from reconstructed CBCTs at different projection views (300, 150, 70, and 39 projections) did not exceed the clinical 1.5 mm acceptable reference measurement for accuracy of tracings. This means that CBCT could be used with significantly reduced radiation in circumstances when only conventional 2D views are needed.

CBCT has proven to be successful for assessment of alveolar bone grafting [41]. According to Pan and Kau [28], the CBCT is a reliable method for evaluation of interradicular bone mass. One of the biggest advantages of CBCTs is that practitioners can get much more accurate information from one scan than from multiple 2D views that are traditionally used with less radiation. When it comes to precision, it seems that this improves with higher-resolution scans, but this possibly should be reserved for research purposes when high precision is desired.

11.4 CBCT Analysis

The introduction of CBCT has improved the quality of radiographic data and has overcome disadvantages of conventional 2D radiographs. The precision of angular and linear measurements on 2D images and 3D volumes, as well as the ease and reliability of landmark identification, and the precision of superimposition of images have been extensively evaluated [37, 42–49].

CBCT analyses may allow the orthodontist to plan the individual torque needs of a specific patient or tooth, as well as to evaluate the alveolar bone widths and heights that may limit orthodontic tooth movement. Posterior cross sections can allow for buccolingual skeletal and dental evaluations of individual teeth or pairs of teeth. Individual or multiple teeth angulations can be observed and assessed from CBCT images as they offer views that are not possible with any other routinely used radiographic technique. The error related to patient positioning that is present in 2D images commonly used for diagnostic purposes in orthodontics is not present in CBCT images as the data acquired represents volume and can be oriented as needed due to the isotropy of voxels.

The 3D data has expanded the possibilities of measurements for evaluation and comparisons.

Previously known 2D linear and angular measurements can now be complemented by volumetric measurements and curvature measurements. This allows for a better understanding of certain conditions, such as a true mandibular asymmetry, that can be better evaluated on a 3D volumetric image rather than on a conventional 2D lateral cephalometric X-ray and/or photographic images. In order to achieve a sound method for analysis of 3D CBCT images, volume segmentation and image reorientation are paramount and need to be done before the identifications of landmarks. Volume segmentation is the allocation and separation of an anatomical structure or area of interest from the three-dimensional volumes to be able to individually view it. Although a relatively simple concept, the actual segmentation of an individual structure or region can be complicated due to biologic and anatomic reasons. Also, the quality of the scan has a large role in this process; low contrast between structures, noise on the images, and movement during image acquisition, all play a role in the ease of identification of boundaries [24].

The current recommendation for assessment of alveolar bone levels via CBCT images is to limit the movement during the scan as much as possible, the use of smallest FOV suitable for the area to be assessed, the use of a small voxel size and the highest gray-scale bit depth, and, when measuring facial or lingual bone, images viewed through the long axis of the tooth to avoid mistakenly measuring interproximal bone [31, 50]. The importance of accurate landmark identification is inversely proportional to the initial thickness of the tissue to be evaluated or compared. For example, a difference of 0.5 mm in the thickness of the lip or the tip of the nose may be negligible. Thus, accuracy of landmark identification for superimposition may not be as critical, but the same difference may be unacceptable for assessment of thicknesses of alveolar bone.

Bone level measurements on CBCTs can be accurate and precise, but the methods are technique sensitive, with very thin alveolar bone missed or underestimated [50, 51]. Also tooth orientation on the multiplanar reconstruction definitely plays a role on circumstances where bone coverage is irregular. Take, for example, in Fig. 11.1, one can imagine that the bone height measurements are going to be significantly different, depending upon where the cuts are made. Also, the presence of alveolar bone on a specific CBCT cut does not necessarily correspond or translate into coverage around the whole facial or lingual surface of a tooth.

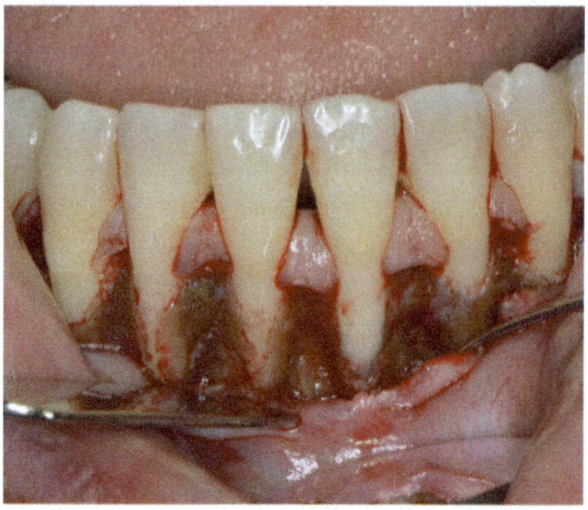

Fig. 11.1 Notice the level of dehiscence present on the teeth. The "relative" presence of alveolar bone coverage around a root on a CBCT image should not be interpreted as coverage around the whole surface

11.5 CBCT Assessment of Alveolar Bone

11.5.1 Tooth-Based Superimposition Method

Until the introduction of three-dimensional (3D) imaging in dentistry, it was not possible for orthodontists to assess facial and lingual alveolar bone by means other than direct observation that required reflecting flaps in the area of interest. Currently, the use of a digital method allows for longitudinal assessment of alveolar bone levels, and the reader can follow a step-by-step guide outlining how to measure and compare alveolar bone levels on the basis of user-defined reference points long. The examples shown were obtained with an ILUMA Ultra (IMTEC, 2401 North Commerce,

Ardmore, OK 73401) with a 16 × 22 cm FOV on 20-s scans at 3.8 mA and 120 kVp.

Once a scan is obtained, and in order to reduce image processing time, only the area of interest is reconstructed to the desired voxel size, e.g., 0.1 mm (Fig. 11.2). The spatial resolution is then measured following manufacturer's instructions. In the examples shown, the spatial resolution was determined to be at 0.1922 mm. Therefore, one can only identify differences greater than 0.19 mm. The DICOM files obtained from the reconstructions were imported to OsiriX software (http://www.osirix-viewer.com/Osirix-64bit.html, Pixmeo SARL, 266 Rue de Bernex, CH-1233 Bernex, Switzerland). OsiriX is an open-source, multidimensional image navigation and display software developed on a Macintosh platform as a stand-alone application for OSX [52, 53].

The 3D multiplanar resolution (MPR) is selected, and the target tooth is oriented through the long axis of the tooth using 3D triangulation, with all three planes set perpendicular to the arch (Fig. 11.3). The images can be viewed at the desired slice thickness, e.g., 0.19 and 0.29 mm slices, using linear table opacity and the color lookup table (CLUT) of choice. These CLUTs help in distinguishing bone and tooth structure by illustrating different color saturations for structures of different density. The numerical range between the white and black levels (window width) and the theoretical midpoint of this range (window level) are selected for best visualization of alveolar bone and tooth structure.

Due to anatomical reasons, distobuccal roots have more bone coverage and might show less bone changes longitudinally; the mesiobuccal roots are usually more prominent and more likely to exhibit changes than the distobuccal roots [54]. Therefore, mandibular molars are bisected through the mesiobuccal and mesiolingual cusps. For maxillary molars, the slices can include the mesiobuccal root and the lingual root if this does not cause a rotation greater than 45° on the slice.

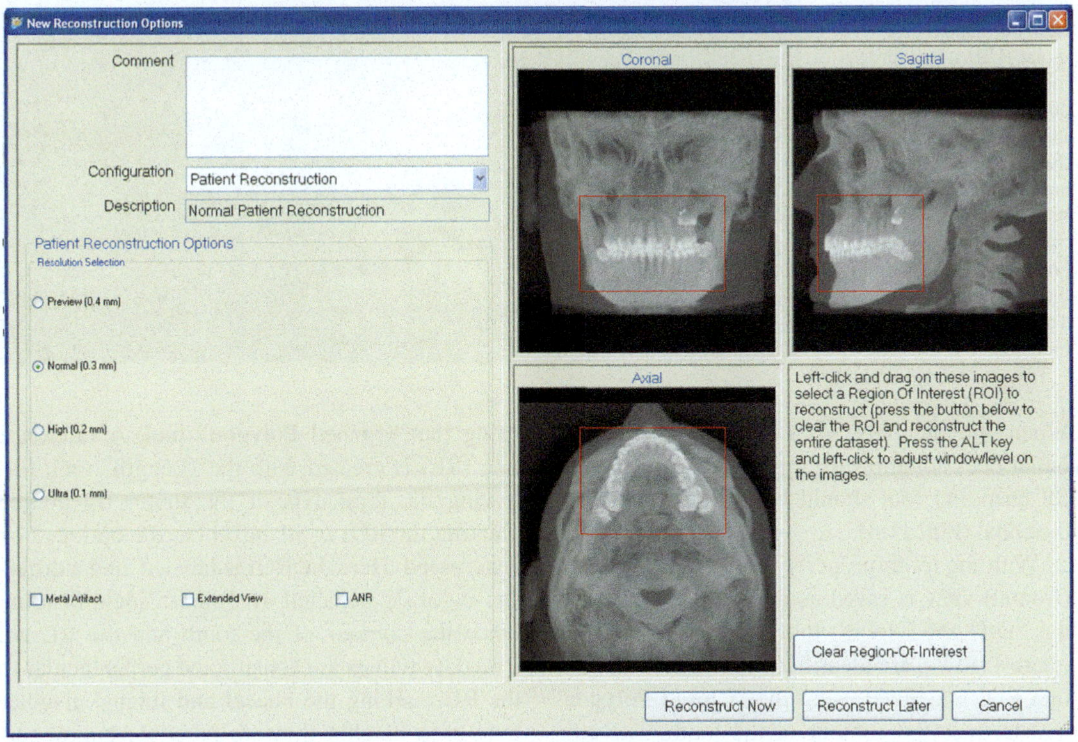

Fig. 11.2 Example showing the selection of the ROI on the ILUMA software. ROIs are reconstructed at the desired spatial resolution

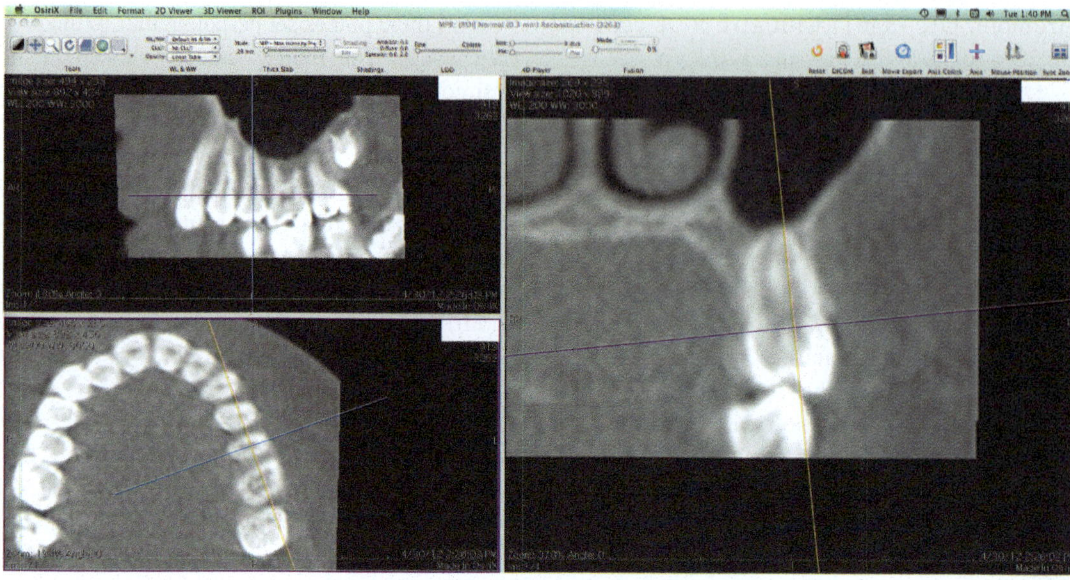

Fig. 11.3 From the 3D Viewer menu on the top, the 3D MPR option is selected. The axial, coronal, and sagittal images can now be oriented following the long axis of the tooth

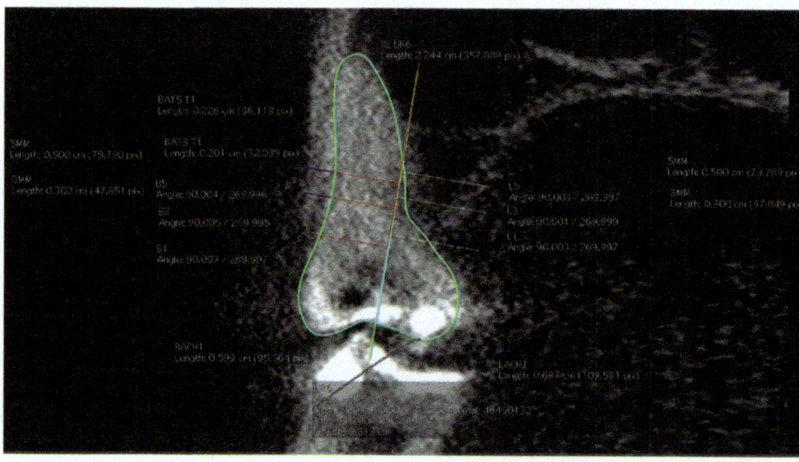

Fig. 11.4 Example of an image for evaluation of bone height and thickness of a maxillary right permanent first molar (UR6). Including the lingual cusp in this cut would have caused the cut to be at a more than 45° angle to the arch circumference, thus, mistakenly capturing interproximal bone as facial bone for the mesiobuccal root. The lingual root was not included

When the anatomy of a particular maxillary molar does not allow for this, only the mesiobuccal cusp and root should be bisected following the canal (Fig. 11.4).

With the tooth properly oriented, the coronal (frontal) view is saved as a DICOM files, using the "DICOM Export" tool, and the regions of interest (ROIs) are identified on this new file. The first ROI is defined using the "Closed Polygon" tool to trace the contour of the tooth. The contours of the buccal and lingual bone are traced using the "Opened Polygon" tool. A reference line (RL) is created with the "Length" tool, following the long axis of the tooth. By double clicking the ROI or its info box, the options box is accessed. Here the ROI is labeled, and a different color is selected for easier identification. Once the contour of the tooth and the RL are traced, two lines are constructed perpendicular to the RL marking the buccal and lingual alveolar bone crest heights; they are called the buccal perpendicular line (BPL) and lingual perpendicular

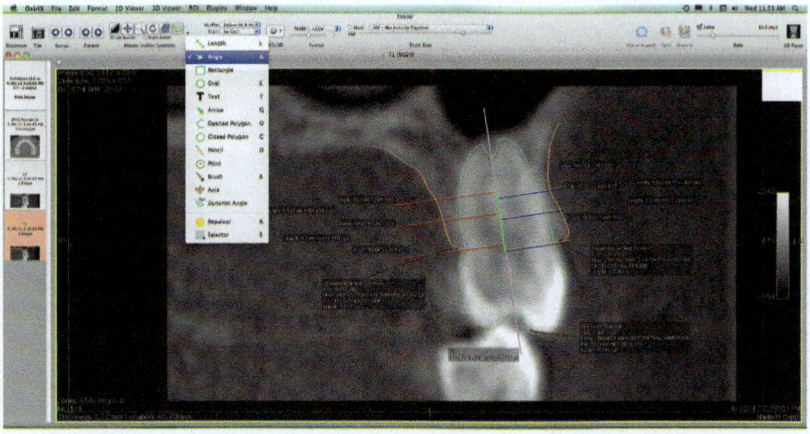

Fig. 11.5 The "Closed Polygon" tool is used to trace the tooth template at the initial time point (T1) (green). The "Opened Polygon" tool is used for tracing the bone contour at T1 (orange). The "Length" tool is used to make a reference line (RL) following the long axis of the tooth (lilac). Four lines overlap each other on top of the RL and are placed at 3 and 5 mm apical to the BPL and LPL. The "Dynamic Angle" and "Length" tool are used to draw lines perpendicular to the RL: BPL, BPL3 and BPL5 (blue), and LPL, LPL3 and LPL5 (red)

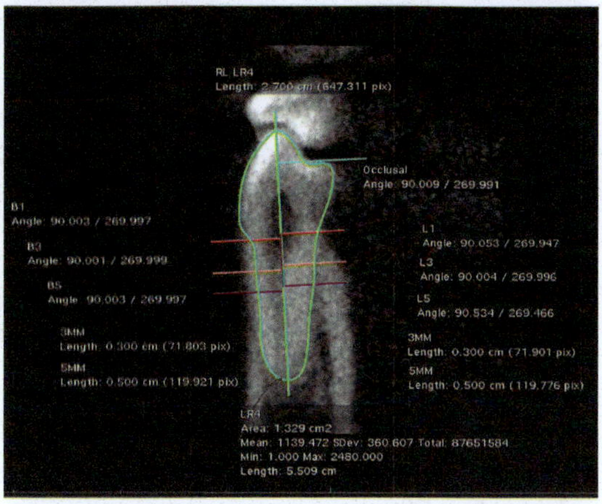

Fig. 11.6 Reference lines are shown on this example on a mandibular right first premolar (LR4). The tooth contour is traced (green). The RL (green line) bisects the tooth following the long axis. B1 and L1 (red) indicate the most coronal points of the alveolar crest heights. B3 and L3 (orange) are drawn 3 mm apical to B1 and L1, perpendicular to RL. B5 and L5 (purple) are drawn 5 mm apical to B1 and L1, perpendicular to RL. Note that for this example, the long axis of the tooth goes from cusp tip to root apex. Therefore, the deepest portion of the central fossa was projected to the RL at 90° (occlusal cyan line)

line (LPL). To make these perpendicular to the RL, first the RL and the tooth template are grouped; this can be done by pressing the <s> key and drawing a selection area, by pressing <COMND-A> (shortcut for "select all"), or by individually selecting the ROIs. The ROIs are then made nonselectable; this step is repeated as needed to make sure that previous ROIs will not be altered. Then, with the "Dynamic Angle" tool, 90° angles are drawn following RL and tangent to the crests (Fig. 11.5).

Following the same procedure used for BPL and LPL, more lines are drawn as desired on each side of RL. These lines can be at any height desired (3 and 5 mm are commonly used). To make sure the lines are at the right height, the "Length" tool is used to mark the desired heights apical to BPL and the LPL on RL. Tracing these lines allows the bone widths to be measured apical from the crest heights (Figs. 11.5 and 11.6). Name the ROIs accordingly.

At this point, the "Length" tool is used to measure heights and thicknesses. The buccal alveolar crest height (BACH) is the distance from the most incisal or occlusal point on the RL to the intersection with BPL. The lingual alveolar crest height (LACH) is the distance from the most incisal or occlusal point on the RL to the intersection with LPL. The buccal alveolar thickness (BAT) and the lingual alveolar thickness (LAT) represent the thicknesses of the buccal and lingual surfaces of the alveolar

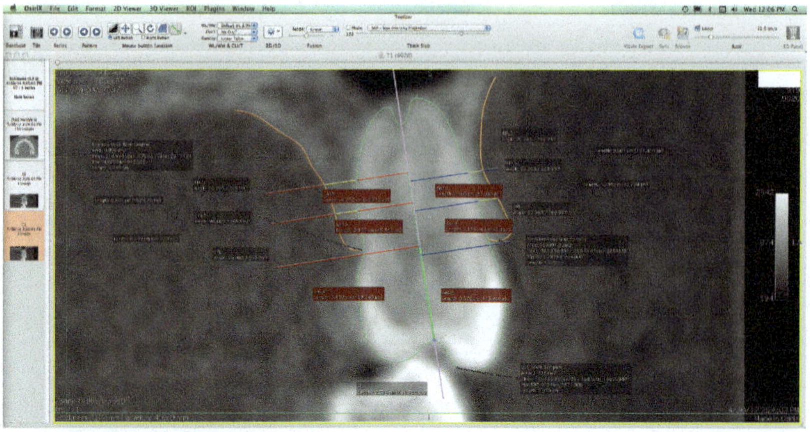

Fig. 11.7 Buccal and lingual crest heights (BACH and LACH) are represented here by green lines on the long axis. The buccal and lingual thicknesses at 3 and 5 mm apical to the crests (BAT3, BAT5, LAT3, and LAT5) are represented by the yellow lines. The measurements in mm are on the highlighted boxes

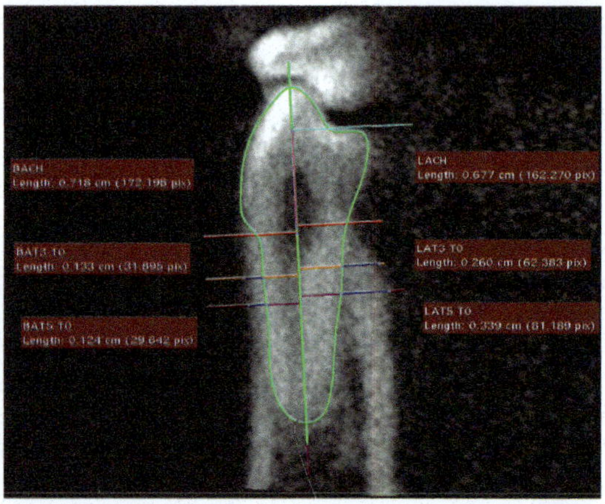

Fig. 11.8 Example images showing measurements for a mandibular right first premolar. BACH (brown), LACH (lilac), and BAT3, BAT5, LAT3, and LAT5 (blue). The ROI information boxes have been selected and highlighted only for demonstration purposes. The ROI information boxes for all the other ROIs have been excluded from this example to make the image and its information easier to read. As shown, this image contains all the elements used in the template for superimposition at T2 or subsequent time points

bone and are measured from the outer contour of the bone to the tooth silhouette on each of the perpendicular accessory lines apical from the crest (Figs. 11.7 and 11.8).

All the structures that are identified on the initial cut are used as a template for a digital tooth-based superimposition on T2 or subsequent images generated following the same guidelines described above. The template for the superimposition is composed by the tooth contour, the RL, and the buccal and lingual perpendicular lines. When the new DICOM file is opened, the ROI template file can either be imported or copied and pasted (Fig. 11.9). The template is then placed over the silhouette of the tooth and moved as needed to find the best fit with the tooth contour. Once the superimposition is completed, the examiner can draw the new bone contour for the new time point if desired. The new perpendicular lines for buccal and lingual crest height are drawn following the same procedure used for BPL and LPL, or the original markers can be moved along the RL as needed if any change had taken place (Fig. 11.10a–c). All the ROIs are then grouped and made nonselectable to measure the new heights and thicknesses, labeling them accordingly, e.g., BACH 2, LAT5, T2, etc. (Figs. 11.11 and 11.12). Once the desired measurements are obtained, the data is extracted to run any calculations needed. OsiriX shows the values for each ROI directly on the screen, and these can be entered manually into a spreadsheet or data processing software.

According to Baumgaertel [35], when looking at alveolar bone levels, the examiner must orient the images the same way at every time point through the long axis of the tooth. If this principle is not followed, bone levels might be mistakenly measured. The tooth template allows one to confirm that the cuts are being made in the same plane and that the alveolar bone thicknesses on the lingual and buccal plates are measured at same level on every time point.

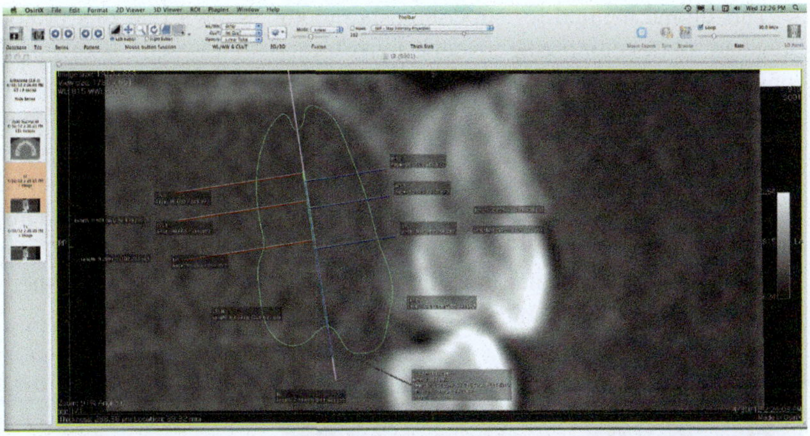

Fig. 11.9 At the subsequent time points, the template is imported from the ROI menu for superimposition of the tooth contour to the best fit. The bone contour and the linear measurements for BACH, LACH, BAT3, BAT5, LAT3, and LAT5 are not necessary for superimposing. Once imported, the ROIs are moved and rotated as needed to find the best fit

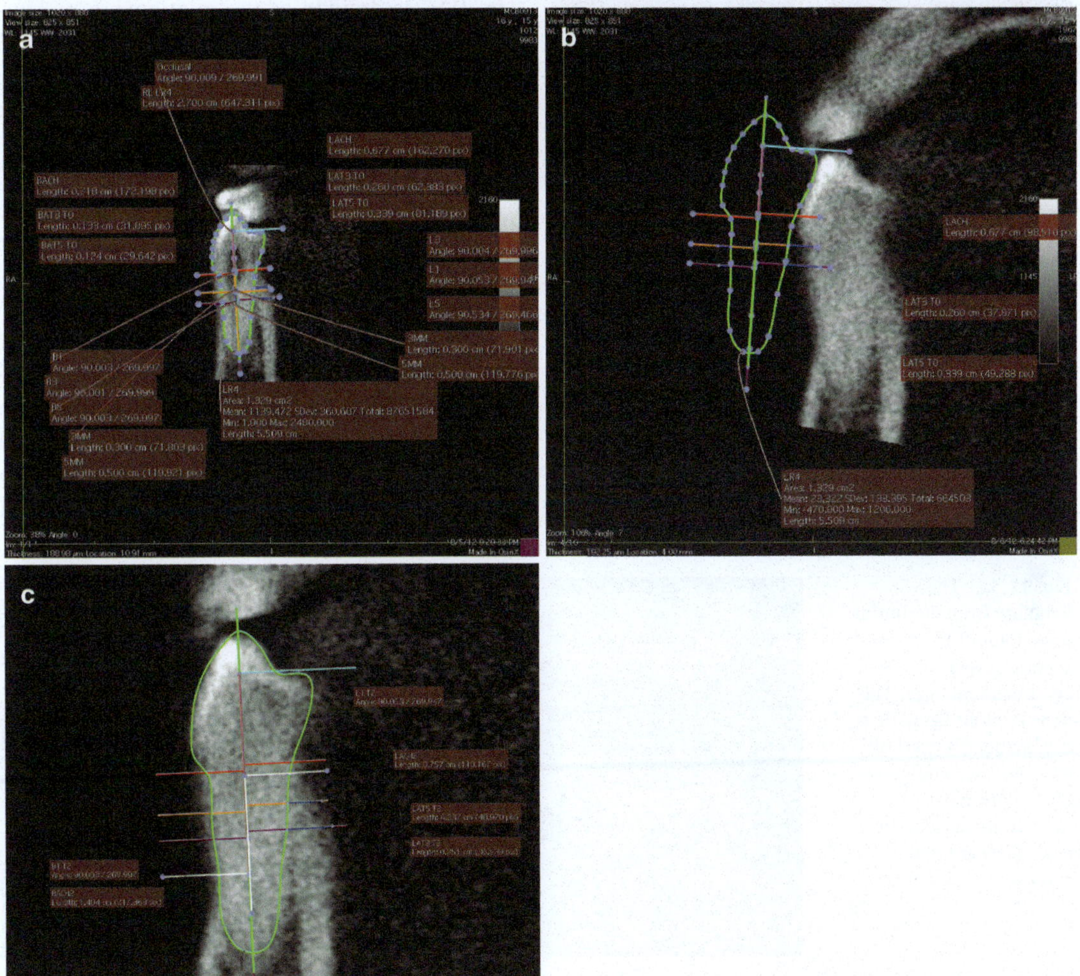

Fig. 11.10 Example images showing all measurements at T1; these ROIs form the template that will be used for superimposition (**a**). Template imported into a T2's cut of LR4 (**b**); the tooth contour and RL are used to superimpose the image at T2 to the best fit. The template at T2 is moved as needed until the best fit is found over the tooth. Once the template has been superimposed, the T2 values could be recorded (**c**). Notice the change in the buccal crest height from initial (red) to retention (white) (the red lines were kept in this example just for illustration purposes). Note that BAT3 and BAT5 were coronal to the new buccal crest

This digital method for superimposing tooth contours allows comparing alveolar bone levels longitudinally. With this method, there is no need to convert DICOM files to other formats such as JPEG or TIFF with no need for printouts, hand superimposition, hand tracings, or hand measurements. Currently, one can use this technique to perform cross-sectional assessments of buccal and lingual alveolar bone levels, as well as longitudinal assessments. Our research has shown that this protocol can be used with low intra- and interexaminer error [33]. A standardization of methods is needed in orthodontics for the assessment of bone levels. Still, clinicians might wonder to what extent it really matters identifying height or width changes within a millimeter, a tenth of a millimeter, or a hundred of a millimeter; this is a question that needs to be answered with further research.

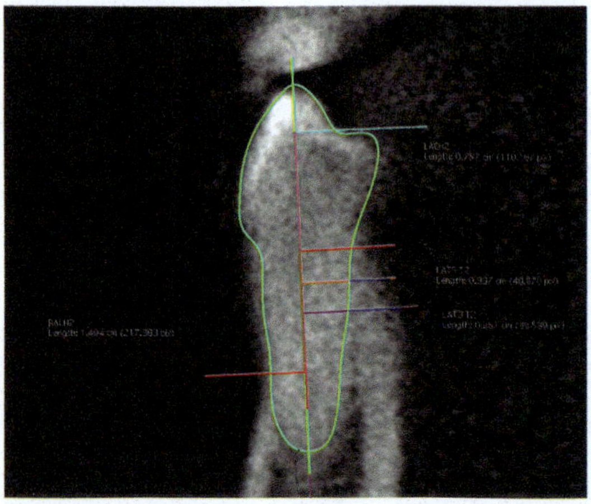

Fig. 11.11 Example of mandibular right first premolar at T2. All measurement values that could be recorded at this time point are shown in the image. As the crest has moved apically past the 3 and 5 mm reference lines, no thickness can be recorded

11.5.2 Evaluation of Interdental Angle (IDA)

The IDA is defined as the relationship of buccolingual axial inclination of contralateral teeth. It is a parameter that can used to evaluate the amount of dental tipping as a consequence of orthodontic therapy. The IDA can be measured using the “Dynamic Angle” tool in OsiriX by drawing two lines through each one of the long axes of contralateral teeth. The IDA can be obtained using the 0.3 mm CBCT scan reconstruction on OsiriX’s 3D MPR viewer. From the axial view, a linear cut bisecting each pair of teeth is obtained. For the mandibular molar, the cut includes the mesial root and cusps; for the maxillary molars, the cut includes the lingual and mesiobuccal roots and cusps. The 0.3 mm reconstruction is used for the IDA as the amount

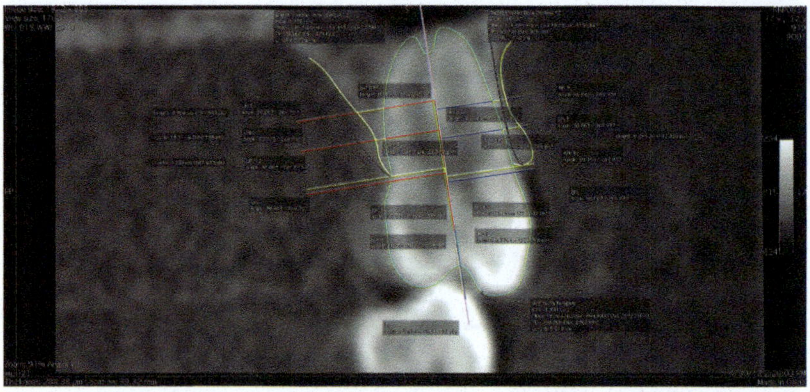

Fig. 11.12 The template from the initial time point (T1) is superimposed posttreatment (T2). The bone contour from T1 is not included as part of the template. The yellow lines represent T2 measurements and the new alveolar bone contour

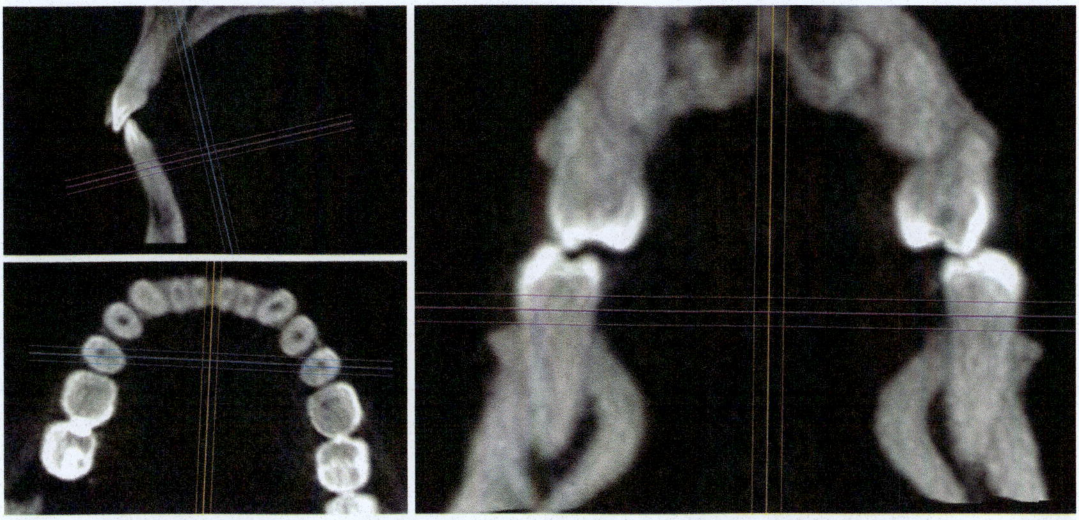

Fig. 11.13 Example of image orientation on 3D MPR. The slice thickness was set at 2.8 mm. Upper left, sagittal view; lower left, axial view; right, coronal view

of scattered is lower than the higher-resolution reconstruction, and it is considered that the rendering speed of lower size files allows for better use of the software. At the same time, the resolution obtained does not sacrifice image quality or important information needed. The slice thicknesses used should allow adequate inclusion of cusp tips and root apices of contralateral teeth (Fig. 11.13). To verify this, the cut is moved from side to side, and it is confirmed on the sagitatal view that the slice encompasses the cusp tips and root apices (Fig. 11.14a, b). The coronal view derived from these slices is saved as a DICOM file and the Dynamic Angle tool is used to calculate the IDA by drawing lines that bisect the teeth following the long axis of each tooth (Fig. 11.15a, b).

OsiriX seems to allow efficient navigation through large sets of data without the need for high-end expensive hardware or software or a highly experienced user. Our method allows superimposing dental contours on the CBCTs, and bone levels are measured directly on the images without sacrificing image quality or resolution. This superimposition method represents a viable way to assess alveolar bone levels longitudinally. The method has been satisfactorily tested with the 32-bit version of the software; and even though the image manipulation was efficient with both versions, we observed a decrease in the rendering speed with the 32-bit version. The 32-bit version is not intended as a commercial medical device for primary diagnostic imaging and is not FDA/CE-1 certified. In the United States, it can only be used as a reviewing, research, or teaching software but not for primary diagnostics or for patient care [52].

This technique is far from infallible, and some limitations do exist. Sometimes, some measurements cannot be recorded on certain teeth due to image quality, stage of eruption, anatomical variances, or greater than 45° dental rotations. The quality of the images will depend on the factors such as contrast, movement during acquisition, presence of metal, signal-to-noise ratio, threshold filters applied, spatial resolution, volume averaging, and properly setting window width and level.

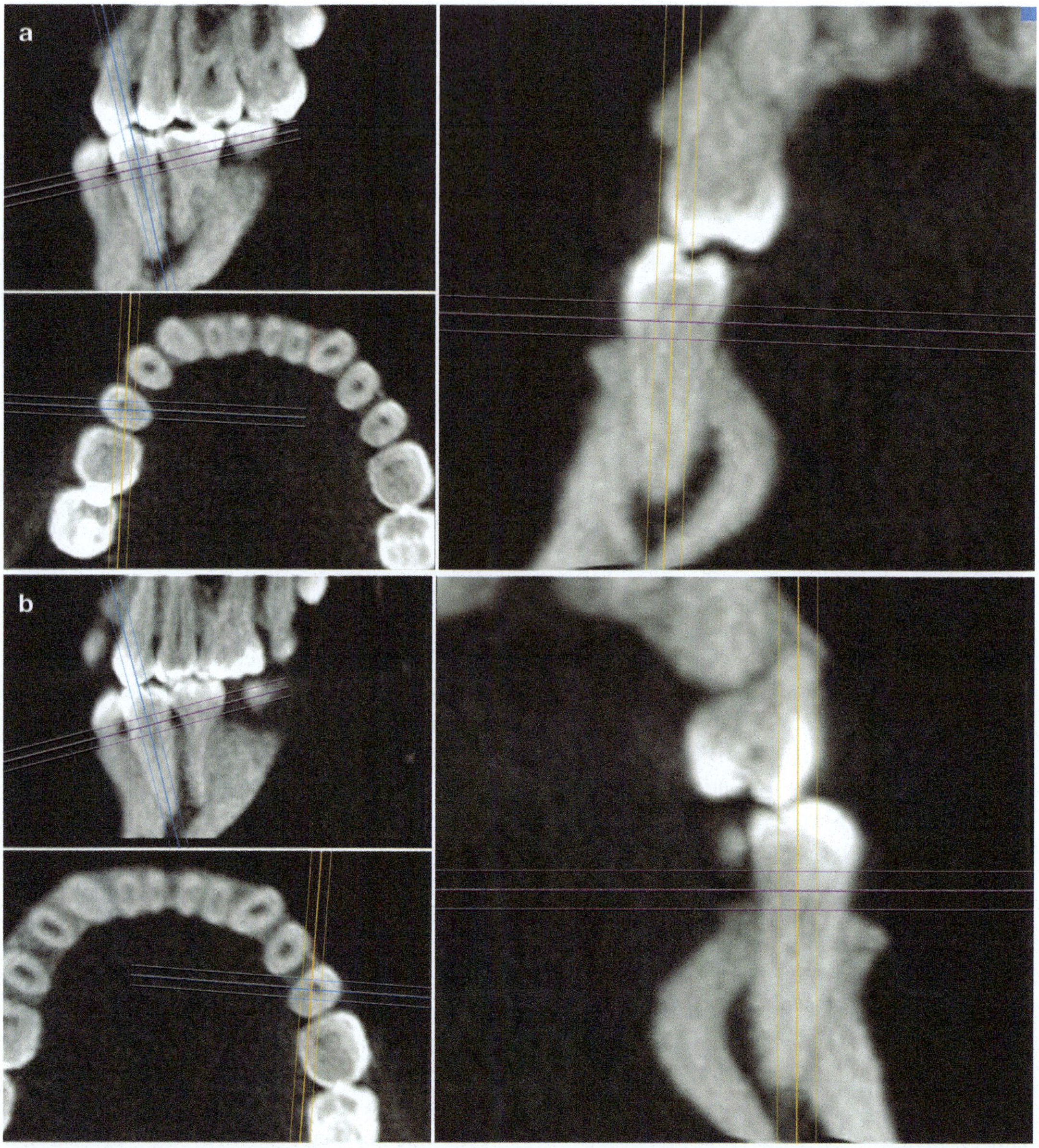

Fig. 11.14 In this example, the image is moved to the mandibular right premolar on the sagittal view to verify that the cusp tips and root apices are included in the cut (**a**). The image is then moved to the mandibular left premolar on the sagittal view to verify that the cusp tips and root apices are included in the cut (**b**)

11.6 The Limits of the Alveolar Bone

The degree to which buccal and lingual tooth movement stimulates new bone formation still appears to be a matter of debate in orthodontics. Orthodontic tooth movement is based on the ability of the bone to adapt and remodel. The stress and strain applied to the periodontal ligament (PDL) and the bone promote bone resorption and deposition [55]. Some clinicians defend the theory that facial movement of teeth with the use of light continuous forces stimulates new bone formation; some claim that bone adaptation takes places and not new bone for-

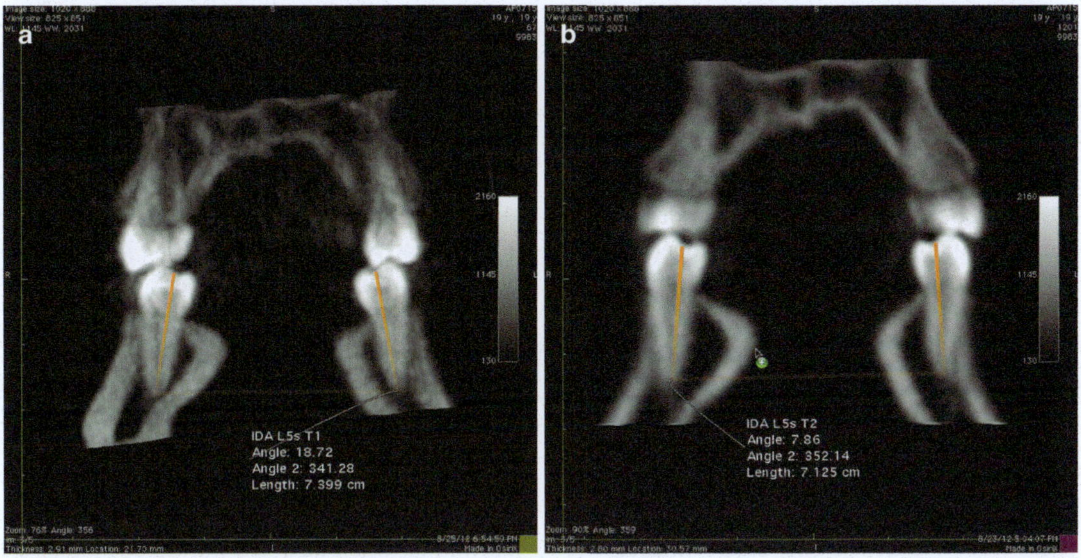

Fig. 11.15 IDA measurements done with the Dynamic Angle tool. An example of mandibular second premolars IDA before orthodontic treatment (**a**) and after (**b**)

mation; others state that buccal or lingual movement of teeth does not generate new bone and that in the absence of bilateral crossbites, expansion of the dental arches is an undesirable and deleterious effect of orthodontic treatment. The evidence currently available seems to favor the use of rapid palatal expanders (RPEs) for orthopedic correction of posterior crossbites of skeletal origin.

It certainly seems very unlikely that the design of brackets or the composition of the brackets and archwires would have a major effect in stimulating formation of new facial or lingual alveolar bone. In any case, if tooth movement was to trigger new bone formation on the facial or lingual aspect of a tooth, then CBCT imaging is currently the best, and least invasive, approach to assess these changes.

11.6.1 Rapid Palatal Expansion (RPE) and CBCT

Over the last 50 years, RPE has become a widely accepted method for overcoming transverse deficiencies in the maxilla. Garib et al. [54] evaluated the periodontal effects of RPE using spiral CT scans. In their comparison of tooth versus tissue-borne expanders, they measured the changes in buccal alveolar heights and thicknesses for maxillary anchor teeth. Their results showed a decrease in buccal bone thickness of 0.6–0.9 mm, an increase in the lingual bone plate thickness of 0.8–1.3 mm, and a loss of alveolar bone height or bone dehiscences on the anchorage teeth's buccal aspect of 7.1 mm (±4.6 mm) at the first premolars and 3.8 mm (±4.4 mm) at the mesiobuccal area of the permanent first molars. Based on studies by Zachrisson [56], a loss of 0.5 mm of interproximal bone height is considered to be clinically significant. Then, a similar loss on facial and lingual aspects should also be clinically significant and maybe even more critical as the bone is considerably thinner than in the interproximal areas.

CT studies have shown the effects of RPE include not only widening of the palatal suture but dental tipping as part of the dental expansion. With this tipping, alveolar bone loss has been observed on the buccal marginal bone, both vertically (height) and horizontally (thickness). Rungcharassaeng et al. [57] used CBCT to determine buccal bone changes of anchorage teeth during RPE and found a loss in thickness of 1 mm and a decrease in bone height of 3–4.4 mm. In 2009 Ballanti et al. [10] used low-dose multislice CT scans to study the periodontal effects of

RPE and observed in their study that soon after active palatal expansion, the buccal plate thicknesses decreased but recovered after 6 months of retention. The authors concluded that this therapy induced a significant increase in the transverse dimension of the maxillary arch in growing subjects without causing permanent injury to the periodontal bony support of anchor teeth.

A study from 2013 [58] evaluated the effects of RPE expansion on the buccal plate of maxillary permanent first molars and premolars via CBCT. Four-tooth expanders (RPE with bands on maxillary first molars and first premolars) or two-tooth expanders (bands on maxillary first molars only) were used in rapid (0.5 mm/day) and slow (0.25 mm/day) expansion protocols on 14 patients. The CBCT images were obtained on an ILUMA Ultra CBCT Scanner (Imtec Corporation, Ardmore, OK) on a 20-s scans at 3.8 mA and then reconstructed into 0.3 and 0.1 mm voxel size on the region of interest. The authors evaluated the variations in the interdental angle, interdental distance, buccal crest height (BCH), and buccal plate thickness at 3 and 5 mm from the most coronal aspect of the alveolar crest. They found the total expansion for molars and premolars to be significant, and the changes in crest height were significant only for the premolars. Their results also showed that the thicknesses at 3 and 5 mm from the alveolar crest were less 6 months after orthodontic treatment than before treatment. A recession of 2.4 mm on buccal alveolar bone on the premolars was reported, and the changes in crest height in the premolars appear to have been related to the significant tipping experienced by the premolar at 14.4°. While taking into consideration that these results are lower than those reported from Garib et al. [54] and Rungcharassaeng [57], the authors concluded that buccal crown tipping of teeth can lead to recessions on buccal alveolar bone.

When measurements are done during active treatment, the thicknesses appear to be less, as opposed to measurements repeated during the retention periods. Ballanti et al. [10] reported a recovery after retention on the alveolar bone thicknesses of the maxillary first molars. These findings could represent a "healing effect" during retention or could be associated with the limitation of the CBCT (or the human eye) to pick up the differences in density between cementum and alveolar bone during active tooth movement.

11.6.2 Extraction and Non-extraction

There is no room for a "one-size-fits-all" approach regarding extraction or non-extraction treatment. The orthodontist needs to consider, among many other things, the individual condition of each patient, the result of past experiences, and the currently available literature. It is known that when teeth are moved outside of the cortical plane, no bone is observed after 1 month, but after 3 months the osteoblast start reforming the cortical plate. Still, encroaching into the cortical plate can result in unintended adverse consequences [59].

In 2012 Lund et al. [17] evaluated mesial, distal, buccal, and lingual alveolar bone levels measured from the CEJ before and after orthodontic treatment with extraction of four premolars. The authors reported a reduction of bone levels following orthodontic treatment and found decreases in bone heights to be larger at the lingual surfaces of the anterior teeth in the mandible and the maxilla. Similarly, a study by Cook et al. [60] on 39 non-extraction patients and 20 four-premolar extraction cases that presented moderate-to-severe crowding demonstrated that both extraction and non-extraction protocols produced statistically significant bone loss in the anterior segments. Both protocols showed similar patterns of alveolar bone loss, and no location with either protocol showed statistically or clinically significant increases in bony support.

In non-extraction treatment, one would expect to see a reduction of alveolar bone levels on the facial surfaces of teeth following expansion and incisor proclination, but the literature currently available seems to suggest that both extraction of four premolars and non-extraction treatments may lead to reduced marginal bone levels on the lingual sides of the anterior teeth. While the loss on the lingual aspect of anterior teeth could be related to the distance the teeth were retracted or to the retroclination of lower incisors on extraction cases, this does not explain the loss seen in non-extraction cases where the dental effects are usually the opposite, e.g., proclination of anterior teeth. Maybe this bone loss is not necessarily correlated to the treatment modality but to the individual traits of certain patients that could present with thinner bone plates or a higher predisposition to alveolar bone loss.

11.6.3 The Importance of a Retention Period for Bone Levels

The forces that create tooth movement have a repercussion on the alveolar bone. As teeth move, osteoblasts, osteoclasts, and their associated activation factors and cofactors take part in bone remodeling and bone turnover. Newly formed bone matrix is deposited by osteoblasts, and after about a week of maturation, it becomes mineralizable osteoid. Osteoblasts deposit approximately 85% of the bone mineral content by primary mineralization [55]. Mineral maturation is completed at approximately 6–8 months, and it is referred to as secondary mineralization [61]. As the bone becomes more mature, it becomes more mineralized, thus more radiopaque. The use of a retention period of about 6–8 months after orthodontics treatment is supposed to allow this process to be completed, thus allowing teeth to be in more mature bone. This enhances the possibilities of detecting thin bone as it is more mineralized than it was during orthodontic treatment [31, 50, 62].

Starnbach et al. [63] studied palatal expansion in monkeys and found that animals that were scarified immediately after the palatal expansion showed remodeled buccal plates and extensive areas of localized bone resorption, whereas the animals sacrificed 3 months after treatment showed progressive improvement of supporting structures and recent alveolar bone formation. In the same study, animals scarified at 6 months after treatment presented near normal alveolar bone and dental supporting structures. This gives credibility to the theory that facial bone may regenerate after active treatment, but it certainly supports the idea that radiographic evaluation of bone levels will be more adequate after at least 6 months post-active orthodontic treatment.

11.7 Conclusions

Orthodontics, as a dental science, is constantly presented with new advancements in technology that aim at expanding the capacities of clinicians and improving patient care. This is why it is important for orthodontists, as well as other dental practitioners, to stay up-to-date on these new advancements and developments and to be aware of the need for scientific validation of claims.

In the years to come, with further reduction of the radiation levels from CBCT units and with continuous improvements in computer hardware and software, 3D imaging will eventually substitute 2D imaging devices in the orthodontic office, just like 2D digital radiography that has slowly, but steadily, replaced conventional 2D films in most parts of the United States. The cost associated with CBCT units and the hardware and software necessary to use these devices will continue to be, without a doubt, the major limitation for a more rapid spread and adoption of this technology worldwide. While 3D imaging technology is continuously improving, practitioners should be striving for excellence, not utopic perfection, and should continue to keep in mind the patient's best interest, respecting their autonomy, and not imposing on them undue burdens but rather letting the patients and/or their families make educated decisions.

The use of CBCT imaging has opened up the doors to new questions in orthodontics as it allows for views that were not possible with conventional two-dimensional radiographic techniques. The advances in the hardware and software, as well as the development of new mathematical algorithms, have allowed for better images to be obtained during shorter exposure times and with lower radiation doses. While the dose of radiation of a CBCT is greater than that of other more conventional radiographic methods, its use is currently the state-of-the-art method for longitudinal or cross-sectional indirect assessment of facial and lingual alveolar bone levels. Even though the limitations of CBCT should be kept in mind, according to Silva et al. [64], its usefulness must not be underestimated as 3D imaging, and its use over time can provide the data necessary to expand our clinical research for evidence-based dentistry.

The effects of orthodontic treatment on the alveolar bone have been a topic for debate for most part of the twentieth and twenty-first century. Research supports the idea that both extraction and non-extraction protocols can have adverse effects on both lingual and facial bone levels if the biologic limits of the alveolar bone are not respected. The decision to treat patients with or without the removal of permanent teeth, then, depends on careful individual considerations that the orthodontist needs to make on every case, e.g., "will the treatment increase or decrease the alveolar bone levels around certain teeth?"

The versatility of CBCTs should not be, by any means, underestimated. It seems unlikely that CBCT will be replaced as the technology of choice for 3D imaging in dentistry any time soon. This will certainly not happen unless a better and safer technology is developed; only then will the profession experience the next paradigm-shift in imaging technology. There is no question in our minds that CBT is here to stay, and we will most likely witness many more improvements to this technology during our professional lifetime, as well as an expansion of its uses and applications in both research and everyday dentistry.

References

1. ADA. 2013. ADA clinical practice guidelines, ADA Center for Evidence-Based Dentistry. Chicago, IL
2. Harrel WE. Three-dimensional diagnosis and treatment planning: the use of 3-D facial imaging and 3-D Cone beam CT in orthodontics and dentistry. Oral presentation. AAO annual session. Chicago. 2011.
3. Kau CH, Richmond S, Palomo JM, Hans MG. Three-dimensional cone beam computerized tomography in orthodontics. J Orthod. 2005;32:282–93.
4. De Vos W, Casselman J, Swennen GR. Cone-beam computerized tomography (CBCT) imaging of the oral and maxillofacial region: a systematic review of the literature. Int J Oral Maxillofac Surg. 2009;38:609–25.
5. Sukovic P, Brooks S, Perez L, Clinthorne N. DentoCATTM – a novel design of a cone-beam CT scanner for dentomaxillofacial imaging: introduction and preliminary results. Proceedings of the 15th international congress and exhibition on computer assisted radiology and surgery. 2001.
6. Sukovic P. Cone beam computed tomography in craniofacial imaging. Orthod Craniofac Res. 2003;6(Suppl 1):31–6. discussion 179-82
7. Timock AM, Cook V, Mcdonald T, Leo MC, Crowe J, Benninger BL, Covell DA Jr. Accuracy and reliability of buccal bone height and thickness measurements from cone-beam computed tomography imaging. Am J Orthod Dentofac Orthop. 2011;140:734–44.
8. Mah J, Hatcher D. Current status and future needs in craniofacial imaging. Orthod Craniofac Res. 2003;6:10–6.
9. Farman AG. Fundamentals of image acquisition and processing in the digital era. Orthod Craniofac Res. 2003;6(Suppl 1):17–22.
10. Ballanti F, Lione R, Fanucci E, Franchi L, Baccetti T, Cozza P. Immediate and post-retention effects of rapid maxillary expansion investigated by computed tomography in growing patients. Angle Orthod. 2009;79:24–9.
11. Grauer D, Cevidanes LS, Proffit WR. Working with DICOM craniofacial images. Am J Orthod Dentofac Orthop. 2009;136:460–70.
12. Hatcher DC, Aboudara CL. Diagnosis goes digital. Am J Orthod Dentofac Orthop. 2004;125:512–5.
13. Farman AG, Scarfe WC. Development of imaging selection criteria and procedures should precede cephalometric assessment with cone-beam computed tomography. Am J Orthod Dentofac Orthop. 2006;130:257–65.
14. Scarfe WC, Farman AG, Sukovic P. Clinical applications of cone-beam computed tomography in dental practice. J Can Dent Assoc. 2006;72:75–80.
15. Swennen GR, Schutyser F. Three-dimensional cephalometry: spiral multi-slice vs cone-beam computed tomography. Am J Orthod Dentofac Orthop. 2006;130:410–6.
16. Mah JK, Yi L, Huang RC, Choo H. Advanced applications of cone beam computed tomography in orthodontics. Semin Orthod. 2011;17:57–71.
17. Lund H, Grondahl K, Grondahl HG. Cone beam computed tomography evaluations of marginal alveolar bone before and after orthodontic treatment combined with premolar extractions. Eur J Oral Sci. 2012;120:201–11.
18. Schulze D, Heiland M, Thurmann H, Adam G. Radiation exposure during midfacial imaging using 4- and 16-slice computed tomography, cone beam computed tomography systems and conventional radiography. Dentomaxillofac Radiol. 2004;33:83–6.

19. Kiefer H, Lambrecht JT, Roth J. Dose exposure from analog and digital full mouth radiography and panoramic radiography. Schweiz Monatsschr Zahnmed. 2004;114:687–93.
20. Mah JK. X-ray imaging and oral healthcare. 2006. http://www.orbitimaging.com/PDF/Xrays-OralHealthcarePatientAAO.pdf. Accessed 20 Mar 2012.
21. Cevidanes LH, Oliveira AE, Grauer D, Styner M, Proffit WR. Clinical application of 3D imaging for assessment of treatment outcomes. Semin Orthod. 2011;17:72–80.
22. Suomalainen A. Cone beam computed tomography in oral radiology. Doctoral dissertation, University of Helsinki. 2010.
23. Ludlow JB, Davies-Ludlow LE, White SC. Patient risk related to common dental radiographic examinations: the impact of 2007 international commission on radiological protection recommendations regarding dose calculation. J Am Dent Assoc. 2008;139: 1237–43.
24. Bayome M, Park JH, Kook Y-A. Computed tomography: new research. Hauppauge, NY: Nova Science Publisher; 2013.
25. Kusnoto B, Kaur P, Salem A, Zhang Z, Galang-Boquiren MT, Viana G, Evans CA, Manasse R, Monahan R, Begole E, Abood A, Han X, Sidky E, Pan X. Implementation of ultra-low-dose CBCT for routine 2D orthodontic diagnostic radiographs: cephalometric landmark identification and image quality assessment. Semin Orthod. 2015;21:233–47.
26. Gamache C, English JD, Salas-Lopez AM, Rong J, Akyalcin S. Assessment of image quality in maxillofacial cone-beam computed tomography imaging. Semin Orthod. 2015;21:248–53.
27. Molen AD. The 3D orthodontist: the modern orthodontist's source for information on 3D technologies. 2011. http://www.3DOrthodontist.com/. Accessed 20 Mar 2012.
28. Pan F, Kau CH. The anatomical evaluation of the dental arches using cone beam computed tomography – a pilot investigation of the availability of bone for placement of mini-implants in class I patients. IFMBE proceedings, 2009, Munich. pp. 2273–2275.
29. Loubele M, Van Assche N, Carpentier K, Maes F, Jacobs R, Van Steenberghe D, Suetens P. Comparative localized linear accuracy of small-field cone-beam CT and multislice CT for alveolar bone measurements. Oral Surg Oral Med Oral Pathol Oral Radiol Endod. 2008;105:512–8.
30. Creed B, Kau CH, English JD, Xia JJ, Lee RP. A comparison of the accuracy of linear measurements obtained from cone beam computerized tomography images and digital models. Semin Orthod. 2011;17:49–56.
31. Molen AD. Considerations in the use of cone-beam computed tomography for buccal bone measurements. Am J Orthod Dentofac Orthop. 2010;137:S130–5.
32. Lund H, Grondahl K, Grondahl HG. Cone beam computed tomography for assessment of root length and marginal bone level during orthodontic treatment. Angle Orthod. 2010;80:466–73.
33. Romero-Delmastro A, Kadioglu O, Currier GF, Cook T. Digital tooth-based superimposition method for assessment of alveolar bone levels on cone-beam computed tomography images. Am J Orthod Dentofac Orthop. 2014;146:255–63.
34. Kobayashi K, Shimoda S, Nakagawa Y, Yamamoto A. Accuracy in measurement of distance using limited cone-beam computerized tomography. Int J Oral Maxillofac Implants. 2004;19:228–31.
35. Baumgaertel S, Palomo JM, Palomo L, Hans MG. Reliability and accuracy of cone-beam computed tomography dental measurements. Am J Orthod Dentofac Orthop. 2009;136:19–25. discussion 25–8
36. Lascala CA, Panella J, Marques MM. Analysis of the accuracy of linear measurements obtained by cone beam computed tomography (CBCT-NewTom). Dentomaxillofac Radiol. 2004;33:291–4.
37. Berco M, Rigali PH Jr, Miner RM, Deluca S, Anderson NK, Will LA. Accuracy and reliability of linear cephalometric measurements from cone-beam computed tomography scans of a dry human skull. Am J Orthod Dentofac Orthop. 2009;136(17):e1–9. discussion 17–8
38. Bagis N, Kolsuz ME, Kursun S, Orhan K. Comparison of intraoral radiography and cone-beam computed tomography for the detection of periodontal defects: an in vitro study. BMC Oral Health. 2015;15:64.
39. Misch KA, Yi ES, Sarment DP. Accuracy of cone beam computed tomography for periodontal defect measurements. J Periodontol. 2006;77:1261–6.
40. Covell DA. Assessing alveolar bone height and thickness using cone beam computed tomography: are the looks deceiving? AAO annual session, San Diego, 2017.
41. Hamada Y, Kondoh T, Noguchi K, Iino M, Isono H, Ishii H, Mishima A, Kobayashi K, Seto K. Application of limited cone beam computed tomography to clinical assessment of alveolar bone grafting: a preliminary report. Cleft Palate Craniofac J. 2005;42:128–37.
42. Gribel BF, Gribel MN, Frazao DC, Mcnamara JA Jr, Manzi FR. Accuracy and reliability of craniometric measurements on lateral cephalometry and 3D measurements on CBCT scans. Angle Orthod. 2011;81:26–35.
43. Lamichane M, Anderson NK, Rigali PH, Seldin EB, Will LA. Accuracy of reconstructed images from cone-beam computed tomography scans. Am J Orthod Dentofac Orthop. 2009;136:156.e1–6. discussion 156–7
44. Moerenhout BA, Gelaude F, Swennen GR, Casselman JW, Van Der Sloten J, Mommaerts MY. Accuracy and repeatability of cone-beam computed tomography (CBCT) measurements used in the determination of facial indices in the laboratory setup. J Craniomaxillofac Surg. 2009;37:18–23.
45. Lagravere MO, Gordon JM, Guedes IH, Flores-Mir C, Carey JP, Heo G, Major PW. Reliability of traditional cephalometric landmarks as seen in three-dimensional

analysis in maxillary expansion treatments. Angle Orthod. 2009;79:1047–56.
46. De Oliveira AE, Cevidanes LH, Phillips C, Motta A, Burke B, Tyndall D. Observer reliability of three-dimensional cephalometric landmark identification on cone-beam computerized tomography. Oral Surg Oral Med Oral Pathol Oral Radiol Endod. 2009;107:256–65.
47. Ballrick JW, Palomo JM, Ruch E, Amberman BD, Hans MG. Image distortion and spatial resolution of a commercially available cone-beam computed tomography machine. Am J Orthod Dentofac Orthop. 2008;134:573–82.
48. Cevidanes LH, Bailey LJ, Tucker GR Jr, Styner MA, Mol A, Phillips CL, Proffit WR, Turvey T. Superimposition of 3D cone-beam CT models of orthognathic surgery patients. Dentomaxillofac Radiol. 2005;34:369–75.
49. Kumar V, Ludlow JB, Mol A, Cevidanes L. Comparison of conventional and cone beam CT synthesized cephalograms. Dentomaxillofac Radiol. 2007;36:263–9.
50. Westerlund A, Oikimoui C, Ransjö M, Ekestubbe A, Bresin A, Lund H. Cone-beam computed tomographic evaluation of the long-term effects of orthodontic retainers on marginal bone levels. Am J Orthod Dentofac Orthop. 2017;151:74–81.
51. Patcas R, Müller L, Ullrich O, Peltomäki T. Accuracy of cone-beam computed tomography at different resolutions assessed on the bony covering of the mandibular anterior teeth. Am J Orthod Dentofac Orthop. 2012;141:41–50.
52. Rosset A, Spadola L, Ratib O. OsiriX: an open-source software for navigating in multidimensional DICOM images. J Digit Imaging. 2004;17:205–16.
53. Osirix. 2012. http://www.OsiriX-viewer.com/. Accessed 10 May 2012.
54. Garib DG, Henriques JF, Janson G, De Freitas MR, Fernandes AY. Periodontal effects of rapid maxillary expansion with tooth-tissue-borne and tooth-borne expanders: a computed tomography evaluation. Am J Orthod Dentofac Orthop. 2006;129:749–58.
55. Roberts WE, Simmons KE, Garetto LP, Decastro RA. Bone physiology and metabolism in dental implantology: risk factors for osteoporosis and other metabolic bone diseases. Implant Dent. 1992;1:11–21.
56. Zachrisson BU. Cause and prevention of injuries to teeth and supporting structures during orthodontic treatment. Am J Orthod. 1976;69:285–300.
57. Rungcharassaeng K, Caruso JM, Kan JYK, Kim J, Taylor G. Factors affecting buccal bone changes of maxillary posterior teeth after rapid maxillary expansion. Am J Orthod Dentofac Orthop. 2007;132:428.e1–8.
58. Nguyen B, Kadioglu O, Currier GF, Olsen J. Cone beam computed tomography evaluation after palatal expansion and orthodontics. J World Fed Orthod. 2013;2:e9–e13.
59. Hsu JT, Chang HW, Huang HL, Yu JH, Li YF, Tu MG. Bone density changes around teeth during orthodontic treatment. Clin Oral Investig. 2011;15:511–9.
60. Cook T, Currier F, Kadioglu O, Griffin T. Comparison of the anterior alveolar bony changes of moderately crowded cases treated either with extraction or non-extraction orthodontic treatment. Semin Orthod. 2015;21:283–90.
61. Roberts WE. Bone physiology, metabolism, and biomechanics in orthodontic practice. In: Graber TM, Vanarsdall RL, KWL V, editors. Orthodontics current principles and techniques, vol. 4. St. Louis, MO: Elsevier Mosby; 2005.
62. Melsen B. Biological reaction of alveolar bone to orthodontic tooth movement. Angle Orthod. 1999;69:151–8.
63. Starnbach H, Bayne D, Cleall J, Subtelny JD. Facioskeletal and dental changes resulting from rapid maxillary expansion. Angle Orthod. 1966;36:152–64.
64. Silva MG, Wolf U, Heinicke F, Bumann A, Visser H, Hirsch E. Cone-beam computed tomography for routine orthodontic treatment planning: a radiation dose evaluation. Am J Orthod Dentofac Orthop. 2008;133:640.e1–5.

Virtual Surgical Planning (VSP)

12

David Sylvester and Steven M. Sullivan

12.1 Introduction

Orthognathic surgery has seen many developments and advances throughout its evolutionary history. Recent trends have seen a shift away from traditional model surgery and have sought to embrace more contemporary methods of virtual surgical planning (VSP) [1]. Since its inception, orthognathic surgery has successfully and predictably been planned and performed by means of traditional, analytical model surgery. Introduction of the cone-beam CT in 1998 in Europe and then in the USA in 2001 provoked the beginnings of a paradigm shift [2, 3]. The literature has since been replete with studies focused on providing evidence that virtual surgical planning is as effective and predictable as traditional model surgery [4–6].

Virtual surgical planning is a process that integrates both computer-aided design (CAD) and computer-aided manufacturing (CAM) with surgery. Xia et al. first published a protocol for transposing the orthognathic surgery workup to a digital platform. This protocol was termed computer-assisted surgical simulation (CASS) [7]. CASS consists of four distinct phases: data acquisition, planning, surgical, and assessment. In recent years, several authors have contributed techniques and proprietary technology to facilitate more accurate digital records [6, 8–10]. While the accuracy of VSP is no longer disputed, many surgeons must now reconcile its practicality and cost with traditional office workflows and expenses during this transformative phase in the history of surgery [1, 11, 12].

Both analytical model surgery and virtual surgical planning begin with a detailed presurgical evaluation. This includes extraoral and intraoral photography; anthropometric measurements; analysis of occlusal discrepancies as they relate to overbite, overjet, canine, and molar relations; two-dimensional and three-dimensional radiographs; dental impressions; and bite registrations. Following acquisition of this information, predictive tracings are performed. The occlusion is evaluated. If necessary, segmentalization is performed to obtain a proper final occlusion. Correction of the patient's unique dentofacial deformity is attempted by proposing calculated jaw movements. The feasibility of these three-dimensional movements is assessed during a virtual surgical planning meeting with a software engineer or during traditional model surgery. Surgical movements are facilitated intraoperatively via splints which are either 3-D printed

D. Sylvester (✉)
Assistant Clinical Professor, Department of Surgical Sciences, University of Oklahoma Health Sciences Center, Oklahoma City, OK, USA
e-mail: david-sylvester@ouhsc.edu

S. M. Sullivan
Professor and Chair, Department of Surgical Sciences, University of Oklahoma Health Sciences Center, Oklahoma City, OK, USA
e-mail: Steven-Sullivan@ouhsc.edu

O. Kadioglu, G. F. Currier (eds.), *Craniofacial 3D Imaging*,
https://doi.org/10.1007/978-3-030-00722-5_12

or manually fabricated. Precision and accuracy of surgical results are assessed during follow-up by obtaining two-dimensional and three-dimensional imaging and clinical photographs.

12.2 Case Presentation

By way of example, this 14-year-old male was referred by his pediatric rheumatologist for evaluation of his marked right temporomandibular joint degeneration with a resulting dentofacial deformity (Fig. 12.1). The patient's chief complaint was an inability to chew as well as pronounced snoring and daytime somnolence. Past medical history was significant for juvenile rheumatoid arthritis with degeneration of the right condyle. At the time of consultation, the patient had been in remission for several months. Past surgical history was significant for third molar removal. Medications included methotrexate and etanercept which were being managed and weened by his pediatric rheumatologist.

Initial evaluation revealed a remarkable anterior open bite on the left side with severe canting of the maxilla and mandible and attendant facial

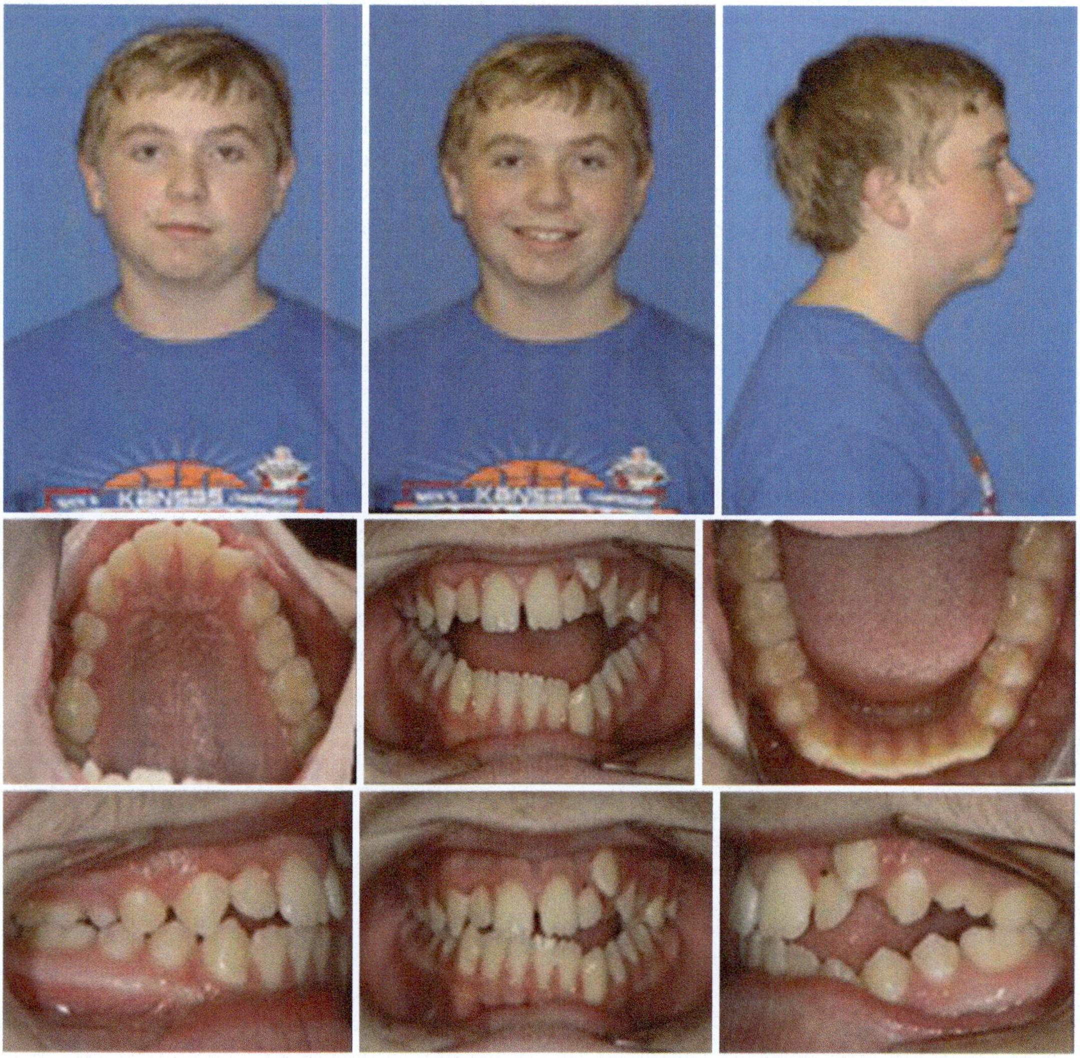

Fig. 12.1 Intraoral and extraoral photographs at the time of consultation

asymmetry. Radiographs were obtained showing significant degenerative changes of the right mandibular condyle, resulting in profound facial asymmetry, mandibular hypoplasia, and a retruded and collapsed oropharyngeal airway. The patient was referred to an orthodontist for full-banded orthodontic treatment and dental decompensation with leveling and alignment of the dental arches. Due to the degree of facial asymmetry as well as degeneration of the right TMJ, the patient was deemed an ideal candidate for virtual surgical planning and right total joint replacement with TMJ Concepts® (Figs. 12.2, 12.3, 12.4, 12.5, 12.6, 12.7 and 12.8).

His presurgical orthodontic phase included digital planning using SureSmile® by OraMetrix. This system permitted very rapid and precise presurgical orthodontics requiring only 10 months. Prior to surgery, standard clinical measurements were obtained as well as extraoral and intraoral photographs, dental impressions, and radiographs which included a medical-grade CT for fabrication of his TMJ Concepts® prosthesis and VSP.

Dental impressions were poured in a type IV dental plaster. Occlusal analysis was performed, and selective enameloplasty was performed where indicated. A stable final occlusion necessitated the maxilla be segmentalized between canines and lateral incisors creating a three-piece maxilla. The maxilla was stabilized in its anticipated final position, and a final bite registration was captured with Blu-Mousse®. Original and altered dental models were provided to MedCad®

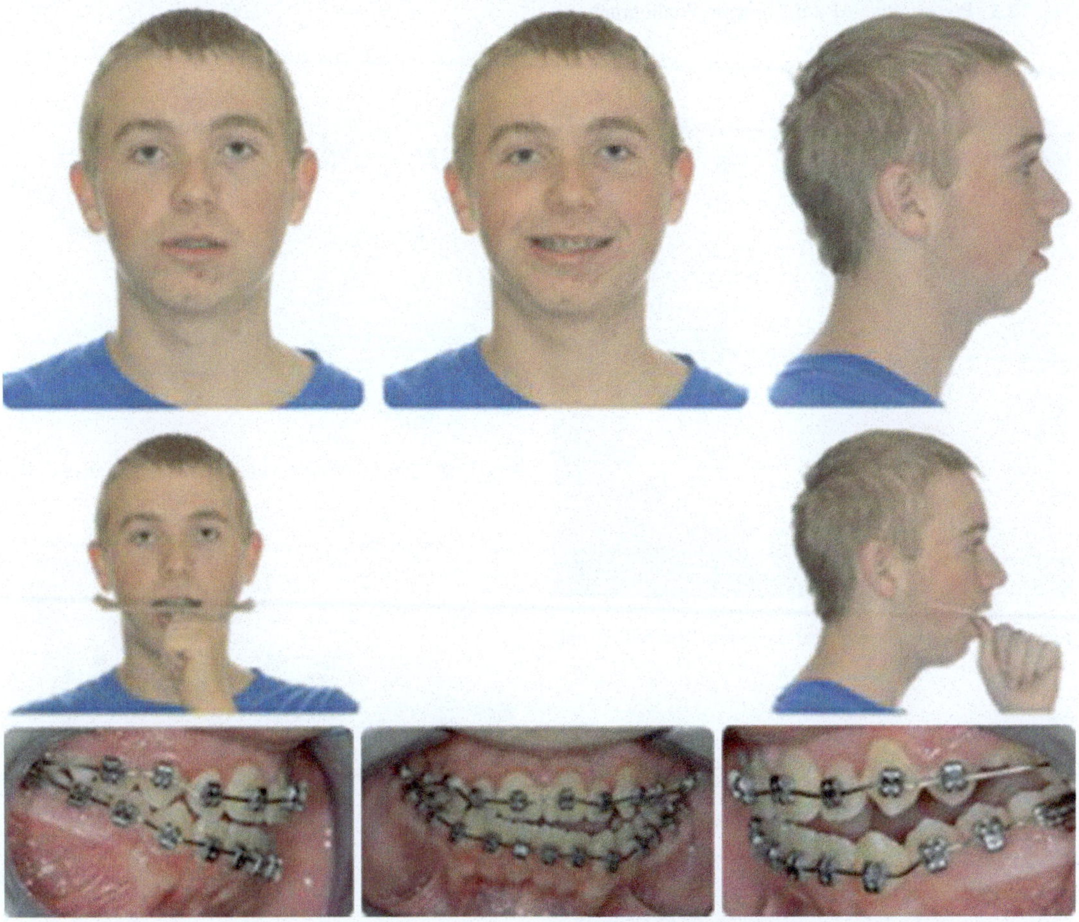

Fig. 12.2 Preoperative intraoral and extraoral photographs following a 10-month period of orthodontics

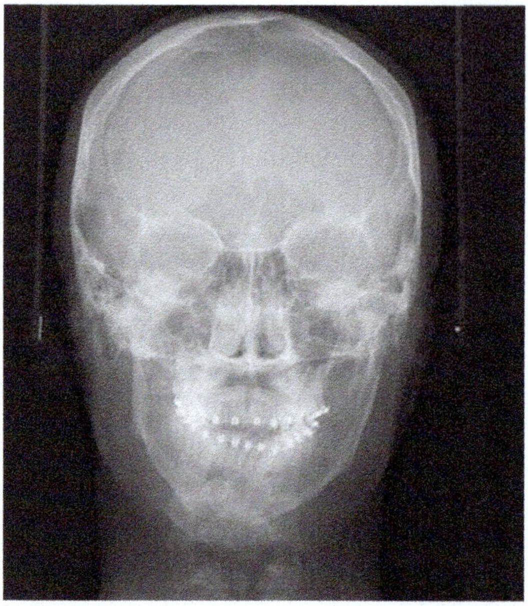

Fig. 12.3 Preoperative AP cephalometric radiograph

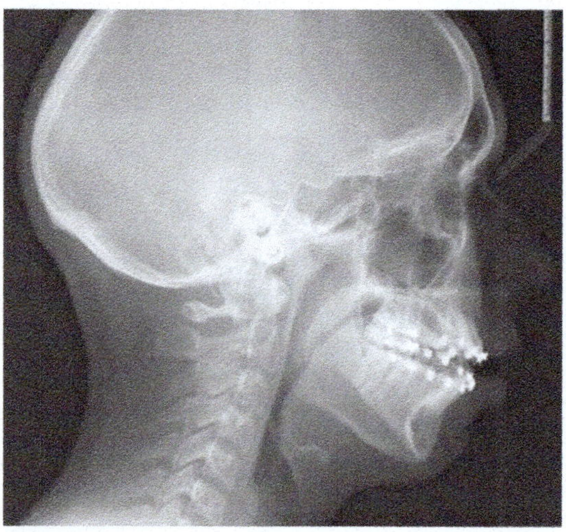

Fig. 12.4 Preoperative lateral cephalometric radiograph

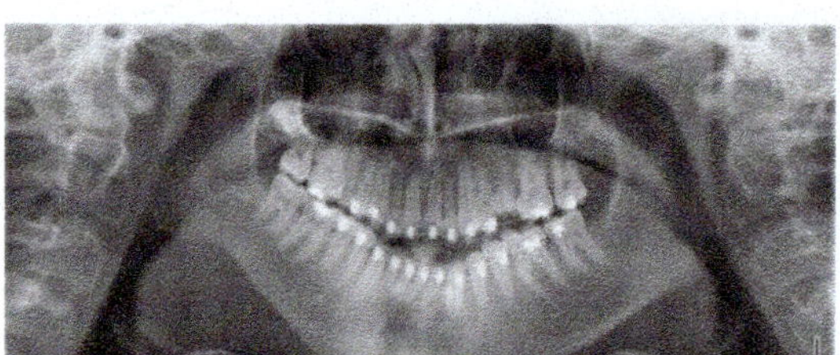

Fig. 12.5 Preoperative orthopantomogram

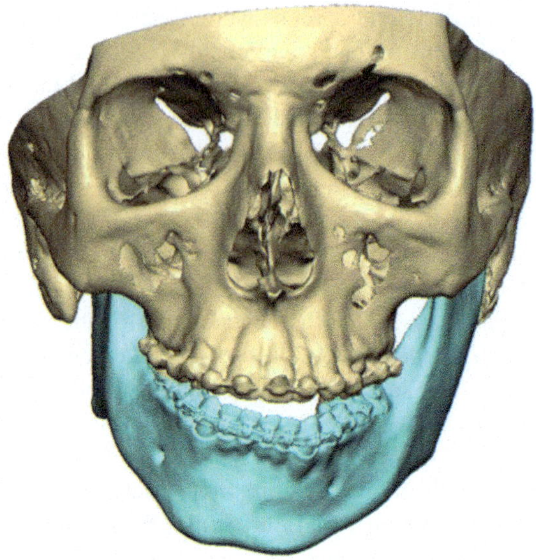

Fig. 12.6 3D rendering obtained from preoperative CBCT

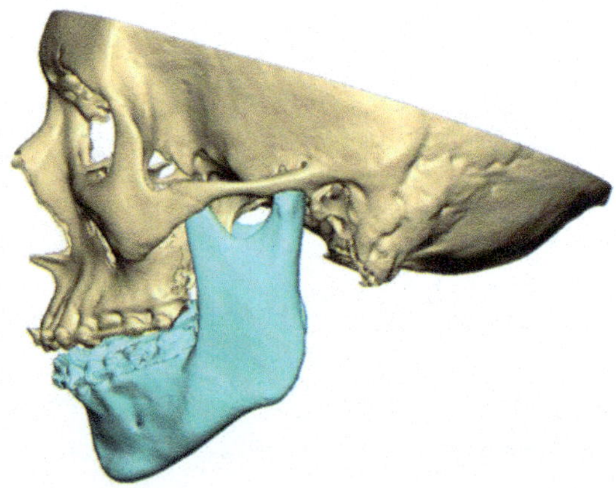

Fig. 12.7 3D rendering obtained from preoperative CBCT

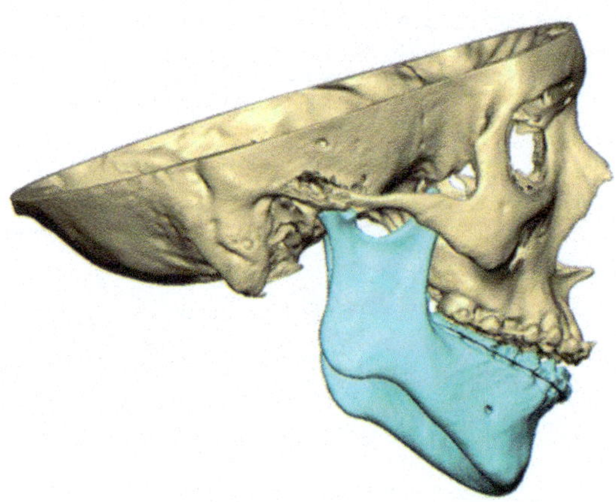

Fig. 12.8 3D rendering obtained from preoperative CBCT

with respective bite registrations. Models were laser scanned and digitally integrated with the patient's CT to obtain accurate dentofacial anatomy. Predictive two-dimensional tracings were performed in standard fashion using the patient's lateral cephalogram.

A MedCad® software engineer aided with virtual surgical planning. Key anatomic landmarks which included A point, B point, anterior nasal spine, posterior nasal spine, pogonion, central incisors, canines, and first molars were referenced as the maxilla was repositioned. Three-dimensional correction necessitated an almost 5 mm cant correction on the right side, counterclockwise rotation of the maxillomandibular complex with a 4° alteration of the occlusal plane and approximately 2° of yaw correction. A left sagittal split osteotomy and right total joint prosthesis were planned for the mandible. The mandible was virtually set to the new maxillary position and evaluated for interferences. A 4 mm genioglossus advancement was planned. A mandible-first surgical sequencing was communicated as it is imperative that this sequence be adhered to when joint replacement is required. Appropriate intermediate and final surgical splints were 3-D printed and provided to the surgeons (Figs. 12.9 and 12.10).

The data sets of the planned surgical movements facilitated the 3-D printing of the stereolithic model necessary for fabrication of the patient's fitted TMJ prosthesis by TMJ Concepts®. Right-sided TMJ ramal and fossa components were waxed into place taking care to avoid vital structures (Figs. 12.11 and 12.12). This custom prosthesis was then verified by the surgeons prior to its final fabrication.

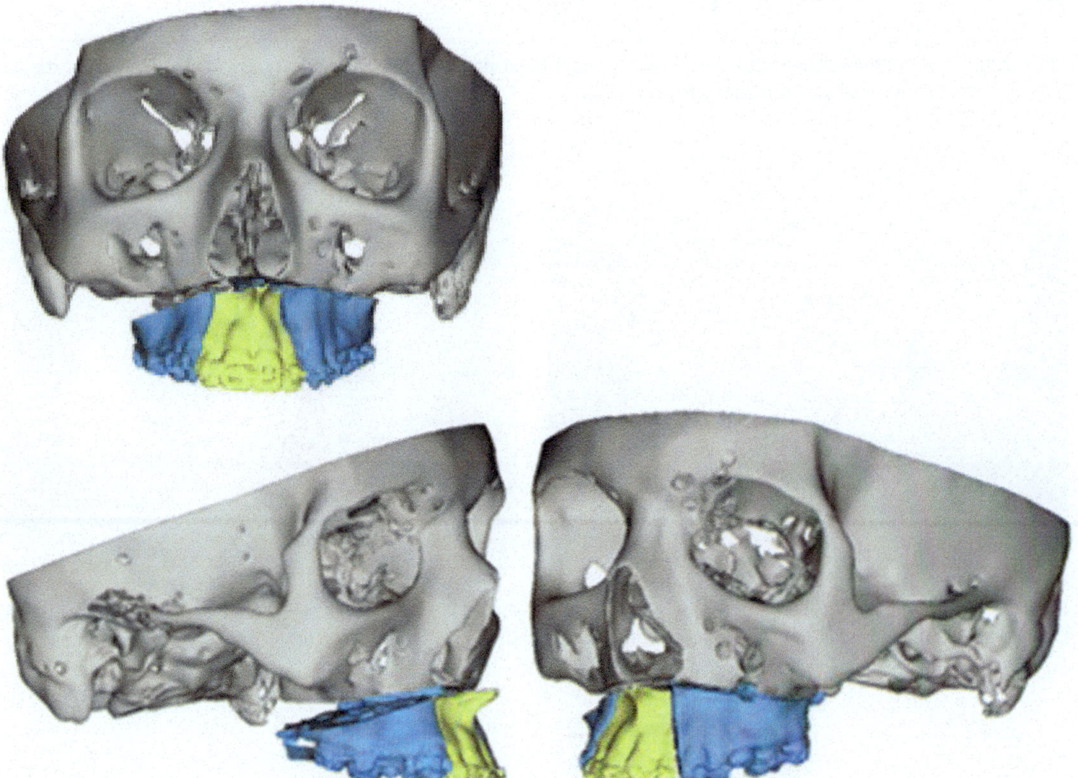

Fig. 12.9 VSP with MedCad® showing the final maxillary position. The maxilla has been segmentalized between the lateral incisors and canines. Note the almost 5 mm cant correction on the right side

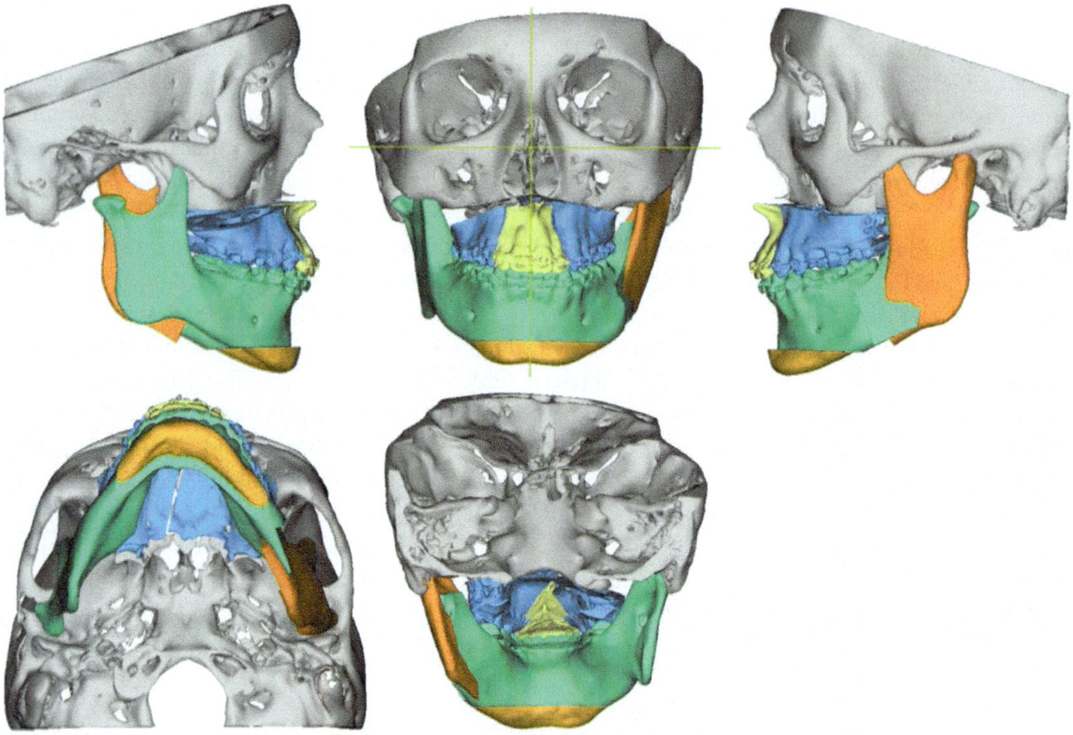

Fig. 12.10 VSP with MedCad® showing the final mandibular position with genioplasty. BSSO is planned for the left side with total joint replacement planned for right side

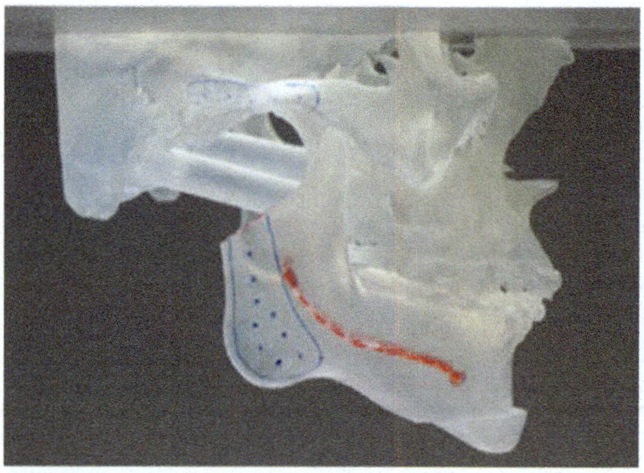

Fig. 12.11 Stereolithic model outlining right TMJ Concepts® ramal and fossa total joint components

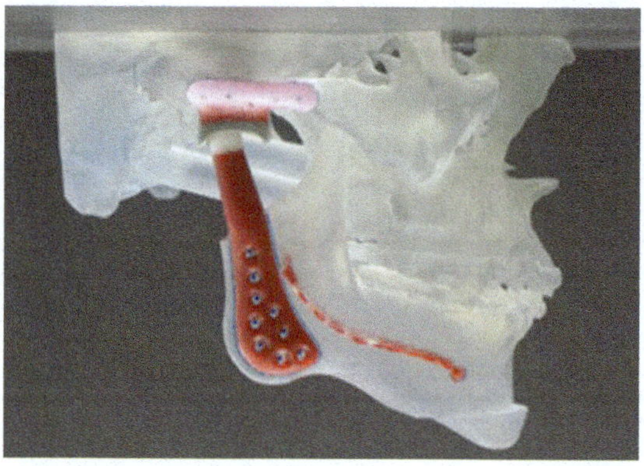

Fig. 12.12 Stereolithic model with wax-up of TMJ Concepts® right ramal and fossa total joint components

Surgery was performed in standard fashion without complications. In the weeks following surgery, the patient reported compliance with all postoperative instructions including a strict non-chew diet. Guiding elastics were transitioned as needed. The patient was followed for a period of 1 year at which time final records were obtained (Figs. 12.13, 12.14, 12.15, 12.16, 12.17, 12.18, and 12.19). The patient was most recently seen for his 36-month follow-up (Fig. 12.20).

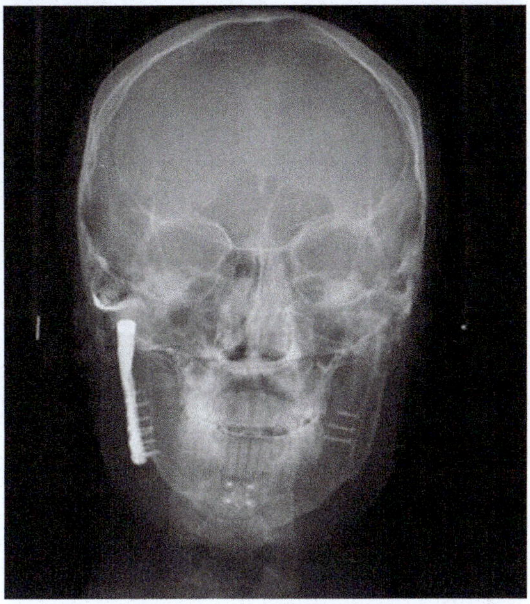

Fig. 12.13 Postoperative AP cephalometric radiograph

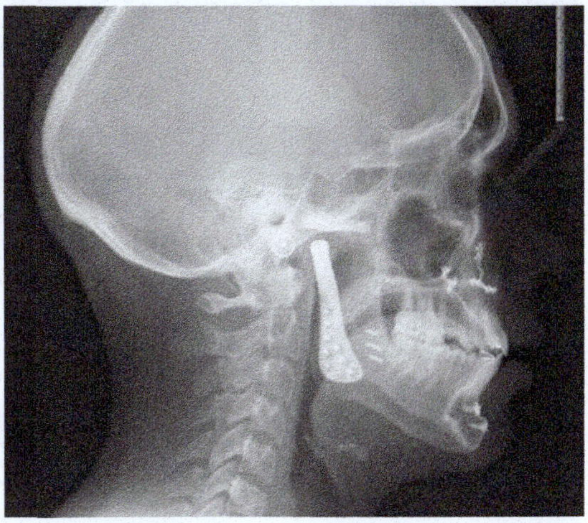

Fig. 12.14 Postoperative lateral cephalometric radiograph

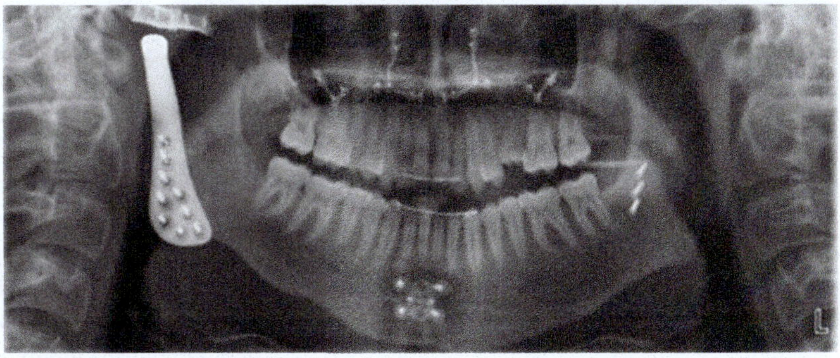

Fig. 12.15 Postoperative orthopantomogram

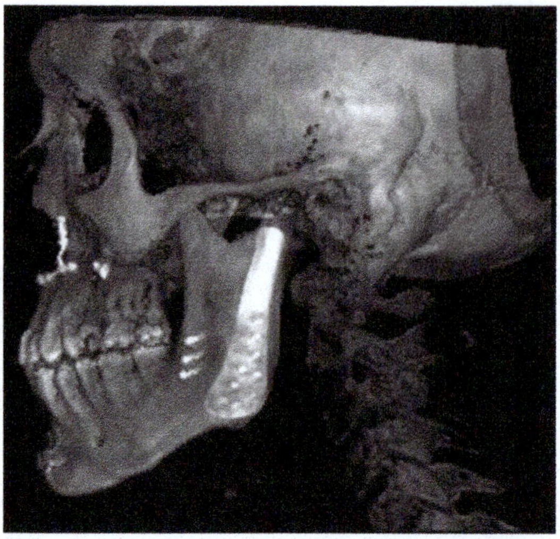

Fig. 12.16 Final CBCT left profile view

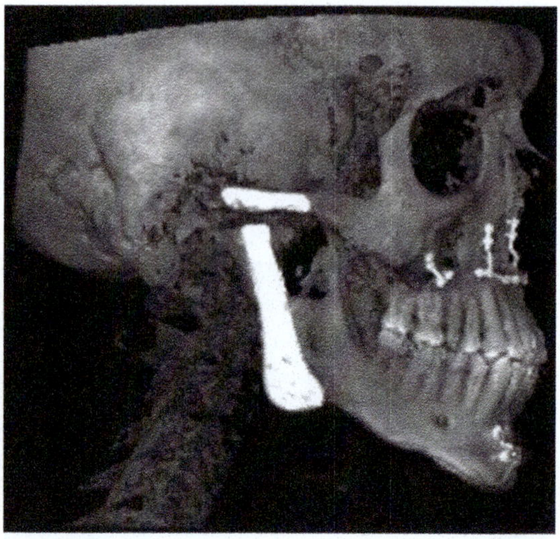

Fig. 12.18 Final CBCT right profile view

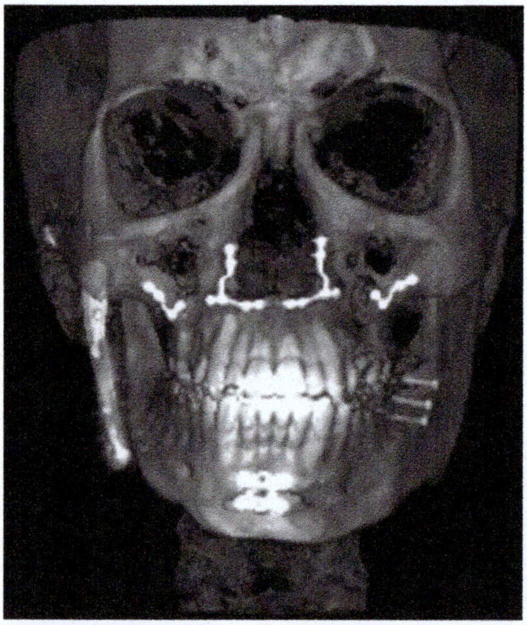

Fig. 12.17 Final CBCT frontal view

12.3 Discussion

The workhorse of traditional orthognathic surgery treatment planning has been the superimposition of two-dimensional predictive tracings in a 1:1 ratio with the patient's cephalometric radiographs. This two dimensionally weighted approach to treating a three-dimensional dentofacial deformity works remarkably well in most cases. However, facial asymmetries of pitch, roll, and particularly yaw can be difficult to accurately predict and may yield themselves more amenable to VSP. At our institution, we continue to draw two-dimensional predictive tracings even for cases treated with VSP. The vast majority of these predictions are remarkably accurate.

There are many other inaccuracies with traditional model surgery. There are inherent inaccuracies with the facebow transfer, dental impression deformities, errors obtaining an accurate bite registration, mounting errors on a semi-adjustable hinge articulator, occlusal wear during splint fabrication, and general operator error at any number of points during the preoperative planning. Additionally, there can be error in identification of cephalometric landmarks as well as tracing errors.

VSP has several limitations as well. Currently, it is not FDA approved for segmental orthognathic surgery. CT scanned images have not reliably captured dental occlusal surfaces. This limitation requires dental impressions and a bite registration be obtained in a traditional fashion. However, new technology is emerging

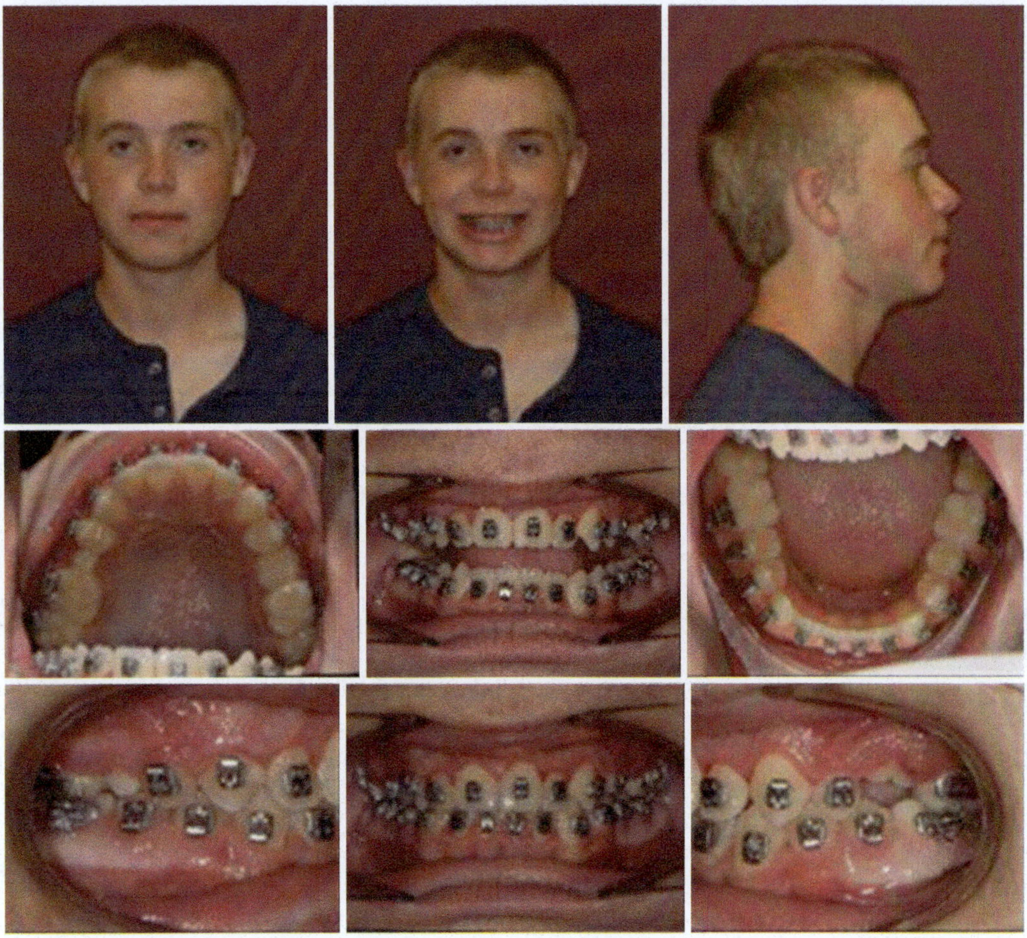

Fig. 12.19 Intraoral and extraoral final records

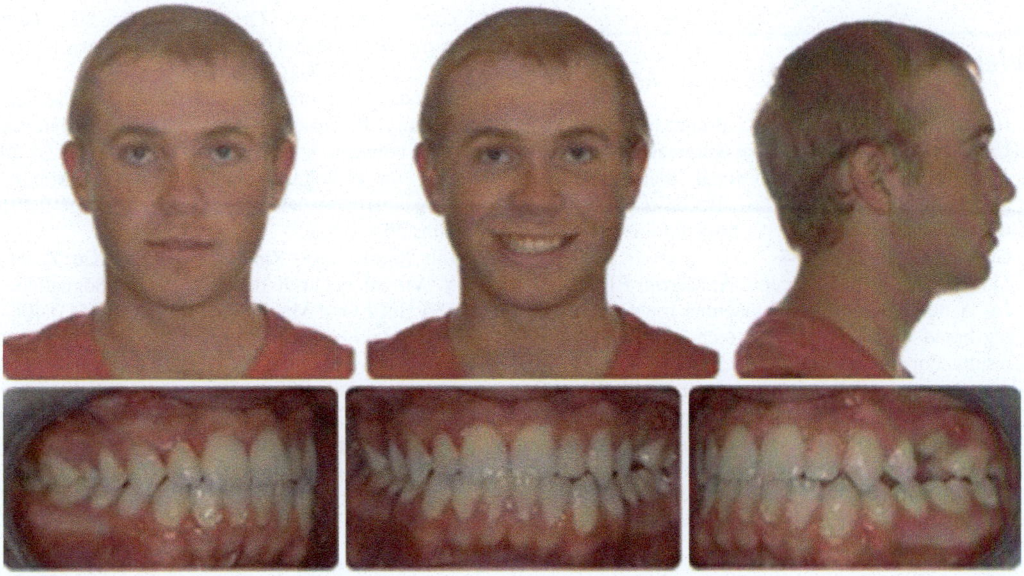

Fig. 12.20 Intraoral and extraoral photographs at 36-month follow-up

obviating the need for laser-scanned models [6, 13, 14]. Currently, VSP must be planned with a software engineer and cannot be done alone. Intraoperative splints are printed at outside facilities and mailed causing a delay between splint fabrication and surgery [1]. Lastly, the use of stereolithic splints incurs additional costs which are not insurance covered benefits. Prices for each splint can be in excess of $800. Recent research comparing the cost differential between standard and virtual surgical planning reported that VSP took significantly less time and was less expensive for all types of cases analyzed. This conclusion was derived by evaluating and comparing the cumulative time and cost for all steps of the orthognathic surgery planning process [12].

Many limitations of VSP are likely to diminish with time. Technology is likely to soon obviate the need for laser-scanned dental models. Routine use of software may empower the surgeon to virtually plan cases without the need for a software engineer. Accessibility to 3-D printing is likely to improve. Costs for VSP and splint fabrication are likely to decrease making VSP an appealing and accessible alternative to traditional model surgery.

Acknowledgement *Special thanks*: The authors would like to thank Dr. J. Peter Kierl for the orthodontic component of this case and photographs used in this chapter.

References

1. Hammoudeh JA, Howell LK, Boutros S, Scott MA, Urata MM. Current status of surgical planning for orthognathic surgery: traditional methods versus 3D surgical planning. Plast Reconstr Surg Glob Open. 2015;3(2):e307. https://doi.org/10.1097/GOX.0000000000000184.
2. Mozzo P, Procacci C, Tacconi A, Martini PT, Andreis IA. A new volumetric CT machine for dental imaging based on the cone-beam technique: preliminary results. Eur Radiol. 1998;8(9):1558–64.
3. American Dental Association Council on Scientific Affairs. The use of cone-beam computed tomography in dentistry: an advisory statement from the American Dental Association Council on Scientific Affairs. J Am Dent Assoc 1939. 2012;143(8):899–902.
4. Bengtsson M, Wall G, Miranda-Burgos P, Rasmusson L. Treatment outcome in orthognathic surgery – a prospective comparison of accuracy in computer assisted two and three-dimensional prediction techniques. J Craniomaxillofac Surg. 2017; https://doi.org/10.1016/j.jcms.2017.01.035.
5. Gelesko S, Markiewicz MR, Weimer K, Bell RB. Computer-aided orthognathic surgery. Atlas Oral Maxillofac Surg Clin North Am. 2012;20(1):107–18. https://doi.org/10.1016/j.cxom.2012.01.002.
6. Yuan P, Mai H, Li J, et al. Design, development and clinical validation of computer-aided surgical simulation system for streamlined orthognathic surgical planning. Int J Comput Assist Radiol Surg. 2017; https://doi.org/10.1007/s11548-017-1585-6.
7. Xia JJ, Gateno J, Teichgraeber JF. New clinical protocol to evaluate craniomaxillofacial deformity and plan surgical correction. J Oral Maxillofac Surg. 2009;67(10):2093–106. https://doi.org/10.1016/j.joms.2009.04.057.
8. Levine JP, Patel A, Saadeh PB, Hirsch DL. Computer-aided design and manufacturing in craniomaxillofacial surgery: the new state of the art. J Craniofac Surg. 2012;23(1):288–93. https://doi.org/10.1097/SCS.0b013e318241ba92.
9. Swennen GRJ, Mollemans W, Schutyser F. Three-dimensional treatment planning of orthognathic surgery in the era of virtual imaging. J Oral Maxillofac Surg. 2009;67(10):2080–92. https://doi.org/10.1016/j.joms.2009.06.007.
10. Bobek S, Farrell B, Choi C, Farrell B, Weimer K, Tucker M. Virtual surgical planning for orthognathic surgery using digital data transfer and an intraoral fiducial marker: the charlotte method. J Oral Maxillofac Surg. 2015;73(6):1143–58. https://doi.org/10.1016/j.joms.2014.12.008.
11. Swennen GRJ. Timing of three-dimensional virtual treatment planning of orthognathic surgery: a prospective single-surgeon evaluation on 350 consecutive cases. Oral Maxillofac Surg Clin N Am. 2014;26(4):475–85. https://doi.org/10.1016/j.coms.2014.08.001.
12. Resnick CM, Inverso G, Wrzosek M, Padwa BL, Kaban LB, Peacock ZS. Is there a difference in cost between standard and virtual surgical planning for orthognathic surgery? J Oral Maxillofac Surg. 2016;74(9):1827–33. https://doi.org/10.1016/j.joms.2016.03.035.
13. Liu XJ, Li QQ, Zhang Z, Li TT, Xie Z, Zhang Y. Virtual occlusal definition for orthognathic surgery. Int J Oral Maxillofac Surg. 2016;45(3):406–11. https://doi.org/10.1016/j.ijom.2015.07.022.
14. Nilsson J, Richards RG, Thor A, Kamer L. Virtual bite registration using intraoral digital scanning, CT and CBCT: In vitro evaluation of a new method and its implication for orthognathic surgery. J Craniomaxillofac Surg. 2016;44(9):1194–200. https://doi.org/10.1016/j.jcms.2016.06.013.

3D Imaging for Craniofacial Anomalies

13

Kevin S. Smith and Myles Davidson

Abstract

Craniofacial anomalies (CFAs) can arise from any type of abnormal growth or deformity of the structures of the craniofacial skeleton. The variations of these anomalies arise from a number of different factors, including genetic factors, environmental factors, and folic acid deficiencies, and can range from very mild to severe, requiring surgery. Some of the most common types of craniofacial anomalies include cleft lip/cleft palate, craniosynostosis, hemifacial microsomia, vascular malformations, hemangioma, and deformational or positional plagiocephaly. Three-dimensional imaging of these CFAs continues to evolve with advances in technology. While cone-beam computed tomography (CBCT) is often considered the workhorse for imaging of CFAs, it is not without its own limitations. Indications for alternative or adjunctive imaging modalities include soft tissue detail, inflammatory processes, and temporomandibular joint morphology. This chapter provides a review of three-dimensional imaging techniques most commonly used for the diagnosis and management of craniofacial anomalies.

K. S. Smith (✉) · M. Davidson
Department of Oral and Maxillofacial Surgery, University of Oklahoma Health Science Center, Oklahoma City, OK, USA
e-mail: kevin-smith@ouhsc.edu; Myles-Davidson@ouhsc.edu

13.1 Computed Tomography

There are a multitude of methods one can utilize when studying craniofacial anomalies. The first is computed tomography (CT) or computerized axial tomography (CAT) imaging. This uses specific X-ray equipment to generate cross-sectional images of the body. CT can be further divided into two groups: fan beam and cone beam. Originally described in the late 1960s by Sir Godfrey Hounsfield, fan beam computed tomography utilizes sequential axial scans obtained from the tissues of interest [1]. These cross-sectional images, or "slices," allow detailed analysis of internal anatomy in an atraumatic fashion (Fig. 13.1). The craniofacial bones are readily visualized using computed tomography, which is considered the gold standard for analysis of the skeletal variations encountered in craniofacial anomalies (Fig. 13.2). Furthermore, the addition of IV or enteral contrast materials to the CT scanning protocol broadens its utility and allows for more accurate study of soft tissue and pathophysiologies which may otherwise be difficult to image.

Cone-beam computed tomography (CBCT) was first introduced in the early 1980s and utilizes X-ray technology in a similar but unique fashion. Unlike the fan-shaped X-ray beam utilized in traditional CT, cone-beam computed tomography relies on a three-dimensional divergent "cone" of X-rays. This allows image capture

O. Kadioglu, G. F. Currier (eds.), *Craniofacial 3D Imaging*,
https://doi.org/10.1007/978-3-030-00722-5_13

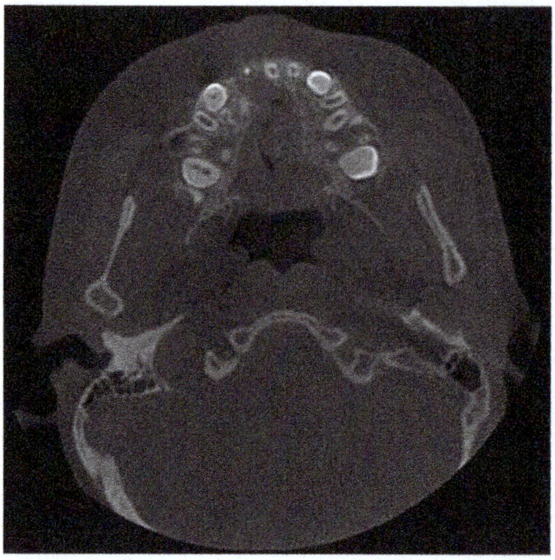

Fig. 13.1 Cone-beam computed tomography axial image visualizing unilateral cleft lip and palate deformity

in a single rotation around the patient without the need for overlapping slices. CBCT machines overcome many limitations of conventional CT scanning devices. Some of the advantages include ease of patient positioning, decreased scan time, improved spatial resolution or sharpness, clinical ease of use, fewer space and equipment requirements, and lower radiation doses [2, 3]. Qu et al. reported CBCT effective radiation doses to be anywhere from several up to ten times smaller than traditional computed tomography [4]. These factors have allowed CBCT machines to become commonplace in outpatient clinics and a vital tool with which craniofacial abnormalities are diagnosed and treated. Figure 13.3 presents the case of an 18-year-old female with bilateral idiopathic condylar resorption. This condition, which typically presents with a retrognathic

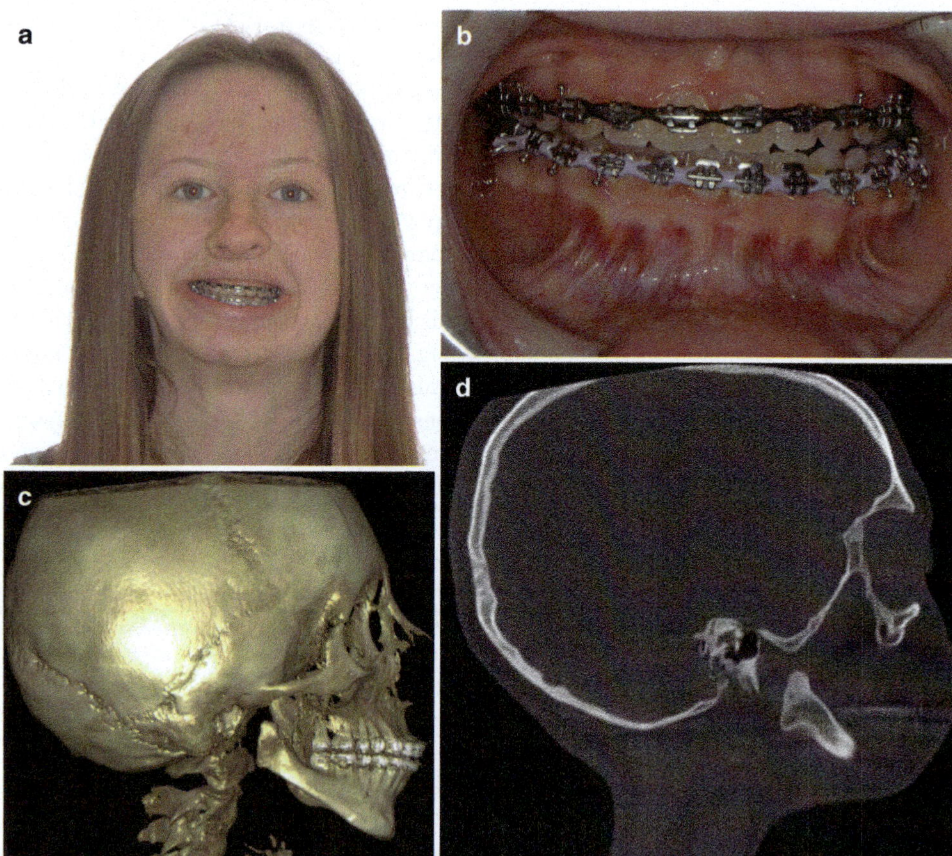

Fig. 13.2 Three-dimensional imaging in a 15-year-old patient with hemifacial microsomia. (**a**) Smile highlighting marked asymmetry of the mandible. (**b**) Intraoral view showing noticeable cant associated with patient's craniofacial anomaly. (**c**, **d**) 3D reconstruction and sagittal slice detailing the extent of hypoplasia in the right mandibular ramus/condyle

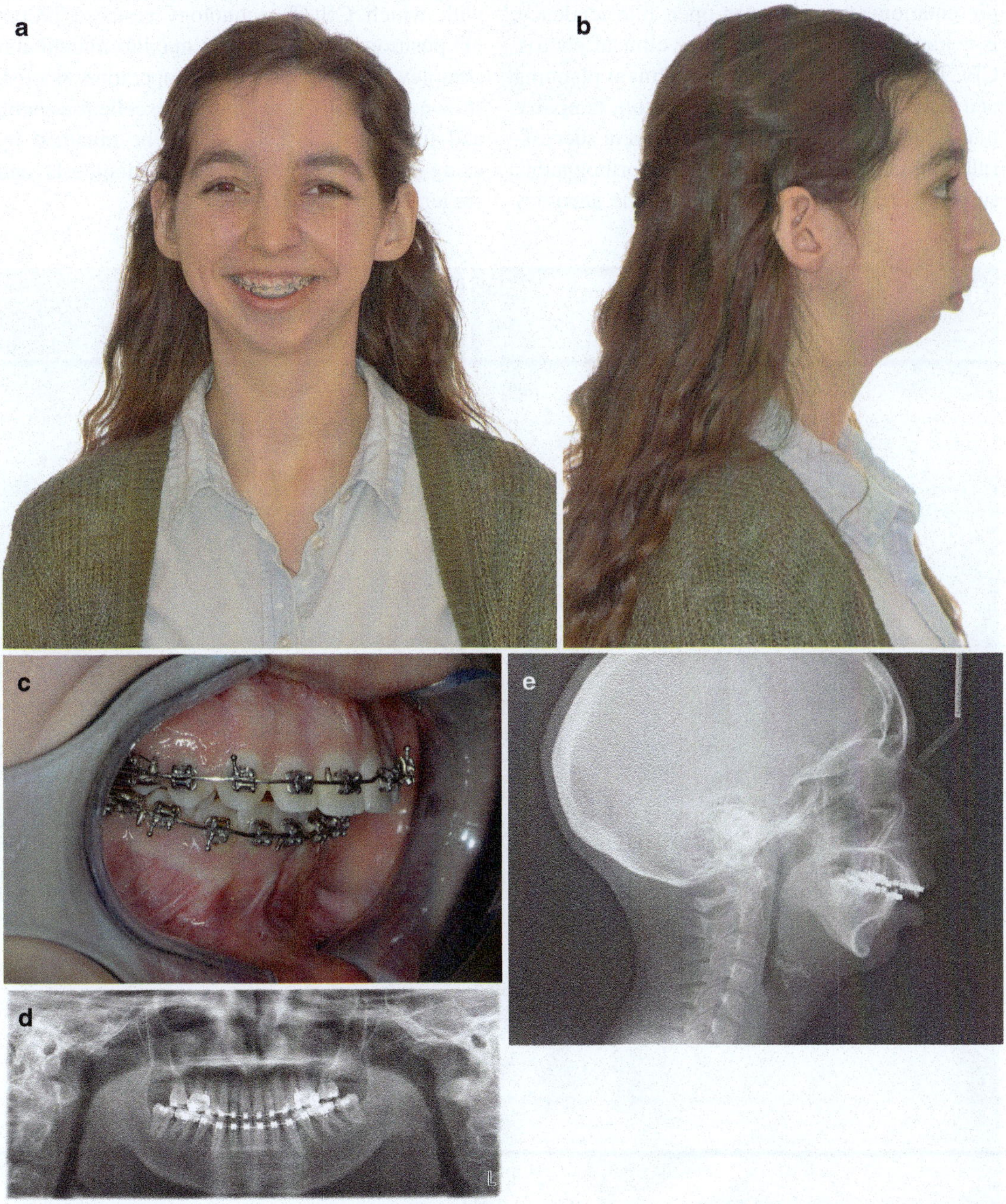

Fig. 13.3 Eighteen-year-old patient presenting with bilateral idiopathic condylar resorption. (**a–c**) Retrognathia, class II skeletal malocclusion, and anterior open bite tendency commonly associated with this condition are demonstrated. (**d, e**) Preoperative lateral cephalometric and panoramic imaging

presentation of an anterior open bite tendency, is evident from the preoperative clinical photos. CBCT was used to guide the treatment planning and for fabrication of bilateral custom temporomandibular joint prostheses which were successfully implanted in conjunction with orthognathic surgery. Figure 13.4 demonstrates the accuracy with which CBCT technology restored proper 3D position of this patient's maxillo-mandibular complex. This patient's postoperative photos, shown in Fig. 13.5, exhibit the excellent esthetic and functional results that can be obtained by using CBCT technology to aid diagnosis and surgery.

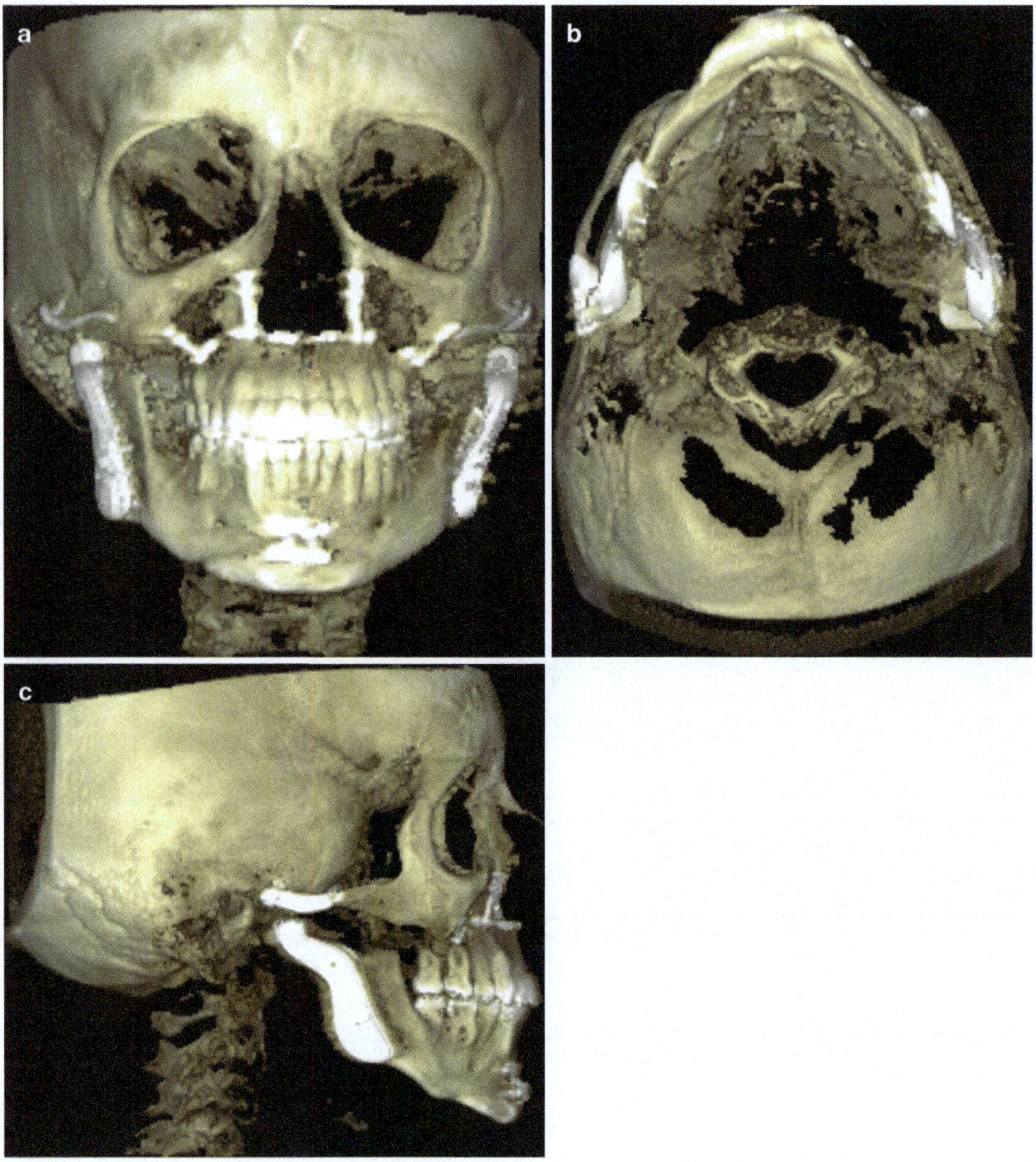

Fig. 13.4 The same eighteen-year-old patient following accurate repositioning of her maxillo-mandibular complex viewed from the (**a**) frontal, (**b**) axial, and (**c**) sagittal planes. CBCT technology was used during the preoperative planning and prosthesis fabrication stages to ensure ideal postoperative outcomes

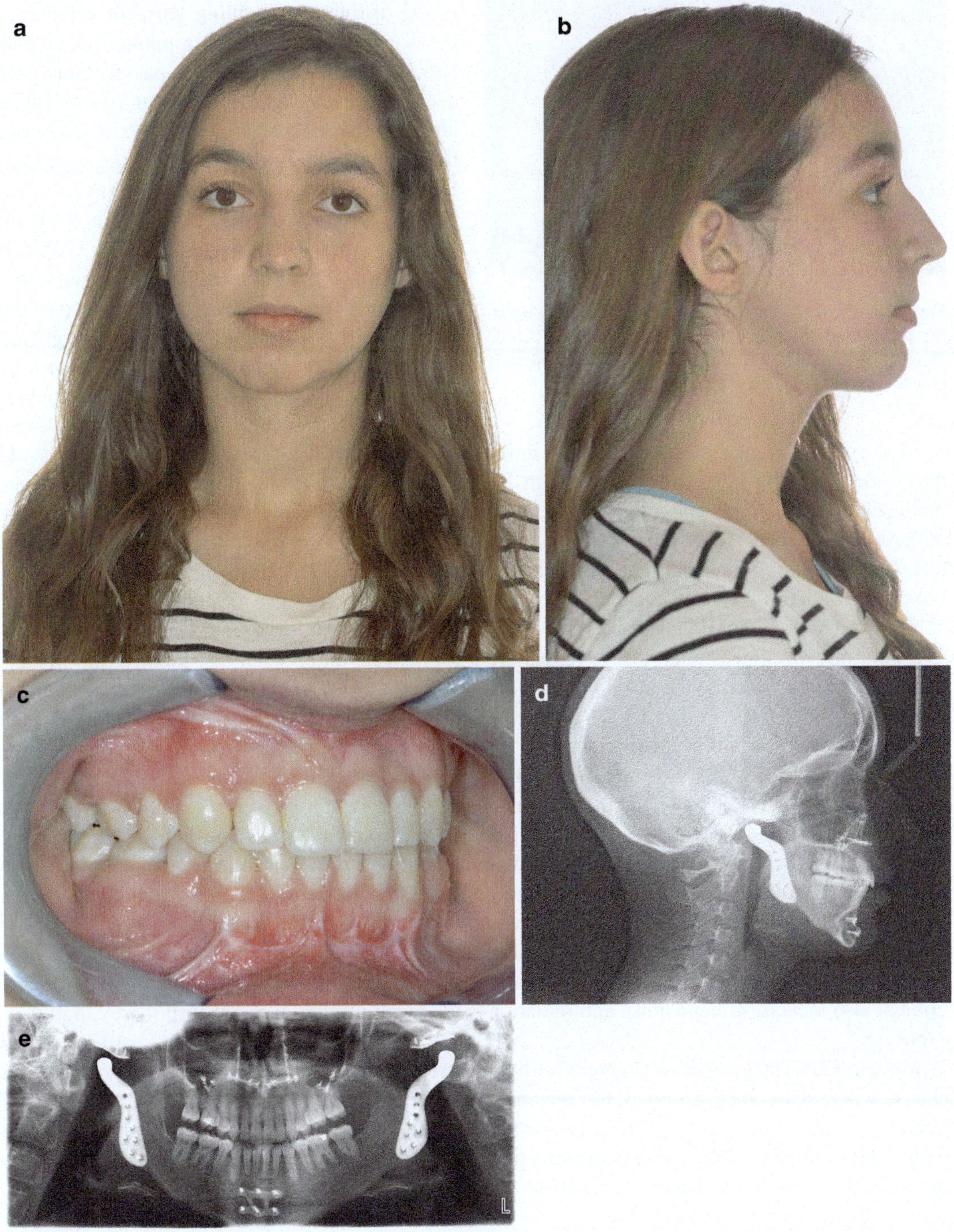

Fig. 13.5 (**a–c**) CBCT-based planning and surgery facilitated excellent cosmetic and functional improvements. (**d**, **e**) Postoperative lateral cephalometric and panoramic imaging

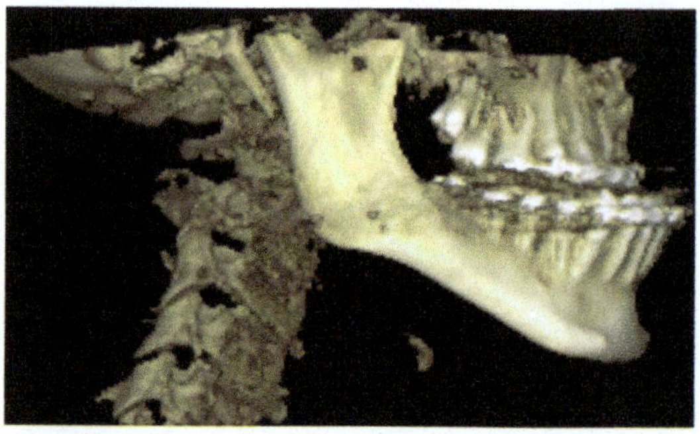

Fig. 13.6 Field of view customization allows 3D imaging of isolated portions of the facial skeleton. Focused imaging on a portion of the facial skeleton can drastically reduce ionizing radiation exposure

Modern CBCT machines continue to introduce new technology that tailors each scan for optimization of treatment planning. One example of these advancements is the increased resolution that contemporary scans are able to achieve. Voxels are values which are assigned a specific point within three-dimensional space. Analogous to a pixel in two-dimensional space, voxels are frequently used to describe the resolution of a 3D image. The voxel sizes used in CBCT have progressively decreased since the introduction of this technology. While CBCT originally relied on voxel sizes of several hundred micrometers, modern CBCT units allow sizes under 100 μm and can be customized from one scan to another [5]. While this advancement seems to be largely beneficial, clinical applications should be evaluated on a patient-to-patient basis, as a smaller voxel size increases not only the spatial resolution but also increases image noise and radiation exposure.

Modern CBCT has provided a better ability to choose where scans are positioned. For instance, you can choose to scan the entire head or only focus on the chin area (Fig. 13.6). Customization of the field of view even allows one to offset the center of rotation, enabling only one rotation of the X-ray detector around a patient rather than two. Eliminating multiple passes not only decreases the chance for patient movement to alter the image reconstruction but also decreases the amount of ionizing radiation exposure. Recent studies have shown reductions in radiation exposure of up to 25% and 60% in the mandible and maxilla, respectively, when CBCT is tailored to a region of interest rather than including the entire head [6, 7]. This feature is particularly useful when treatment planning growth abnormalities or pathology that affects specific areas of the craniofacial skeleton. The case presented in Fig. 13.7 illustrates the selective capabilities of CBCT technology in a 24-year-old female with right mandibular ameloblastoma. Much less exposure to ionizing radiation was achieved by limiting the field of view on the patient's CBCT exam to her lower facial third. This feature becomes particularly useful in situations where multiple CBCT volumes are necessary over time, such as monitoring of this patient's pre- and postoperative progression.

Cited disadvantages of CBCT include increased presence of artifacts, more image scatter, and decreased ability to differentiate low contrast visibility tissues [2, 3, 8]. Furthermore, when compared to traditional 2D imaging modalities, CBCT is still more expensive, emits more radiation, and is very susceptible to motion and metal artifacts. Metal artifacts in particular are a common problem associated with CBCT imaging in dentistry. These metal artifacts are the result of high X-ray absorption by high-density metallic elements. Beam hardening and streak artifacts around dental restorations can render diagnosis and treatment planning a challenge. Figure 13.8 demonstrates just how problematic these artifacts can be to postoperative evaluation.

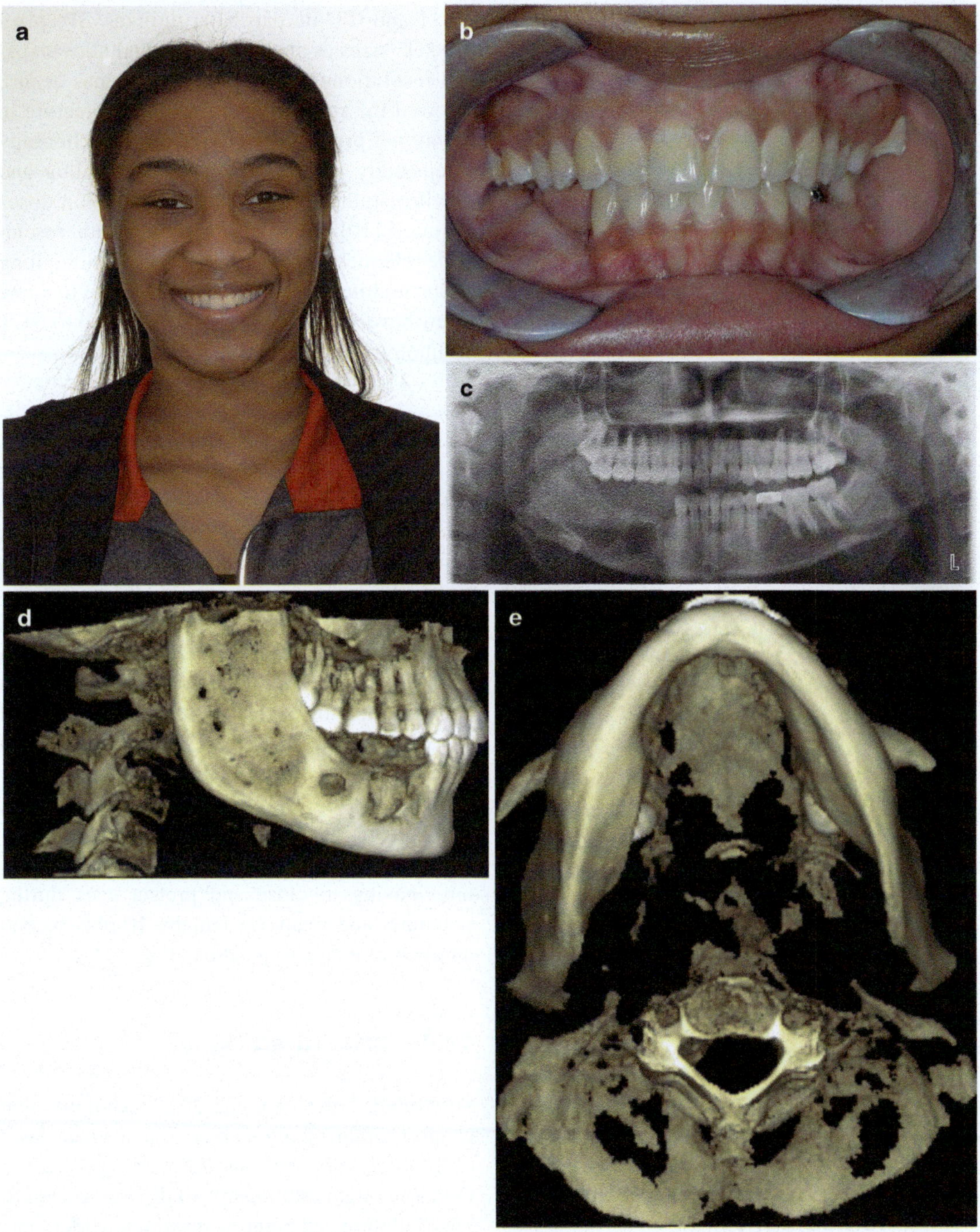

Fig. 13.7 (**a**) Patient with right mandibular ameloblastoma. (**b**) Intraorally, notable expansion of right posterior mandible with confirmation of bony lesion on (**c**) panoramic radiograph. (**d** and **e**) Lower levels of ionizing radiation were achieved by customizing the CBCT field of view to include only the lower facial third

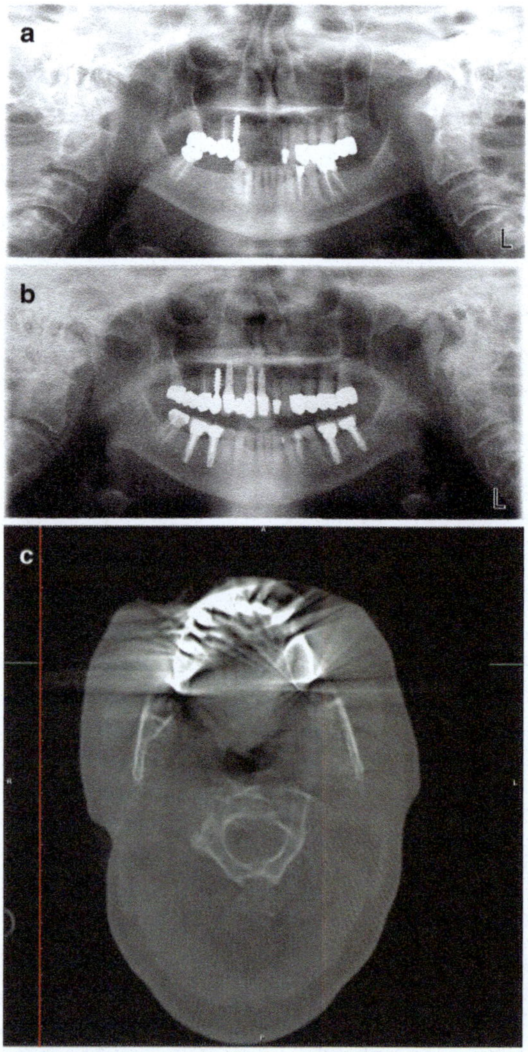

Fig. 13.8 (**a**) Preoperative and (**b**) postoperative panoramic radiographs showing successful implant restoration of multiple edentulous spaces in a 67-year-old patient. (**c**) Postoperative axial CBCT cut illustrating the image-degrading effect of streak artifacts and beam hardening around this patient's metallic restorations

13.2 Micro- and Nano-computed Tomography

A more recently developed method of imaging is micro- and nano-computed tomography (MCT/NCT). This is essentially the same as CT scanning except for the fact that the reconstructed cross sections are bounded to a much smaller area. These devices can have 10,000 times more resolution than medical CT scanners do. Nano-CT in particular utilizes 3D pixel "voxel" sizes as small as 25 nm and allows spatial resolutions of 400 nm [9]. This has in turn allowed the study of tissue nano-architecture in a way not previously possible. This technology is currently restricted to animal modeling and in vitro studies due to the high radiation doses required [10]. However, imaging of atherosclerotic plaques, cerebral microcirculation, lung tissue architecture, and trabecular bone to name a few have all been successfully applied to NCT technology [9, 10].

13.3 3D Laser Scanning

Three-dimensional laser scanning is a less invasive method of capturing the face in comparison to many imaging modalities. Perhaps most importantly, 3D laser scanning avoids the potentially harmful ionizing radiation encountered with traditional X-ray techniques. Laser scanning can supply 3D images for treatment planning or evaluating effects of many orthodontic and surgical procedures [11, 12]. In particular, 3D laser scans have been successfully applied to analyze soft tissue changes following orthognathic surgery [13, 14]. Drawbacks of this method can include difficulty in capturing fine soft tissue detail and texture, having to close and protect eyes during scanning, and relatively lengthy 10-plus second duration of image acquisition [15].

13.4 Structured Light

Structured light imaging is a technique that enables capturing of the three-dimensional shape of the face without the use of ionizing radiation. Following light illumination of the face, positioning of illuminated points is integrated with points on 3D cephalometric tracings. The result is a 3D shape of the patient's face, viewable on a computer screen. The aim of this technique is to combine the facial shape and underlying radiographic data from other sources to study 3D structures for diagnosis, treatment planning, and evaluation of treatment results. This technique has traditionally been applied to intraoral imaging in orthodontics, cleft lip analysis, and both cosmetic and orthognathic surgery. Cited limitations have included the inability of some patients to remain still for the necessary length of time, the need for multiple light sources and patterns to ensure accuracy, and the difficulty in reconstructing anatomy in areas of undercut profile. The use of multiple cameras, or a single camera using multiple views at different angles, is often necessary for accurate assessment of the entire face. The emergence of handheld structured light devices has improved accuracy, particularly for intraoral scanning purposes [16–18].

13.5 Stereophotogrammetry

Stereophotogrammetry is the process of photographing a 3D object from two different coplanar planes in order to acquire a 3D reconstruction of the images. While clinical application of stereophotogrammetry was initially used to analyze the effects from orthodontic therapy, the technology is now applicable to imaging in many areas of dentistry and craniofacial surgery [19]. Namely, pre-surgical rendering of 3D facial soft tissue movements can have great benefit to craniomaxillofacial surgeons. Photogrammetric scanning in combination with CBCT has recently been shown by multiple authors to accurately simulate 3D soft tissue changes after orthognathic surgery [20, 21]. In addition, comparison of soft tissue morphologic variations in different patterns of cleft lip and palate deformity has been successfully performed [22]. Because of its ability to help surgeons visualize predicted soft tissue movements, this modality can be valuable for patient acceptance and understanding of possible postoperative results.

Facial cosmetic surgery is often complicated by differing expectations between the surgeon and patient. Interpretation of surgical results can be quantified in a more predictable and reproducible manner through the use of stereophotogrammetric technology. One recent example uses stereophotogrammetry as a way to classify changes in volume distribution following fat grafting and other cosmetic rejuvenation procedures [23]. As this technology has improved over the years, 3D surface scanning is now able to be integrated into modern CBCT units (Fig. 13.9). This allows providers the ability to capture both 3D photos and a CBCT volume in the same imaging session. Should CBCT and its radiation be unnecessary to the treatment plan, 3D soft tissue scans can be acquired separately in a radiation-free manner for use in dedicated orthodontic, maxillofacial, or cosmetic surgery planning [24–26]. Structured light, for example, can be used in combination with stereophotogrammetry for this kind of three-dimensional analysis (Fig. 13.10).

Unfortunately, the ability to obtain noncontact 3D imaging is still often limited by patient compliance with younger age groups. Modern 3D digital stereophotogrammetry, with acquisition times as short as 0.0015 s, does require much less patient compliance to capture a successful image [27]. Stereophotogrammetry, like any maxillofacial imaging technology, is not without its own disadvantages. Postural changes between images, hair intervention, soft tissue glare or reflection, and poor detail on curved surfaces such as the submental and subnasal regions will continue to challenge the accuracy and ease with which this technology can be implemented into craniofacial treatment plans (Fig. 13.11).

13.6 Three-Dimensional Facial Morphometry

Three-dimensional facial morphometry (3DFM) is a process that can be used following other imaging procedures, particularly as a supplement to the lateral cephalometric analysis. Landmarks are positioned on the face and later covered with 2 mm semispherical reflective markers. An ultraviolet stroboscope is then used to light up the projective markers. Analysis of multiple facial angles is generally needed to acquire a complete mapping of the face. While simplistic in concept, the landmarks for this method are often located through palpation and hand tracing [15, 28]. This can be an elaborate and time-consuming process that fails to match the accuracy found in other modern craniofacial imaging technologies.

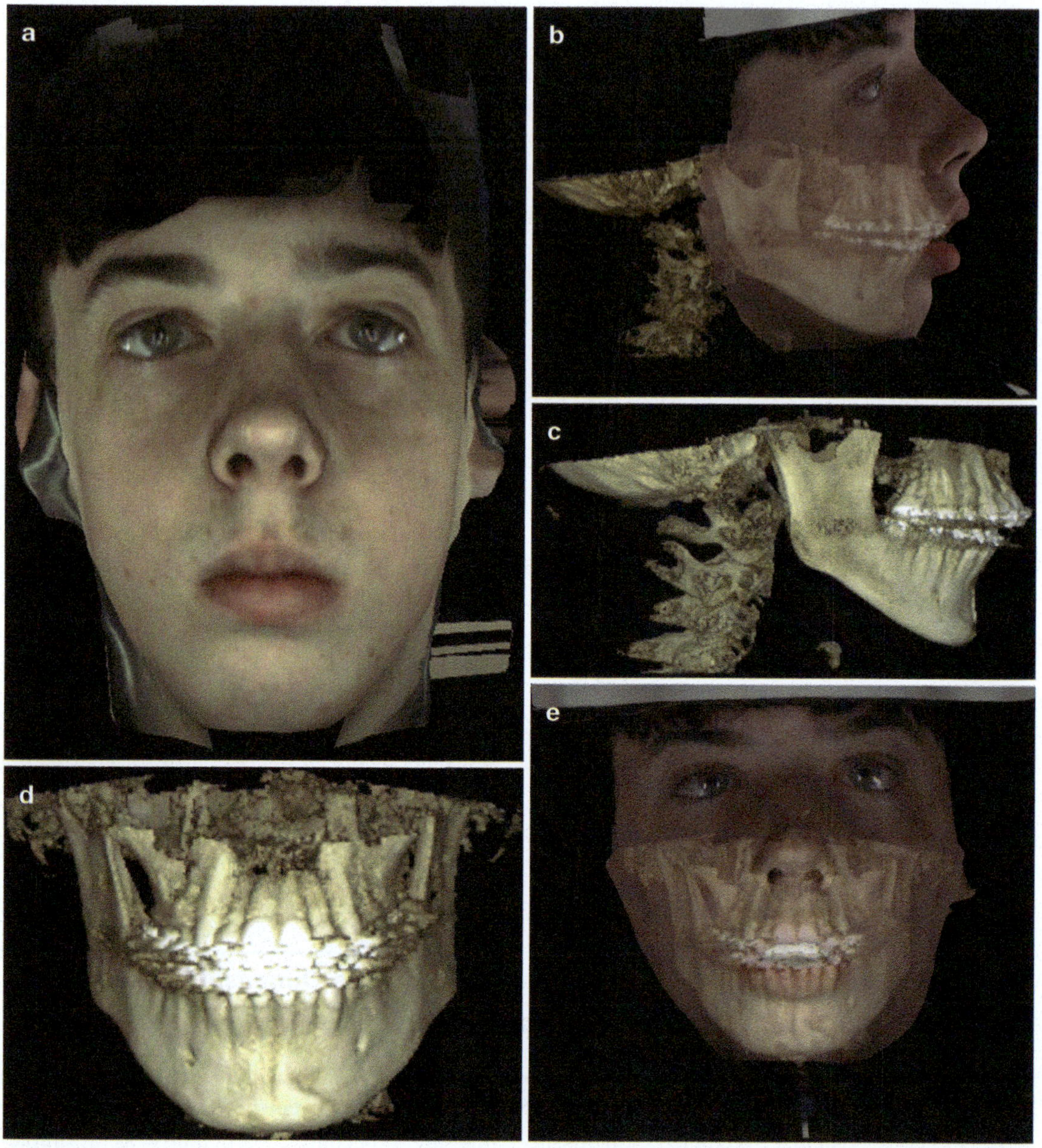

Fig. 13.9 (**a**) 3D image created from a CBCT study showing soft tissue contours of a 17-year-old patient. (**b–e**) There are multiple degrees of adjustable soft tissue overlay which can be viewed on the patient's underlying facial skeleton

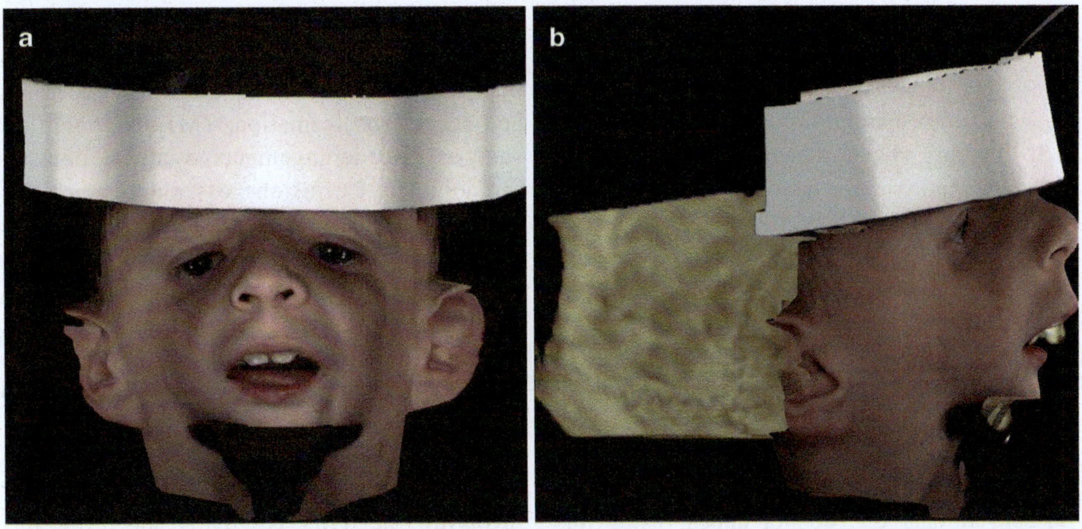

Fig. 13.10 Eight-year-old patient with Treacher Collins syndrome showing several limitations of 3D stereophotogrammetric imaging. (**a**) Motion artifacts (particularly on the patient's right), poor soft tissue detail around the chin, and obstructive forehead brace all contributing to diminished image quality. (**b**) Obstructive forehead brace and poor detail around the chin are visualized. Additional lack of detail is often noted around the ears with this imaging modality

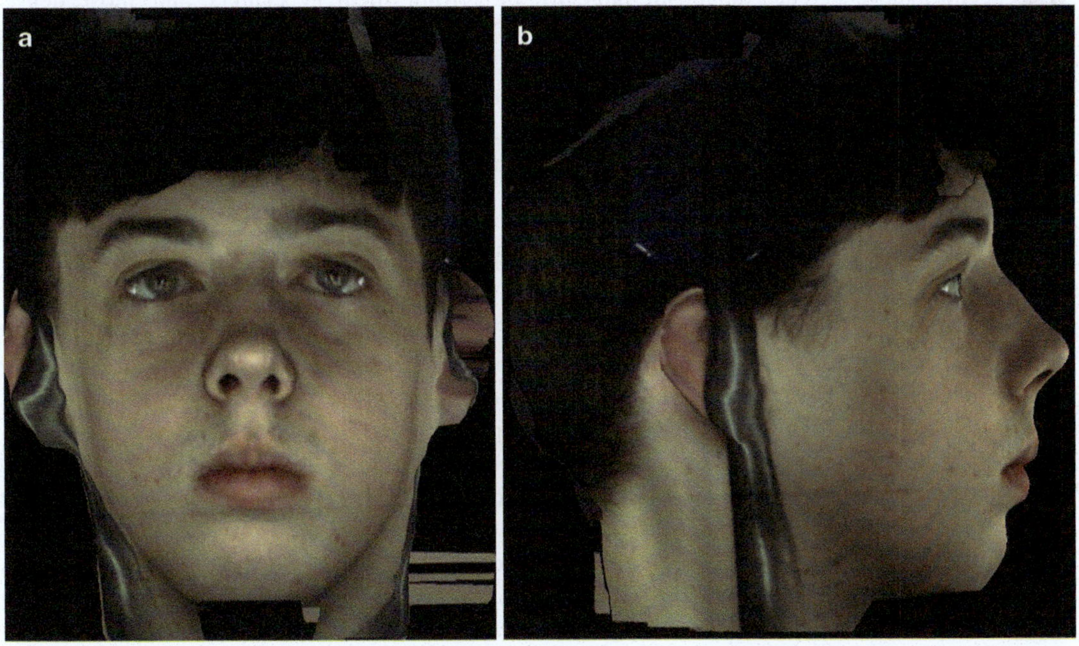

Fig. 13.11 The same 17-year-old patient further demonstrating limitations in image quality using combined CBCT/stereophotogrammetry. (**a, b**) Both images illustrate the lack of detail common to the forehead and auricular regions. Additional interference is noted from the bilateral sidebar appliances used to discourage patient motion and maintain proper head positioning

13.7 Tuned-Aperture Computed Tomography

Tuned-aperture computed tomography (TACT) is a 3D imaging system that uses a calibration or reference marker in the area of interest to permit synthetic reconstruction of the desired image. TACT utilizes a series of 2D periapical radiographs taken from different angles to retrospectively generate tomographic slices [29, 30]. 3D data is then produced from these collections of two-dimensional images. Interestingly, overall radiation dose is no more than two times that of a conventional PA film, and artifacts such as starburst patterns seen around metallic restorations on conventional CT, do not appear using this technique. TACT has been suggested as a diagnostic tool for evaluation of healing in both mandibular and calvarial defects. However, studies supporting this technique have only been performed in vitro at this time [31–33]. Currently, TACT seems to have a greater diagnostic value in its ability to detect dental caries, visualize vertical root fractures, and assist in pre-implant treatment planning [34–36].

13.8 Magnetic Resonance Imaging

Magnetic resonance imaging (MRI) is a process that works by detecting a resonance signal from the hydrogen nucleus. Therefore, it can essentially be considered as imaging of water within the tissue. MRI is the highest contrast resolution medical imaging technique. Radio waves are sent to the desired location for examination in a magnetic field. The energy produced from hydrogen atoms in the cells stimulated by radio waves is converted to numerical values which are processed on a computer and then converted to an image. MRIs are helpful for the study of skeletal physiology, tumors, and healing of grafts. Advantages of MRI include its ability to provide valuable information about the position and morphology of the temporomandibular joint (TMJ) disc, detailed soft and hard tissue resolution, and radiation-free imaging. It also provides an opportunity to examine inflammatory processes and scar tissue formation. TMJ abnormalities, such as disc derangements, synovial thickening, joint space effusions, osseous degenerative changes, and bone marrow edema, can all be accurately diagnosed using MRI [37–39] (Fig. 13.12). Furthermore, it can be safely used in patients who are allergic to contrast agents, and the images can be obtained without repositioning the patient. Because of these numerous advantages, MRI is considered as the gold standard for imaging of the TMJ. Treatment planning for craniofacial anomalies commonly involving the TMJ is greatly aided by this modality, in particular, hemifacial microsomia and other variations of the oculo-auriculo-vertebral spectrum [38]. Several disadvantages of MRI include its reliance on expensive and advanced medical equipment, its unavailability in certain medical centers and most dental offices, and the extended length of time it takes to evaluate joints such as the TMJ. Additionally, it can be contraindicated in patients with claustrophobia and those with ferromagnetic implants [37].

13.9 Cine Magnetic Resonance Imaging

Another unique way of observing and studying processes within the skull is phase contrast cine MRI. A cine MRI is used to observe cerebrospinal fluid (CSF) flow in order to demonstrate if there is an abnormality present. With each heartbeat CSF is forced out of the ventricles of the brain into the subarachnoid cisterns and down the spinal canal. When the heart relaxes, the CSF flow reverses. The movie-like cine MRI captures this fluid movement. Phase-contrast cine MRIs can be used to study various types of hydrocephalus, arachnoid cysts, and other cystic lesions and to provide information for evaluation of Chiari 1 malformations [40, 41].

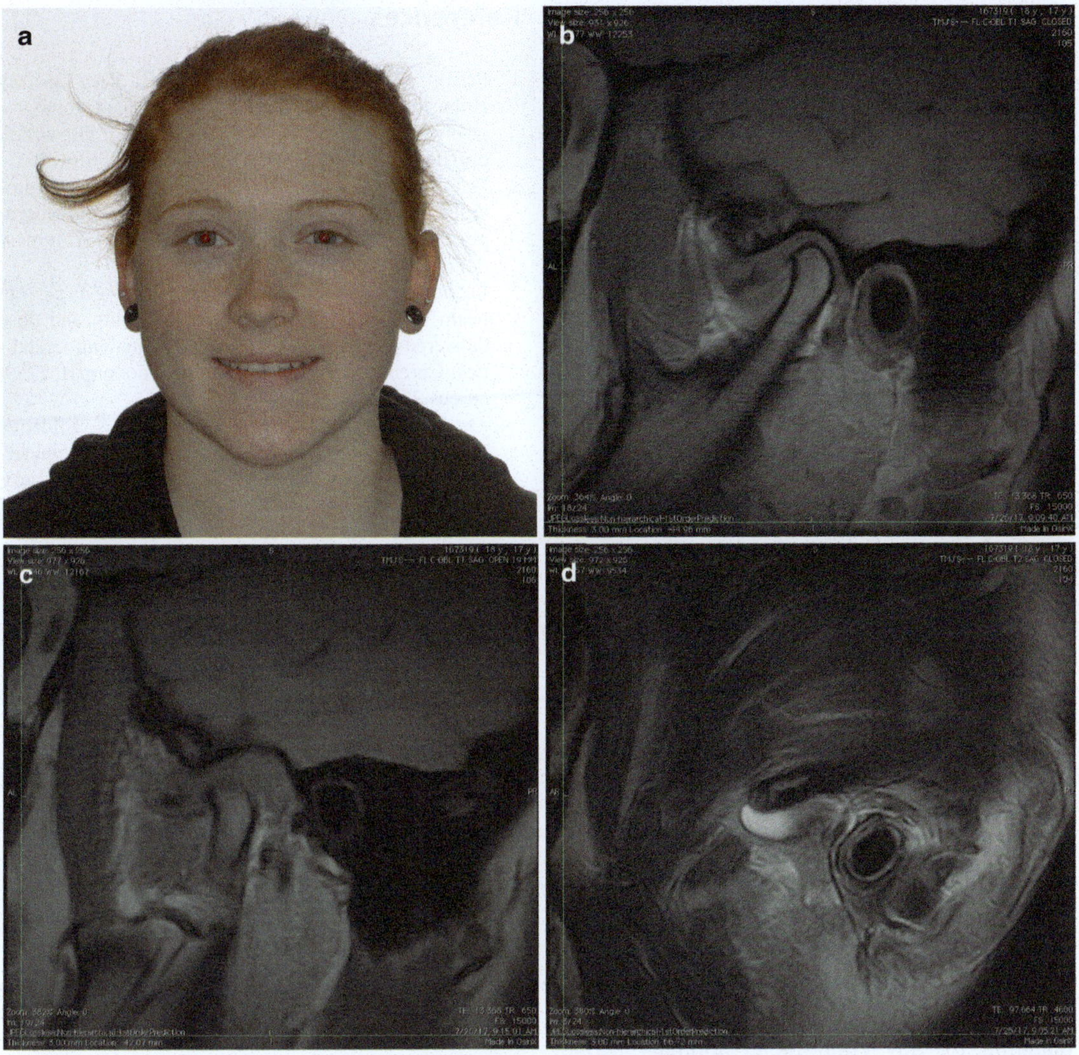

Fig. 13.12 Magnetic resonance imaging (MRI) being used for evaluation of the TMJ. (**a**) An 18-year-old female presenting with bilateral temporomandibular joint pain and dysfunction including periodic closed locking. (**b**) T1-weighted image showing anterior disc displacement in a closed mandibular position. (**c**) T1-weighted image confirming anterior disc displacement without reduction as the mandible transitions into an open position. (**d**) Radiopaque signal in a T2-weighted image corresponding to an inflammatory TMJ effusion

13.10 Angiography

A final important craniofacial imaging technique is called an arteriogram. Essentially this is an "X-ray" of the arterial blood vessels. It is performed to evaluate various vascular conditions, such as an aneurysm, a blockage, or a malformation. This process is also called an angiogram or arteriography [42]. Fluoroscopy is often used during arteriograms. Fluoroscopy is the radiographic study of moving body structures, so you can compare it to an "X-ray movie" of sorts. A continuous X-ray beam is passed through the body part being examined and is transmitted to a TV-like monitor so that the body part and its motion can be seen in detail. Generally, a dye is injected into an artery so that the arteries in question may be more visible on the X-ray. While this modality still has a valuable place in medicine, computed tomographic angiography (CTA) and magnetic resonance angiography (MRA) have begun to offer an alternative to past techniques. CTA, for example, particularly in combination with ultrasound, offers similar accuracy and is less invasive and stressful than traditional angiography for diagnosing carotid artery stenosis [43, 44]. MRA offers the advantage of not using ionizing radiation and in certain applications can be diagnostic without the use of nephrotoxic contrast agents [45]. Diagnoses of facial vascular anomalies, such as hemangiomas and arteriovenous malformations, are just several examples of the benefits of this technique [46]. Case-by-case selection should obviously have a large role in choosing this modality for vascular imaging between patients.

13.11 Conclusions

Three-dimensional imaging for craniofacial anomalies remains an evolving field. Applications and indications for various techniques continue to be driven by technological advances to each technique. With multiple methods now being applicable in most clinical scenarios, accuracy, cost, time, risks/benefits, and optimal resource allocation are only several of the factors that one should consider in diagnosing and treatment planning for any craniofacial anomaly. Cone-beam computed tomography remains the gold standard for the imaging of hard tissue-based, craniofacial abnormalities. The improvements in CBCT acquisition speed, resolution, and field of view customization have largely eliminated the day-to-day use of many of the previously introduced modalities discussed above. Furthermore, because newer CBCT machines allow 3D surface scans to be obtained simultaneously, clinicians are able to evaluate soft tissue landmarks with much more accuracy than in the past. While CBCT can be considered the imaging method of choice for most scenarios, MRI still has an important role in providing imaging free of ionizing radiation and for detailed evaluation of soft tissues such as the TMJ disc.

References

1. Petrik V, Apok V, Britton JA, Bell BA. Godfrey Hounsfield and the dawn of computed tomography. Neurosurgery. 2006;58:780–7. https://doi.org/10.1227/01.NEU.0000204309.91666.06.
2. Garayoa J, Castro P. A study on image quality provided by a kilovoltage cone-beam computed tomography. J Appl Clin Med Phys. 2013;14:239–57. https://doi.org/10.1120/jacmp.v14i1.3888.
3. Lechuga L, Weidlich GA. Cone beam CT vs. fan beam CT: a comparison of image quality and dose delivered between two differing CT imaging modalities. Cureus. 2016;8(9):1–13. https://doi.org/10.7759/cureus.778.
4. Qu X, Li G, Ludlow JB, Zhang Z, Ma X. Effective radiation dose of ProMax 3D cone-beam computerized tomography scanner with different dental protocols. Oral Surg Oral Med Oral Pathol Oral Radiol Endod. 2010;110(6):770–6. https://doi.org/10.1016/j.tripleo.2010.06.013.
5. Pauwels R, Araki K, Siewerdsen JH, Thongvigitmanee SS. Technical aspects of dental CBCT: state of the art. Dentomaxillofac Radiol. 2015;44:1–20. https://doi.org/10.1259/dmfr.20140224.
6. Davies J, Johnson B, Drage NA. Effective doses from cone beam CT investigation of the jaws. Dentomaxillofac Radiol. 2012;41:30–6. https://doi.org/10.1259/dmfr/30177908.
7. Ludlow JB, Ivanovic M. Comparative dosimetry of dental CBCT devices and 64-slice CT for oral and maxillofacial radiology. Oral Surg Oral Med Oral Pathol Oral Radiol Endod. 2008;106:106–14. https://doi.org/10.1016/j.tripleo.2008.03.018.
8. Elstrøm UV, Muren LP, Petersen JBB, Grau C. Evaluation of image quality for different kV cone-beam CT acquisition and reconstruction methods in the head and neck region. Acta Oncol. 2011;50(6):908–17. https://doi.org/10.3109/0284186X.2011.590525.
9. Peyrin F, Dong P, Pacureanu A, Langer M. Micro and nano CT for the study of bone ultrastructure. Curr Osteoporos Rep. 2014;12:465–74. https://doi.org/10.1007/s11914-014-0233-0.
10. Kampschulte M, Langheinirch AC, Sender J, Litzlbauer HD, Althohn U, Schwab JD, et al. Nano-computed tomography: technique and applications. Fortschr Röntgenstr. 2016;188:146–54. https://doi.org/10.1055/s-0041-106541.
11. Day CJ, Lee RT. Three-dimensional assessment of the facial soft tissue changes that occur postoperatively in orthognathic patients. World J Orthod. 2006;7:15–26.
12. Yu Z, Mu X, Feng S, Han J, Chang T. Flip-registration procedure of three-dimensional laser surface scanning images on quantitative evaluation of facial asymmetries. J Craniofac Surg. 2009;20(1):157–60. https://doi.org/10.1097/SCS.0b013e318191ce88.
13. Baik HS, Kim SY. Facial soft-tissue changes in skeletal class III orthognathic surgery patients analyzed

with 3-dimensional laser scanning. Am J Orthod Dentofac Orthop. 2010;138:167–78. https://doi.org/10.1016/j.ajodo.2010.02.022.
14. Soncul M, Bamber MA. Evaluation of facial soft tissue changes with optical surface scan after surgical correction of class III deformities. J Oral Maxillofac Surg. 2004;62:1331–40. https://doi.org/10.1016/j.joms.2004.04.019.
15. Langdon J, Patel M, Ord R, Brennan P. Operative oral and maxillofacial surgery. 3rd ed. Boca Raton, FL: CRC Press: Taylor & Francis Group; 2010.
16. Nguyen C, Nissanov J, Ozturk C, Nuveen M, Tuncay OC. Three-dimensional imaging of the craniofacial complex. Clin Orthod Res. 2000;3:46–50. https://doi.org/10.1034/j.1600-0544.2000.030108.x.
17. Chan B, Auyeung J, Rudan JF, Ellis RE, Kunz M. Intraoperative application of hand-held structured light scanning: a feasibility study. Int J CARS. 2016;11:1101–8. https://doi.org/10.1007/s11548-016-1381-8.
18. Hajeer MY, Millett DT, Ayoub AF, Siebert JP. Applications of 3D imaging in orthodontics: part II. J Orthod. 2004;31(2):154–62. https://doi.org/10.1179/146531204225020472.
19. Burke PH, Beard FH. Stereophotogrammetry of the face: a preliminary investigation into the accuracy of a simplified system evolved for contour mapping by photography. Am J Orthod. 1967;53(10):769–82.
20. Schendel SA, Jacobson R, Khalessi S. 3-Dimensional facial simulation in orthognathic surgery: is it accurate? J Oral Maxillofac Surg. 2013;71:1406–14. https://doi.org/10.1016/j.joms.2013.02.010.
21. Khambay B, Nebel JC, Bowman J, Walker F, Hadley DM, Ayoub A. A pilot study: 3D stereophotogrammetric image superimposition onto 3D CT scan images – the future of orthognathic surgery. Int J Orthodon Orthognath Surg. 2002;17:331–41.
22. Bugaighis I, Mattick CR, Orth F, Tiddeman B, Hobson R. 3D facial morphometry in children with oral clefts. Cleft Palate Craniofac J. 2014;51(4):452–61. https://doi.org/10.1597/12-217.
23. Mailey B, Baker JL, Hosseini A, Collins J, Suliman A, Wallace AM, Cohen SR. Evaluation of facial volume changes after rejuvenation surgery using a 3-dimensional camera. Aesthet Surg J. 2016;36(4):379–87. https://doi.org/10.1093/asj/sjv226.
24. Knoops PGM, Beaumont CAA, Borghi A, Rodriguez-Florez N, Breakey RWF, Rodgers W, et al. Comparison of three-dimensional scanner systems for craniomaxillofacial imaging. J Plast Reconstr Aesthet Surg. 2017;70:441–9. https://doi.org/10.1016/j.bjps.2016.12.015.
25. Tzou CHJ, Artner NM, Pona I, Hold A, Placheta E, Kropatsch WG, Frey M. Comparison of three-dimensional surface-imaging systems. J Plast Reconstr Aesth Surg. 2014;67:489–97. https://doi.org/10.1016/j.bjps.2014.01.003.
26. Maal TJ, Van Loon B, Plooij JM, Rangel F, Ettema AM, Borstlap WA, Berge SJ. Registration of 3-dimensional facial photographs for clinical use. J Oral Maxillofac Surg. 2010;68:2391–401. https://doi.org/10.1016/j.joms.2009.10.017.
27. Heike CL, Upson K, Stuhaug E, Weinberg SM. 3D digital stereophotogrammetry: a practical guide to facial image acquisition. Head Face Med. 2010;6:18. https://doi.org/10.1186/1746-160X-6-18.
28. Ferrario VF, Sforza C, Poggio CE, Serrao G. Facial three-dimensional morphometry. Am J Orthod Dentofacial Orthop. 1996;109(1):86–93.
29. Webber RL, Horton RA, Tyndall DA, Ludlow JB. Tuned-aperture computed tomography (TACT). Theory and application for three-dimensional dentoalveolar imaging. Dentomaxillofac Radiol. 1997;26(1):53–62. https://doi.org/10.1038/sj.dmfr.4600201.
30. Shah N, Bansal N, Logani A. Recent advances in imaging technologies in dentistry. World J Radiol. 2014;6(10):794–807. https://doi.org/10.4329/wjr.v6.i10.794.
31. Nair MK, Seyedain A, Webber RL, Nair UP, Piesco NP, Agarwal S, et al. Fractal analyses of osseous healing using tuned aperture computed tomography images. Eur Radiol. 2001;11(8):1510–5. https://doi.org/10.1007/s003300000773.
32. Nair MK, Nair UP, Seyedain A, Gassner R, Piesco N, Mooney M, et al. Correlation of tuned aperture computed tomography with conventional computed tomography for evaluation of osseous healing in calvarial defects. Oral Surg Oral Med Oral Pathol Oral Radiol Endod. 2007;103(2):267–73. https://doi.org/10.1016/j.tripleo.2006.02.006.
33. Nair MK, Seyedain A, Agarwall S, Webber RL, Nair UP, Piesco NP, et al. Tuned aperture computed tomography to evaluate osseous healing. J Dent Res. 2001;80(7):1621–4. https://doi.org/10.1177/00220345010800070501.
34. Nair MK, Tyndall DA, Ludlow JB, May K. Tuned aperture computed tomography and detection of recurrent caries. Caries Res. 1998;32:23–30. https://doi.org/10.1159/000016426.
35. Nance R, Tyndall D, Levin LG, Trope M. Identification of root canals in molars by tuned-aperture computed tomography. Int Endod J. 2000;33(4):392–6. https://doi.org/10.1046/j.1365-2591.2000.00330.x.
36. Liang H, Tyndall DA, Ludlow JB, Lang LA. Cross-sectional presurgical implant imaging using tuned aperture computed tomography. Dentomaxillofac Radiol. 1999;28(4):232–7. https://doi.org/10.1038/sj/dmfr/4600451.
37. Ferreira LA, Grossmann E, Januzzi E, de Paula MVQ, Carvalho ACP. Diagnosis of temporomandibular joint disorders: indication of imaging exams. Braz J Otorhinolaryngol. 2016;82:341–52. https://doi.org/10.1016/j.bjorl.2015.06.010.
38. Navallas M, Inarejos EJ, Iglesias E, Cho Lee GY, Rodriguez N, Anton J. MR imaging of the temporomandibular joint in juvenile idiopathic arthritis: technique and findings. Radiographics. 2017;37:595–612. https://doi.org/10.1148/rg.2017160078.

39. Hechler BL, Phero JA, Van Mater H, Matthews NS. Ultrasound versus magnetic resonance imaging of the temporomandibular joint in juvenile idiopathic arthritis: a systematic review. Int J Oral Maxillofac Surg. 2018;47:83–9. https://doi.org/10.1016/j.ijom.2017.07.014.
40. Karatas O, Toy E. Three-dimensional imaging techniques: a literature review. Eur J Dent. 2014;8(1):132. https://doi.org/10.4103/1305-7456.126269.
41. Battal B, Kocaoglu M, Bulakbasi N, Husmen G, Sanal HT, Tayfun C. Cerebrospinal fluid flow imaging by using phase-contrast MR technique. Br J Radiol. 2011;84:758–65. https://doi.org/10.1259/bjr/66206791.
42. The Johns Hopkins University. Arteriogram. http://www.hopkinsmedicine.org/healthlibrary/conditions/radiology/arteriogram_85,p01274/. Accessed 5 Jun 2017.
43. Herzig R, Burval S, Krupka B, Vlachova I, Urbanek K, Mares J. Comparison of ultrasonography, CT angiography, and digital subtraction angiography in severe carotid stenosis. Eur J Neurol. 2004;11:774–81. https://doi.org/10.1111/j.1468-1331.2004.00878.x.
44. Newton C. Comparing CTA and MRA. Diagn Invasive Cardiol. 2010;50(3):22–3.
45. Hartung MP, Grist TM, Francois CJ. Magnetic resonance angiography: current status and future directions. J Cardiovasc Magn Reson. 2011;13:19. https://doi.org/10.1186/1532-429X-13-19.
46. Hassanien OA, Ghieda UE, Younes RL, Shaban EA. Facial vascular anomalies; MRI and TRICKS-MR angiography diagnostic approach. Egypt J Radiol Nucl Med. 2017;48:885–95. https://doi.org/10.1016/j.ejrnm.2017.08.013.

The manufacturer's authorised representative in the EU is Springer Nature Customer Service Centre GmbH, Europaplatz 3, 69115 Heidelberg, Germany. If you have any concerns regarding our products, please contact ProductSafety@springernature.com

Printed and bound by CPI Group (UK) Ltd, Croydon, CR0 4YY
22/09/2026
02221056-0006